T0205993

Data-Driven Model-Free Controllers

Data-Driven Model-Free Controllers

Radu-Emil Precup, Raul-Cristian Roman, and
Ali Safaei

CRC Press
Taylor & Francis Group
Boca Raton London New York

CRC Press is an imprint of the
Taylor & Francis Group, an **informa** business

MATLAB® is a trademark of The MathWorks, Inc. and is used with permission. The MathWorks does not warrant the accuracy of the text or exercises in this book. This book's use or discussion of MATLAB® software or related products does not constitute endorsement or sponsorship by The MathWorks of a particular pedagogical approach or particular use of the MATLAB® software.

First edition published 2022
by CRC Press
6000 Broken Sound Parkway NW, Suite 300, Boca Raton, FL 33487-2742

and by CRC Press
2 Park Square, Milton Park, Abingdon, Oxon, OX14 4RN

CRC Press is an imprint of Taylor & Francis Group, LLC

© 2022 Radu-Emil Precup, Raul-Cristian Roman, Ali Safaei

Library of Congress Cataloging-in-Publication Data
Names: Precup, Radu-Emil, 1963- author. | Roman, Raul-Cristian, author. | Safaei, Ali, author.
Title: Data-driven model-free controllers / Radu-Emil Precup, Raul-Cristian Roman, Ali Safaei.
Description: First edition. | Boca Raton, FL : CRC Press, 2022. | Includes bibliographical references and index. | Summary: "This book categorizes the wide area of data-driven model-free controllers, reveals the exact benefits of such controllers, gives the in-depth theory and mathematical proofs behind them, and finally discusses their applications"– Provided by publisher.
Identifiers: LCCN 2021042242 (print) | LCCN 2021042243 (ebook) |
ISBN 9780367697303 (hbk) | ISBN 9780367698287 (pbk) | ISBN 9781003143444 (ebk)
Subjects: LCSH: Programmable controllers. | Automatic control.
Classification: LCC TJ223.P76 P68 2022 (print) | LCC TJ223.P76 (ebook) |
DDC 629.8/95–dc23/eng/20211027
LC record available at https://lccn.loc.gov/2021042242
LC ebook record available at https://lccn.loc.gov/2021042243

ISBN: 978-0-367-69730-3 (hbk)
ISBN: 978-0-367-69828-7 (pbk)
ISBN: 978-1-003-14344-4 (ebk)

DOI: 10.1201/9781003143444

Typeset in Times
by codeMantra

Access the Support Material: http://www.routledge.com/9780367697303

Contents

Authors

Radu-Emil Precup received a Dipl.Ing. (Hons.) degree in automation and computers from the "Traian Vuia" Polytechnic Institute of Timisoara, Timisoara, Romania, in 1987, a Diploma in mathematics from the West University of Timisoara, Timisoara, in 1993, and a Ph.D. degree in automatic systems from the "Politehnica" University of Timisoara, Timisoara, in 1996. He was a Professor in the Department of Automation and Applied Informatics in the Politehnica University of Timisoara, Romania since 2000, and currently, he is a Doctoral Supervisor of automation and systems engineering in the same institution. He is also an Adjunct Professor within the School of Engineering, Edith Cowan University, Joondalup, WA, Australia, and an Honorary Professor and a Member of the Doctoral School of Applied Informatics with the Óbuda University (previously named Budapest Tech Polytechnical Institution), Budapest, Hungary. He is currently the Director of the Automatic Systems Engineering Research Centre in the Politehnica University of Timisoara, Romania.

Raul-Cristian Roman received a Bachelor's degree in engineering in automation and computers in 2012, a Master's degree in automation and computers in 2014 and a Ph.D. degree in systems engineering in 2018 from the Politehnica University of Timisoara, Timisoara, Romania. From 2012 to 2015, he was a software engineer at S.C. Elster Rometrics S.R.L., Romania. He is currently an Assistant Professor in the Department of Automation and Applied Informatics in the Politehnica University of Timisoara, Romania.

Ali Safaei received his B.Sc. and M.Sc. degrees in mechanical engineering from Isfahan University of Technology, Isfahan, Iran in 2009 and University of Tehran, Tehran, Iran in 2012, respectively. After gaining an experience of 5 years in automotive, oil-gas and power plant industries, he won a TWAS-USM postgraduate fellowship and started his Ph.D. studies in 2017 at the School of Electrical and Electronic Engineering, Universiti Sains Malaysia, Penang, Malaysia. In 2019, he finished his PhD studies and received a MITACS Accelerate grant for a postdoctoral period at McGill University and HumanITAS start-up company in Montreal, QC.

1 Introduction

1.1 THE MOTIVATION OF DATA-DRIVEN MODEL-FREE CONTROL

Model reference control has evolved in time as a popular approach to controller design and tuning because the control system engineer must usually develop control systems based on performance specifications. One representative control solution in this regard is the model reference adaptive control (MRAC). In many situations, these specifications are formulated as constraints on the time response of the controlled output of the control system, such as overshoot, settling time, rise time, and steady-state error. This has always been the most direct and easily understandable way of imposing and assessing the behavior and the quality of control systems. Other performance specifications can be formulated in the frequency domain, as, for example, bandwidth, phase and gain margins and initial slope of open-loop magnitude plot (for linear systems).

As shown in [Rad11a], a usual way to deal with model reference control is formulated as follows: given a set performance specifications that can be often described by the response in time of a reference model response with respect to a reference input, design a control system that makes the controlled output(s) of the process track the reference model in terms of time response when driven by the same reference input. The reference model is typically a low-order system. In many applications, the difference of the outputs of the control system and the reference model, referred to as tracking error, defines a measure of the quality of the match between the two outputs and, accordingly, the two systems (i.e., the control system and the reference model). The cost function or the objective function defined based on the tracking error depends on the parameters of the controller. The model reference problem requires finding the suitable set of controller tuning parameters that minimize the objective function, which brings close the two outputs that form the aforementioned error. Thus, the model reference control problem becomes an optimization problem that can be solved analytically or numerically making use of optimization techniques. The vector variable of the optimization problem is the parameter vector of the controller whose elements are the controller tuning parameters.

However, this way hides important aspects in the general context of the fact that the control designs usually boil down to tradeoff. The opponents in this tradeoff are the process and the controller, and their relationship determines those aspects, namely the disturbance rejection and the parameter sensitivity, which in turn determine the robust stability and the robust performance.

The controller is subjected to the designer's choice in both structure complexity and parameter values, whereas the process can only be known partially. The control system designers use process models, which are inherently simplifications of the reality. Therefore, with models available, the controller design can be carried out and

DOI: 10.1201/9781003143444-1

next tested but with no guarantee that the implementation of the proposed solution would give satisfactory results in the real-world operation of the control systems.

The control system designers always have modeling tools at hand such as first principles modeling or system identification, which lead to mathematical models of the processes. So, one aspect that has always been fixed with respect to the controller design is the classical and also modern model-based approach. Whether the control system designers operate on nominal models or simplified models, the knowledge of the process model has been essential to the design and analysis of the control systems. Knowing the model allows for a better insight into the limitations of the controller design and allows to conduct simulation using as detailed as possible process modes before controller implementation.

As shown in [Rad11a], on the shaky grounds of model uncertainty and parameter variations, tools had to be developed such that the specifications would be met under a broad range of conditions; this is the reason for development of adaptive control and robust control. Adaptive control has emerged as a solution to cope with parameter variations by means of automatically redesigning and retuning the controller. Generally, the difficulty associated with this approach is mainly the analysis that has to be carried out in a time-varying nonlinear context and possibly the cost of implementation. The most popular versions of adaptive control are gain scheduling, MRAC (both direct and indirect) and self-tuning regulators. Even when using approaches like the "MIT rule" or the Lyapunov redesign, use is still made of the process model. Therefore, bringing together and treating the issues of stability analysis, convergence and robustness, all of them in the time-varying nonlinear framework, is not an easy task for the majority of control engineers.

Model-free controllers, also referred to as **data-driven controllers**, were investigated in the literature of control engineering over more than two decades. These controllers do not rely on a priori known mathematical model of the dynamic system (i.e., the process) that the controller is going to be implemented on. In other words, few or even no information on the process that is controlled is used in controller parameter tuning, conducting experiments on the real-world control system replacing it. Instead, a more generic model is considered, and the unknown parameters in that model are estimated online, based on the measured input-output data sets of the dynamic system. This feature of model-free controllers makes them attractive to practitioners. This book uses the data-driven model-free control (MFC) name in order to highlight that use is made of measured input-output data sets; therefore, the controllers are model-free in the tuning. However, these techniques make use of system responses in time or frequency domain, which are nonparametric models.

The control system performance improvement offered by data-driven control techniques is achieved by simple specifications and relatively easily interpretable performance indices. These indices are usually specified in the time domain (e.g., overshoot, settling time, rise time), and they are aggregated in integral-type or sum-type objective functions such as the linear-quadratic Gaussian (LQG) ones. The minimization of the objective functions in constrained optimization problems can fulfill various objectives as reference trajectory tracking (including model reference tracking), control input (or control signal) penalty, disturbance rejection, etc.

The logic behind data-driven control techniques is apparently simple: since the classical two-step design approach (modeling and next control) could ignore the performance specifications because of the mismatch between the model and the real process, the designer should skip the modeling step and try to make a design and tuning without using a process model. This idea has been advocated in [Gev96] showing that the modeling and identification should not be seen as a purpose in itself, but rather a mean to help just the purpose of control design. Only the relevant aspects for control should be captured within the process model in the model-based design techniques.

The **data-driven MFC** name should be used carefully. It is valid only in certain situations, under certain assumptions, which cannot be made unless some insight is available about the process that is to be controlled. As shown in [Rad11a], several open questions remain on how much information on the process is sufficient in order to meet the performance specification, how may the empirical observations be used to infer limitations in the design beforehand, and if a model is strictly needed, how detailed should that model be in the framework of the cost of identification versus the model quality.

From the point of view of the tradeoff to model quality, design complexity and performance specifications, data-driven MFC ensures a good tradeoff. Data-driven control techniques are focused on achieving performance with no process model or little information on process model at hand and meeting the performance specifications with simple and relatively easily interpretable controllers such as the most popular controllers in the industry, i.e., the PI and PID ones. However, many data-driven control techniques are constrained to be used with pre-parameterized controllers, in that they can only deal with parameter optimization. In addition, if the controller specification is left free, the user would encounter the same model selection problem of system identification, which is typical to standard model-based control techniques.

The main shortcoming of data-driven MFC is the **difficult systematic stability and robustness analyses**. In other words, the tuning to ensure reference trajectory tracking does not guarantee robust stability and robust performance. This is normal because these analyses require detailed mathematical models of the controlled process. Nevertheless, the term "robustness", which in model-based control refers to the property of a controller of having low sensitivity to modeling errors, is usually avoided in data-driven MFC. Since here the tuning process is model-free, the use of the term "robustness" sounds ambiguous and should be clarified.

The above benefits and advantages of model-free controllers over the model-based controllers are proved and reviewed in several publications that will be discussed in the next section. Moreover, their representative applications on practical benchmarks will also be briefly described.

1.2   A CONCISE OVERVIEW OF THE MAIN DATA-DRIVEN MFC TECHNIQUES

The only book on data-driven control already available "off-the-shelf" is [Baz12]. This book covers the optimal type of data-driven model-free controllers and focuses on the description of the theory of data-driven techniques. A part of the present book partially overlaps with [Baz12], in particular, the chapters on iterative feedback tuning (IFT) and virtual reference feedback tuning (VRFT).

The book [Hou13a] discusses an important part of data-driven model-free controllers, namely those based on model-free adaptive control (MFAC). MFAC was first proposed for a class of discrete-time nonlinear systems, which can be equivalently transformed to dynamic linearization data models.

The book [Din14] treats the fault-tolerant control systems in the context of data-driven control. This is motivated by the current industry development, which calls for more system reliability, dependability and safety.

The books [Xu03] and [Owe15] deal with an important data-driven technique, namely iterative learning control (ILC). Although ILC can be seen in the model-based context as far as the control loop is concerned at least in the initial part of its development, ILC is a relevant data-driven model-free technique that acts rather on the control system structure than on the controller itself. A general theoretical analysis of ILC covering various applications is given in [Xu03]. The main theoretical approaches to ILC based on either signal or parameter optimization are presented in [Owe15].

As pointed out in [Pre20b], all successful applications and performance proved by data-driven MFC justify considering, in accordance with the statement given in [Fli20], that MFC, as an important part of these controllers, is a new tool for machine learning (ML). The authors support this statement and highlight that it is also valid for other data-driven model-free techniques, especially the iterative ones, where the parameter-updated law can be viewed as a learning algorithm. In this context, the recent books [Kam18] and [Bru19] give theoretical approaches to ML-based control techniques focusing on optimal and stable control systems. The present book will prove that data-driven MFC is an efficient tool for ML.

A useful discussion on model-based versus data-driven control, which inspired future research directions, is presented in the survey paper [Hou13b]. This survey is continued by other ones that treat industrial applications of data-driven techniques [Yin15b], classification and analysis of observer-integrated control [Hua20], compare model-based and data-driven control in the context of adaptive control [Ben18a], and two special issues on data-driven control and process monitoring for industrial applications [Yin14, Yin15a]. An overview of model-based control and data-driven control methods is carried out in [Hou17a] treating the data-driven equivalent dynamic linearization as a useful tool of data-driven control methods of discrete-time nonlinear systems. The link of data-driven control to fault diagnosis and nature-inspired optimal control of industrial applications is pointed out in [Pre15]. The advantages of data-driven MFC with respect to model-based controller tuning are outlined in the comparison conducted in [For14].

As specified in [Pre20a], unlike much of the existing work, the approach proposed in [VanW20] does not make the a priori assumption of persistency of excitation on the input. Instead, it studies necessary and sufficient conditions on the given data under which different analysis and control problems can be solved. Therefore, it reveals situations in which a controller can be tuned from data, even though unique system identification is impossible.

The most successful data-driven MFC techniques in the authors' opinion are briefly discussed as follows backing them up with classical and recent results. Different points of view are used in the above-discussed literature to classify the data-driven MFC techniques. For example, one point of view applied in [Hou13c] concerns the control system structure. The first category supposed that the controller structure with one or more unknown parameters contains the optimal controller, which is obtained from experimental knowledge on the process structure; the controller design is next transformed into a direct identification problem for the controller parameters. The second category includes the controllers designed based on certain function approximations or equivalent descriptions of the process, such as neural networks, fuzzy models or Taylor approximation; the controller parameters are next tuned by minimizing a specified performance criterion using the input-output data, including offline and online data. In this regard, since the present book does not have the aim to categorize the wide area of data-driven model-free controllers, but to reveal the exact benefits of such controllers, the techniques associated with these controllers will be divided into two categories that aim their implementation, iterative ones and noniterative or one-shot ones.

IFT [Hja94, Hja98, Hja02] is a well-acknowledged iterative data-driven technique that tunes the controller parameters iteratively along the gradient direction of an objective function. IFT is applicable if an initial appropriately parameterized controller that ensures a finite value of the objective function is assumed to be known [Pre07 Jun21].

MFAC has attractive features for control applications because online input-output data collected from the process according to [Hou97, Yu20] are used. The MFAC structures are based on dynamic linearized models of the process, and the control algorithms are formulated in a similar manner to predictive control. Three types of dynamic linearization data models are included in MFAC structures: compacted form dynamic linearization (CFDL) [Hou11a, Hou11b], partial form dynamic linearization (PFDL) [Hou11b] and full-form dynamic linearization (FFDL) [Hou99].

A different approach to MFAC is proposed and exemplified in continuous-time applications in [Saf18a] and [Saf18b]. This approach solves the major challenge for designing and implementing the data-driven model-free controllers represented by the online method for estimating the unknown parameters of the system. Use is made in [Saf18a] and [Saf18b] of parameter estimators.

Simultaneous Perturbation Stochastic Approximation (SPSA) [Spa98, Wan98] is based on the fact that unlike with the deterministic steepest descent, the gradient-based stochastic approximation algorithms employed in IFT and SPSA use estimated gradients of the objective function. If the gradients cannot be calculated using data from real-time experiments, they have to be estimated on the basis of the noisy measurements of the objective function using finite difference approximations around the current operating point. SPSA is advantageous as it reduces the implementation cost by means of only two evaluations of the objective function per iteration [Rad11b, Zam20].

Iterative Correlation-based Tuning (CbT) [Kar04, Miš07] operates in the model reference control framework. The relation between the reference input and the tracking error is highlighted in the correlation function of the two signals if the quasi-stationary framework is realistically assumed [Lju99]. A decorrelation procedure is applied to make the tracking error converge to zero. The objective function depends on the correlation function of the two signals (i.e., the reference input and the tracking error), and it is minimized in an iterative manner. The initial controller in iterative CbT is usually linear and appropriately parameterized, and it should ensure a finite value of the objective function [Sat20].

Frequency-domain Tuning (FdT) [deB99, Kam99, Kam00] makes use of a linear-quadratic (LQ) objective function that penalizes the tracking error in the model reference control framework, and it is minimized in terms of a different version of the classical gradient-based stochastic approximation algorithm, which is based on the estimate of the gradient of the objective function. The objective function is expressed in the frequency domain using Parseval's theorem and next spectral analysis techniques that allow the calculation of different auto- and cross-correlation sequences of the closed-loop control system signals. Spectral estimates are obtained for the transfer functions involved in the FdT algorithm. The advantage of FdT is the use of frequency response functions, which are nonparametric models. The derivatives of the objective function with respect to the controller parameters are thus obtained in the frequency domain [Kha14, daS18]. The stability is ensured between two consecutive iterations by the calculation of the Vinnicombe metric and the generalized stability margin using nonparametric models obtained via spectral analysis.

Iterative Regression Tuning (IRT) [Hal04, Hal06] minimizes an objective function that depends on the controller tuning parameters. A specific feature of IRT is the aggregation in the objective function of performance specifications (and indices) of different nature, which need only to be smooth functions of the controller tuning parameters. A gradient-based search is employed in the minimization of the objective function using either lot of simulations of the control system behavior [Hal04 Hal06] or local linear models derived by finite difference approximations assisted by real-world experiments conducted on the control system [Pre10].

Adaptive online IFT [McD12] tunes the controller parameters online using experimental data during normal system operation. This feature, which is different from the classical IFT, is important because the control system is keeping its reliable performance over a long period of time without the need of any offline tuning or system identification.

As pointed out in [Pre07], **ILC** states that control system performance executing repetitively the same tasks can be improved using the experience gained from previous experiments in control system operation. This is important because of two reasons. First, ILC can be formulated in terms of the iterative solving of a parametric optimization problem, which ensures the minimization of an objective function to meet the performance indices imposed to the control system; this can be viewed in the general context of learning. Second, since ILC does not generally act on the controller parameter but on the control system structure out of the controller, it can be applied to the reference input tuning in two-degree-of-freedom control system structures, with beneficial effects on the control system behavior with respect to both reference and disturbance inputs. As discussed in [Rad15], where constraints are considered and appropriately handled, the ILC-based solving of optimal control problems is formulated in [Gun01] and [Owe05], the specific time and frequency-domain convergence analyses are performed in [Nor02], the stochastic approximation in ILC is treated in [But08], the output tracking of nonlinear stochastic systems is discussed in [Che04], affine constraints are considered in [Mis11] by the transformation of ILC problems with quadratic objective functions into convex quadratic programs, and the system impulse response is estimated in [Jan13] using input/output measurements from previous iterations and next used in a norm-optimal ILC structure that accounts for actuator limitations by means of linear inequality constraints.

ILC has been treated recently in the framework of data-driven control. Some relevant results in this regard are optimal data-driven ILC [Chi15], constrained data-driven optimal ILC [Rad15, Chi18], multi input-multi output (MIMO) ILC [Bol18] and the analysis of ILC in the condition of incomplete information [She18].

Reinforcement Learning (RL) is a data-driven and also ML technique whose specific feature is the use of information gathered from interactions with the environment. The RL problem is formulated in the Markov decision process framework using dynamic programming to solve the optimization problem that ensures optimal reference tracking. An RL agent executes actions in the environment and based on the received reward, adjusts its knowledge about itself and the environment. By applying this process incrementally, the RL agent will become better at picking actions, which will maximize or minimize the rewards [Sut17]. As shown in [Sut92], RL is a viable technique that solves optimal reference tracking problems and fills the gap between ML and control. In this context, the RL agent is the controller that automatically learns how to modify its parameters and how to control a process based on the feedback (reward) it receives from it [Bus18]. Some recent applications of RL that solve optimal control problems are given in [Che18, Ngo18 and Qi19] and combined with other data-driven techniques in [Rad17, Rad18 and Wan17], as they are quite relevant to each other when performing intelligent optimization.

Unfalsified Control (UC) [Saf97, Tsa01] is a data-driven MFC approach that, given measured input-output data, recursively falsifies the controller tuning parameters that fail to satisfy the performance specifications. The controllers whose abilities to meet the performance specifications are not contradicted by the available data are referred to as unfalsified, and one controller that is unfalsified is implemented. A recurrent learning algorithm specific to UC is expressed if this unfalsification procedure is repeated over time [vanH07, VanH08, Bia15, San15, Jia16].

Lazy learning (LL) [Bon99, Bon05] combines an LL approximator, which performs the online local linearization of the system, with conventional linear control techniques as minimum variance, pole placement and optimal control. The advantages of LL as a data-driven model-free technique in control are proved in [Hou17b] and [Dai21].

One of the most popular data-driven model-free noniterative techniques is **VRFT** [Lec01, Cam02, Cam03]. The objective function specific to VRFT penalizes the difference between the outputs of the control system and the reference model and makes use of the 2-norm of discrete-time transfer functions, and it can also be expressed in the frequency domain after the application of Parseval's theorem. VRFT searches for the global minimum of the objective function optimum, being reduced to an identification problem as far as the controller and not the process is concerned. In the case of restricted complexity controller design, the achieved controller is a good approximation of the restricted complexity global optimal controller. VRFT is a one-shot algorithm, i.e., no model identification of the process, and also one-shot, namely it can be applied using a single set of input data generated from the process, with no need for additional specific experiments nor iterations [Car19, For19].

Noniterative CbT [Kar07, vanH11] is also very well known, and it is based on the generation of an estimate of the tracking error in a model reference control problem, which is next employed in the minimization of the objective function. The correlation approach deals with the influence of measurement noise, and a convex optimization problem is obtained for linearly parameterized controllers. It has been proven in [vanH11] that noniterative CbT statistically outperforms other data-driven MFC techniques.

MFC [Fli09, Fli13] combines the well-known and widely used PI and PID controller structure with an intelligent term that compensates for the effects of nonlinear dynamics, disturbances and uncertain parameters. An integral local model of the process is used in this regard, and its identification is not needed [Rom18, Pol19, Joi20, Ols20, Vil20].

Active Disturbance Rejection Control (ADRC) [Gao06] uses an extension of the system model with an additional and fictitious state variable, which models the nonmodeled dynamics of the process. This virtual state is estimated online by an extended state observer and next used in the control signal (or control input) in order to decouple the system from the actual disturbance that acts on the process. The ADRC structure and design approach offers robustness and adaptive features to the control systems [Ben18b, Mad19, Sun19, Rom21].

Data-driven predictive control is based on several system identification approaches [Kad03, Wan07, Luc20], which avoid the inclusion of several predictive control features. This is important in practical applications because data-driven predictive control can deal with systems without complete online measurement of all output variables and work with relatively simple models to tune the predictive control with little additional experimental effort.

Markov data-based LQG control [Shi00] estimates the Markov parameters (i.e., impulse response coefficients) from given input-output data, and a finite-horizon LQG controller is next designed. A receding horizon algorithm can also be designed to track changes in process parameters and/or structure over time [Aan05].

Extremum seeking control [Krs00a] uses a probing signal and demodulation to recover the gradient of the objective function to the controller parameters. The stability of this adaptive control system structure is proven with the averaging method in terms of showing that the system converges to a small neighborhood of the extremum of the objective function [Krs00b]. Some relevant recent applications of extremum seeking control are given in [Gua17, Mar17, Bag18, Gua18, Zho18].

Pulse response-based control [Ben93] uses a nonparametric model represented by the measured response with respect to pulses in control signal (or control input). Minimum time optimal control problems are solved in terms of computing the minimum number of time steps for which a control history exists that consists of a train of pulses and satisfies control signal bounds. The desired outputs are obtained at the end of the control task [Gon00].

LQ data-driven control [Fav99] uses subspace system identification techniques to design finite-horizon LQG controllers. The three steps of the classical model-based LQG controller design, namely system identification, Kalman filter and LQ control design, are replaced by a QR- and an SV-decomposition [daS19].

Data-Driven Inversion-Based Control (D^2-IBC) [Nov15] is based on a two-degree-of-control system structure, which operates with a linear controller and a nonlinear controller in parallel, both designed using experimental data. Attractive automotive applications of D2-IBC are reported in [Gal17] and [Gal19].

The frequency-domain data-driven approach in the Loewner framework [Ker17] solves a model reference tracking problem in two steps. First, based on the available input-output data and the reference model, the frequency response of an ideal controller is computed at several frequencies. Second, a reduced order model of the ideal controller is identified on the basis of its frequency-domain data. Two frequency-domain identification methods are used in the second step, and results of the application of this data-driven model-free approach are presented in [Ker19] and [Ker20].

This book will report the approaches that are known better by the authors, i.e., the ones where the authors have experience, results and more publications about. Since the aim of this book is to illuminate different aspects and features of the data-driven model-free controllers such that to become a very beneficial tool and guide for the students and researchers in the field of control engineering, this book mixes the authors' mix model reference and disturbance rejection controls, and iterative and noniterative tuning. Although this looks more fuzzy and less well-motivated, the authors' approach is useful as the in-depth theory and mathematical proofs behind the data-driven model-free controllers are presented at the minimum understandable level for the readers, and finally their applications are organized straightforward in a transparent manner and accompanied by Matlab & Simulink programs and simulation schemes. Other important data-driven model-free controllers, including those based on ILC and RL, will be treated in the second edition of the book.

1.3 DYNAMIC SYSTEMS USED IN IMPLEMENTATIONS

1.3.1 TOWER CRANE SYSTEM

The dynamic system considered as a controlled process in order to exemplify the single input-singe output (SISO) and MIMO data-driven MFC systems in Chapters 2–7 is the tower crane system. The implementation and real-time testing of MIMO control systems for cart position, arm angular position and payload position can be carried out, and three separate SISO control systems can be implemented and tested separately for each position control.

The tower crane system laboratory equipment in the Intelligent Control Systems Laboratory of the Politehnica University of Timisoara, Romania is organized upon [Int12]. The block diagram of the principle of the experimental setup is given in Figure 1.1.

As shown in [Rom21], the communication between the TCS and the personal computer is made through a RT-DAC/PCI multipurpose digital input-output board that communicates with a power interface. The personal computer is based around an Intel microprocessor and a Windows 10 operating system. The board that ensures the communication of the personal computer with the tower crane systems is accessed from the equipment toolbox, which operates using Matlab & Simulink [Int12].

The experimental setup is illustrated in Figure 1.2. It indicates the axes and the measurement of the three controlled outputs, namely the cart position y_1 (m) = x_3 (m), the arm angular position y_2 (rad) = x_4 (rad) and the payload position y_3 (m) = x_9 (m).

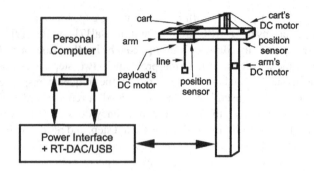

FIGURE 1.1 Block diagram of principle of tower crane system laboratory equipment and experimental setup. (Adapted from [Hed21, Rom21].)

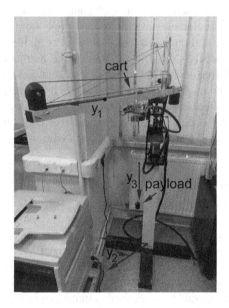

FIGURE 1.2 Experimental setup.

The state variables involved in the system model are as follows: x_1 (rad) and x_2 (rad) – the angles that describe the payload position in xz plane, x_3 (m) $\in [-0.25, 0.25]$ – the first output of the tower crane system, i.e., the cart position, x_4 (rad) $\in [-1.57, 1.57]$ – the second output of the tower crane system, i.e., the arm angular position, x_5 (rad/s) and x_6 (rad/s) – the angular velocities that describe the payload position in xz plane, x_7 (m/s) – the cart velocity and x_8 (rad/s) – the arm angular velocity. Since no payload swing damping is considered in this book, it is assumed, as in [Hed21] and [Rom21], that the line is rigid, the payload angles are small, and the mass and stiffness of the line are neglected. Other two state variables are defined in the conditions of these three assumptions, namely x_9 (m) $\in [-0.4, 0.4]$ – the lift line position and x_{10} (m/s) – lift line velocity.

The three actuators shown in Figures 1.1 and 1.2 are direct current (DC) motors. The variable $m_1 \in [-1, 1]$ is the output of the saturation and dead zone static nonlinearity specific to the first actuator

$$m_1(t) = \begin{cases} -1, & \text{if } u_1(t) \leq -u_{b1}, \\ \dfrac{u_1(t) + u_{c1}}{u_{b1} - u_{c1}}, & \text{if } -u_{b1} < u_1(t) < -u_{c1}, \\ 0, & \text{if } -u_{c1} \leq |u_1(t)| \leq u_{a1}, \\ \dfrac{u_1(t) - u_{a1}}{u_{b1} - u_{a1}}, & \text{if } u_{a1} < u_1(t) < u_{b1}, \\ 1, & \text{if } u_1(t) \geq u_{b1}, \end{cases} \tag{1.1}$$

where t is the independent continuous-time argument, $t \in \Re$, $t \geq 0$, $u_1(\%) \in$, $[-100, 100]$ and next $u_1 \in [-1, 1]$ is the first control signal (or control input), i.e., the pulse width modulation (PWM) duty cycle of the DC motor for cart position control, and the values of the parameters in (1.1) are $u_{a1} = 0.1925$, $u_{b1} = 1$ and $u_{c1} = 0.2$. The variable $m_2 \in [-1, 1]$ is the output of the saturation and dead zone static nonlinearity specific to the second actuator

$$m_2(t) = \begin{cases} -1, & \text{if } u_2(t) \leq -u_{b2}, \\ \dfrac{u_2(t) + u_{c2}}{u_{b2} - u_{c2}}, & \text{if } -u_{b2} < u_2(t) < -u_{c2}, \\ 0, & \text{if } -u_{c2} \leq |u_2(t)| \leq u_{a2}, \\ \dfrac{u_2(t) - u_{a2}}{u_{b2} - u_{a2}}, & \text{if } u_{a2} < u_2(t) < u_{b2}, \\ 1, & \text{if } u_2(t) \geq u_{b2}, \end{cases} \tag{1.2}$$

where $u_2(\%) \in [-100, 100]$, and next $u_2 \in [-1, 1]$ is the second control input, i.e., the PWM duty cycle of the DC motor for arm angular position control, and the values of the parameters in (1.2) are $u_{a2} = 0.18$, $u_{b2} = 1$ and $u_{c2} = 0.1538$. The variable $m_3 \in [-1, 1]$ is the output of the saturation and dead zone static nonlinearity specific to the second actuator

$$m_3(t) = \begin{cases} -1, & \text{if } u_3(t) \leq -u_{b3}, \\ \dfrac{u_3(t) + u_{c3}}{u_{b3} - u_{c3}}, & \text{if } -u_{b3} < u_3(t) < -u_{c3}, \\ 0, & \text{if } -u_{c3} \leq |u_3(t)| \leq u_{a3}, \\ \dfrac{u_3(t) - u_{a3}}{u_{b3} - u_{a3}}, & \text{if } u_{a3} < u_3(t) < u_{b3}, \\ 1, & \text{if } u_3(t) \geq u_{b3}, \end{cases} \tag{1.3}$$

where $u_3(\%) \in [-100, 100]$, and next $u_3 \in [-1, 1]$ is the third control input, i.e., the PWM duty cycle of the DC motor for payload position control, and the values of the parameters in (1.3) are $u_{a3} = 0.1$, $u_{b3} = 1$ and $u_{c3} = 0.13$.

Omitting the time argument for the sake of simplicity, since the lift line position is not constant as in [AlM00] and [Int12],

$$\dot{x}_9 = x_{10}, \tag{1.4}$$

the dynamics of the first two DC motors (for y_1 and y_2) are inserted in terms of appropriate gains and small time constants, and the first eight state equations in [AlM00] and [Int12] are modified and processed leading to

$$\dot{x}_1 = x_5,$$

$$\dot{x}_2 = x_6,$$

$$\dot{x}_3 = x_7,$$

$$\dot{x}_4 = x_8,$$

$$\dot{x}_5 = f_5(\Pi), \tag{1.5}$$

$$\dot{x}_6 = f_6(\text{II}),$$

$$\dot{x}_7 = -\frac{1}{T_{\Sigma 1}} x_7 + \frac{k_{P1}}{T_{\Sigma 1}} m_1,$$

$$\dot{x}_8 = -\frac{1}{T_{\Sigma 2}} x_8 + \frac{k_{P2}}{T_{\Sigma 2}} m_2,$$

where the expressions of the functions f_5 and f_6 are

$$f_5(\Pi) = f_5(x_1, x_2, x_3, x_5, \ldots, x_{10}, m_1, m_2) = \frac{1}{2x_9}(-4x_5 x_{10} + 4x_6 x_8 x_9 \cos^2 x_1 \cos x_2$$

$$-x_6^2 x_9 \sin 2x_1 + x_8^2 x_9 \sin 2x_1 \cos^2 x_2 - 2g \sin x_1 \cos x_2 - 4x_7 x_8 \cos x_1 + 2x_3 x_8^2 \sin x_1 \sin x_2$$

$$+2\frac{x_7}{T_{\Sigma 1}} \sin x_1 \sin x_2 - 2\frac{k_{P1}}{T_{\Sigma 1}} m_1 \sin x_1 \sin x_2 + 4x_8 x_{10} \sin x_2 + 2\frac{x_3 x_8}{T_{\Sigma 2}} \cos x_1$$

$$-2\frac{k_{P2}}{T_{\Sigma 2}} x_3 m_2 \cos x_1 - 2\frac{x_8 x_9}{T_{\Sigma 2}} \sin x_2 + 2\frac{k_{P2}}{T_{\Sigma 2}} x_9 m_2 \sin x_2), \tag{1.6}$$

$$f_6(\text{II}) = f_6(x_1, x_2, x_3, x_5, \ldots, x_{10}, m_1, m_2) = \frac{1}{x_9 \cos x_1}(-2x_6 x_{10} \cos x_1 - 2x_5 x_8 x_9 \cos x_1 \cos x_2$$

$$+2x_5 x_6 x_9 \sin x_1 - g \sin x_2 - 2x_8 x_{10} \sin x_1 \cos x_2 - x_3 x_8^2 \cos x_2 + x_8^2 x_9 \cos x_1 \cos x_2 \sin x_2$$

$$-\frac{x_7}{T_{\Sigma 1}} \cos x_2 + \frac{k_{P1}}{T_{\Sigma 1}} m_1 \cos x_2 + \frac{x_8 x_9}{T_{\Sigma 2}} \sin x_1 \cos x_2 - \frac{k_{P2}}{T_{\Sigma 2}} x_9 m_2 \sin x_1 \cos x_2), \tag{1.7}$$

$k_{P1} = 0.188$ m/s and $k_{P2} = 0.871$ rad/s are the process gains (of the first two DC motors), and $T_{\Sigma 1} = 0.1$ s and $T_{\Sigma 2} = 0.1$ s are the small time constants of the first two DC motors.

The simplified model of payload dynamics in three-dimensional cranes given in [Váz13] is adapted to the tower crane system in terms of

$$m_L \ddot{x}_9 + \mu_L \dot{x}_9 - m_L(g + \ddot{z}_c) = k_{P3}m_3, \tag{1.8}$$

where $m_L = 0.33$ kg is the payload mass, $\mu_L = 1,600$ kg/s is the viscous coefficient associated with the payload motion, $g = 9.81$ m/s^2 is the gravitational acceleration, $k_{P3} = 200$ kg $\cdot$ m/s^2 is process gains of the third DC motor, and z_c (m) is the z coordinate of the payload.

The definitions of the coordinates in the systems of axes and the trigonometric transformations lead to the following equation that gives the z coordinate of the payload in the system of axes defined in Figure 1.2:

$$z_c = -x_9 \cos x_1 \cos x_2. \tag{1.9}$$

Equation (1.9) is differentiated twice and next substituted along with (1.4) and (1.5) in (1.8), leading to the tenth state equation of the tower crane system:

$$\dot{x}_{10} = f_{10}(\aleph), \tag{1.10}$$

where the expression of the function f_{10} is

$$f_{10}(\aleph) = f_{10}(x_1, x_2, x_3, x_5, \ldots, x_{10}, m_1, m_2, m_3) = \frac{1}{1 + \cos x_1 \cos x_2}$$

$$\times \left(\begin{array}{l} -\dfrac{\mu_L}{m_L} x_{10} + g + 2x_6 x_{10} \cos x_1 \sin x_2 + 2x_5 x_{10} \sin x_1 \cos x_2 - 2x_5 x_6 x_9 \sin x_1 \sin x_2 \\[2mm] +x_9 f_6(\amalg) \sin x_2 \cos x_1 + x_9 f_5(\prod) \sin x_1 \cos x_2 + x_9(x_5^2 + x_6^2) \cos x_1 \cos x_2 + \dfrac{k_{P3}m_3}{m_L} \end{array} \right).$$

$$\tag{1.11}$$

Concluding, the nonlinear state-space model of the tower crane system viewed as the controlled process is obtained as follows by merging equations (1.4), (1.5) and (1.10) and adding the three output equations:

$$\dot{x}_1 = x_5,$$

$$\dot{x}_2 = x_6,$$

$$\dot{x}_3 = x_7,$$

$$\dot{x}_4 = x_8,$$

$$\dot{x}_5 = f_5(\Pi),$$

$$\dot{x}_6 = f_6(\amalg),$$

$$\dot{x}_7 = -\frac{1}{T_{\Sigma 1}}x_7 + \frac{k_{P1}}{T_{\Sigma 1}}m_1,$$

$$\dot{x}_8 = -\frac{1}{T_{\Sigma 2}}x_8 + \frac{k_{P2}}{T_{\Sigma 2}}m_2,$$

$$\dot{x}_9 = x_{10},$$

$$\dot{x}_{10} = f_{10}(\aleph),$$

$$y_1 = x_3,$$

$$y_2 = x_4,$$

$$y_3 = x_9, \qquad\qquad\qquad (1.12)$$

where the expression of the functions f_5, f_6 and f_{10} is given in (1.6), (1.7) and (1.11), respectively, and the values of all parameters are presented in this section. This model is implemented in the process_model.m S-function and the Process.mdl Simulink diagram, which are included in the accompanying Matlab & Simulink programs.

The Process.mdl Simulink diagram is built around the process_model.m S-function included in the tower crane system subsystem. This subsystem, which models the controlled process in the control system structures developed in the next chapters, also includes three zero-order hold (ZOH) blocks applied to the control signals (or control inputs), in order to prepare the process for digital control. Each ZOH block operates with the sampling period of 0.01 s, which is useful as it accommodates for quasi-continuous digital control.

The tower crane system subsystem is replaced by a real-time block in the Simulink diagrams of the control systems that are used in the experiments conducted in the next chapters. These real-time Simulink diagrams are not inserted as accompanying Matlab & Simulink programs because they require additional hardware and software resources [Int12] and the laboratory equipment as well.

1.3.2 Nonholonomic Autonomous Ground Rover

The nonholonomic autonomous ground rover is a ground-wheeled mobile robot moving on a plane environment with the nonholonomic constraint; i.e., a constraint on its velocity vector, where the rover can have only velocity in its forward direction and the component of velocity in a perpendicular direction to its centerline would be always zero. The dynamic system of such robot (Figure 1.3), which is also known as a *differential-drive* robot, is considered as a double-integrator MIMO system and defined as follows [Saf18a]:

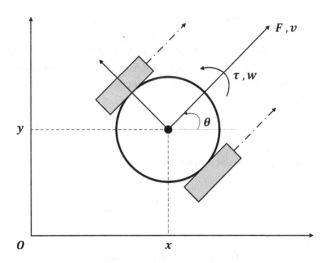

FIGURE 1.3 The schematic of a nonholonomic autonomous ground mobile robot [Saf18a].

$$\dot{x} = v\cos\psi,$$

$$\dot{y} = v\sin\psi,$$

$$\dot{\psi} = \omega, \tag{1.13}$$

$$\dot{v} = \frac{1}{m}(f - k_f v + f_d),$$

$$\dot{\omega} = \frac{1}{J}(\tau - k_\tau \omega + \tau_d).$$

where x and y are the 2D positions of the rover, v is the forward velocity of the rover, ψ is its heading (or yaw) angle, ω is the angular velocity of the rover, and f and τ are the input force and torque of the rover, respectively, while f_d and τ_d are the corresponding external translational and rotational external disturbances, respectively. In addition, m is the mass of the rover, J is its moment of inertia, and k_f and k_τ are the constant coefficients for resistive drag force and torque acting on the rover, respectively. As it can be observed, the dynamic system in (1.13) has three outputs (i.e., x, y and θ) and two inputs (i.e., f and τ). Moreover, by assuming that the rover has two drive wheels, the expressions of the inputs become

$$f = k_v(\omega_l + \omega_r)/r,$$

$$\tau = d\,k_v(\omega_r - \omega_l)/r, \tag{1.14}$$

where ω_l and ω_r are the angular speed of the left and right wheels of the rover, respectively, d is the constant distance between the two wheels, r is the radius of each wheel, and k_v is a coefficient for motors attached to the wheels that relates each motor angular speed to the generated traction force at the wheel. The presented model for a wheeled ground rover robot is utilized as for implementation of the model-free adaptive controller in Chapter 5 and inserted in the Appendix.

To control the position and heading of the ground rover robot, two decoupled subsystems are considered for the heading of rover and its translational forward-direction displacement. In this regard, we define a tracking error for the heading angle of the rover as follows [Saf18a]:

$$e_{\text{hed}} = \psi_{\text{des}} - \psi, \tag{1.15}$$

where the heading angle reference input (or set point) is

$$\psi_{\text{des}} = \tan^{-1}\left(\frac{y_{\text{des}} - y}{x_{\text{des}} - x}\right), \quad -\pi \leq \psi_{\text{des}} \leq \pi. \tag{1.16}$$

In addition, a displacement tracking error is defined as

$$e_{\text{dis}} = \sqrt{(x_{\text{des}} - x)^2 + (y_{\text{des}} - y)^2}. \tag{1.17}$$

The above-mentioned controllers are acting in a completely decoupled way. It means that first the heading controller regulates the heading tracking error, and then the displacement controller would operate and move the rover toward the destination.

1.3.3 UNDERACTUATED AUTONOMOUS QUADROTOR

The underactuated autonomous quadrotor (i.e., a flying mobile robot) has four propellers. This dynamic system is a double-integrator MIMO system, and it is defined as follows [Saf18a]:

$$\dot{\vec{p}} = \vec{v},$$

$$\dot{\Phi} = \mathbf{R}_t^{-1}\vec{\omega},$$

$$\dot{\vec{v}} = \frac{1}{m}(\mathbf{R}_r \vec{f} - k_f \vec{v} - m\vec{f}_g + \vec{f}_d), \tag{1.18}$$

$$\dot{\vec{\omega}} = \mathbf{J}^{-1}(\vec{\tau} - k_\tau \vec{\omega} - \vec{\omega} \times \mathbf{J}\vec{\omega} + \vec{\tau}_d)$$

where $\vec{p} = \begin{bmatrix} x & y & z \end{bmatrix}^T$ and $\vec{v} = \begin{bmatrix} v_x & v_y & v_z \end{bmatrix}^T$ are the position and velocity vectors of the quadrotor in an inertial frame, $\vec{\Phi} = \begin{bmatrix} \phi & \theta & \psi \end{bmatrix}^T$ is the attitude vector of the quadrotor in the form of Euler angles (i.e., roll, pitch and yaw), and $\vec{\omega} = \begin{bmatrix} \omega_x & \omega_y & \omega_z \end{bmatrix}^T$ is the vector of angular velocities of the quadrotor defined in its body frame. In this regard, this dynamic system has 12 states, while only the elements of $\vec{p}$ and $\vec{\Phi}$ should be considered as outputs of the system (i.e., six output variables). Moreover, $\vec{f} = \begin{bmatrix} 0 & 0 & f_T \end{bmatrix}^T$

and $\vec{\tau} = \begin{bmatrix} \tau_x & \tau_y & \tau_z \end{bmatrix}^T$ are the vectors of input thrust force and torques, respectively, in the body frame of quadrotor controlling it in 3D space. Evidently, we have four input variables, and hence, the dynamic system is an underactuated one, since the number of control inputs is lower than the number of system outputs. Here, m is the mass, and $\mathbf{J} \in \Re^{3 \times 3}$ is the inertia matrix of the quadrotor. The parameters k_f and k_τ are the coefficients for the drag forces and torques acting against the motion of quadrotor. Also, $\vec{f}_g = \begin{bmatrix} 0 & 0 & g_e \end{bmatrix}^T$, for $g_e = -9.81$ m/s^2. In addition, $\vec{f}_d$ and $\vec{\tau}_d$ are the vectors of external disturbance forces and torques acting on the system (Figure 1.4). In (1.18), two transformation matrices defined as follows are employed [Saf18a]:

$$\mathbf{R}_t = \begin{bmatrix} 1 & 0 & -\sin\theta \\ 0 & \cos\phi & \cos\theta \cdot \sin\phi \\ 0 & -\sin\phi & \cos\theta \cdot \cos\phi \end{bmatrix}, \tag{1.19}$$

$$\mathbf{R}_r = \begin{bmatrix} \cos\psi \cdot \cos\theta & \cos\psi \cdot \sin\theta \cdot \sin\phi - \sin\psi \cdot \cos\phi & \cos\psi \cdot \sin\theta \cdot \cos\phi + \sin\psi \cdot \sin\phi \\ \sin\psi \cdot \cos\theta & \sin\psi \cdot \sin\theta \cdot \sin\phi + \cos\psi \cdot \cos\phi & \sin\psi \cdot \sin\theta \cdot \cos\phi - \cos\psi \cdot \sin\phi \\ -\sin\theta & \cos\theta \cdot \sin\phi & \cos\theta \cdot \cos\phi \end{bmatrix} \tag{1.20}$$

In the literature, $\mathbf{R}_r$ is mostly called the rotation matrix of quadrotor's body frame into an inertial frame. Besides, recalling the four propellers accommodated on the quadrotor, the following matrix relationship is defined [Saf18a]:

$$\begin{bmatrix} f_T \\ \tau_x \\ \tau_y \\ \tau_z \end{bmatrix} = \begin{bmatrix} k_l & k_l & k_l & k_l \\ -k_l l & 0 & k_l l & 0 \\ 0 & k_l l & 0 & k_l l \\ k_t & -k_t & k_t & -k_t \end{bmatrix} \begin{bmatrix} \omega_1^2 \\ \omega_2^2 \\ \omega_3^2 \\ \omega_4^2 \end{bmatrix}, \tag{1.21}$$

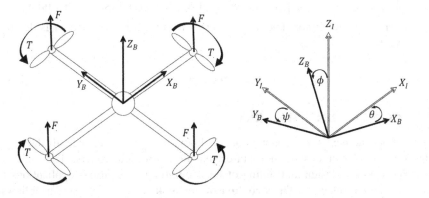

FIGURE 1.4 The schematic of an underactuated autonomous quadrotor [Saf18a].

where ω_i for $i \in \{1,2,3,4\}$ is the angular speed of ith propeller, l is the arm length of the quadrotor, and k_l and k_t are the propellers' coefficients for generating the lift force and torque, respectively. The dynamic model of quadrotor is used for implementing the cooperative adaptive MFC algorithms in Chapter 8.

For controlling the position of a quadrotor in 3D space, a cascade structure is defined in which there is an outer loop for controlling the translational motion of the quadrotor, and an inner loop for controlling its rotational motion. In this regard, the position tracking error is defined as the vector

$$\mathbf{e}_{\text{pos}} = \vec{p}_{\text{des}} - \vec{p}, \tag{1.22}$$

where $\vec{p}_{\text{des}}$ is the desired position of the quadrotor in 3D space. A position controller should be designed to generate the desired velocity of the quadrotor $\vec{v}_{\text{des}}$, by operating on the tracking error

$$e_{\text{vel}} = \vec{v}_{\text{des}} - \vec{v}. \tag{1.23}$$

The output of the velocity controller would be the desired translational forces on the quadrotor $\vec{f}_{\text{des}}$, as follows:

$$\vec{f}_{\text{des}} = \begin{bmatrix} f_x & f_y & f_z \end{bmatrix}^T = \mathbf{R}_r \vec{f}. \tag{1.24}$$

Then, by recalling the definition of $\mathbf{R}_r$ in (1.20), this leads to the following [Saf16]:

$$f_T = \sqrt{f_x^2 + f_y^2 + f_z^2},$$

$$\phi_{\text{des}} = \sin^{-1}\left(\frac{\sin\psi_{\text{des}} f_x - \cos\psi_{\text{des}} f_y}{f_T} \right), \tag{1.25}$$

$$\theta_{\text{des}} = \tan^{-1}\left(\frac{\cos\psi_{\text{des}} f_x + \sin\psi_{\text{des}} f_y}{f_z} \right),$$

where ψ_{des} is the desired yaw angle of the quadrotor ($\psi_{\text{des}} = 0$ in most applications). Here, $\vec{\Phi}_{des} = \begin{bmatrix} \phi_{\text{des}} & \theta_{\text{des}} & \psi_{\text{des}} \end{bmatrix}^T$ is the vector of desired attitude sent to the inner-loop controller of the quadrotor. The tracking error for the attitude controller is defined as the vector

$$\mathbf{e}_{\text{att}} = \vec{\Phi}_{\text{des}} - \vec{\Phi}. \tag{1.26}$$

1.4 CONCLUDING REMARKS

Throughout the next seven chapters of this book, it is tried to include all the main algorithms and methods of the data-driven model-free controllers. The authors' experience in the design and tuning of these controllers in various applications is essential in their selection. **The structure of the book** is briefly discussed as follows. First, in Chapter 2, the IFT algorithm is presented which updates the PI/PID gains by minimizing a cost function (or an objective function), which includes the output tracking error. The cost function may also include the control efforts. The cost function is minimized by computing its gradient. Then, in Chapter 3, the intelligent Pi/PID (i.e., iPI/iPID) controllers are proposed, which are not updating the gains online; instead, they are updating the unknown parameters of the generic ultra-local model of the corresponding dynamic system. After that, Chapter 4 treats the combination of MFC and sliding mode control. Two control system structures are given.

In Chapter 5, the model-free adaptive controllers for discrete-time and continuous-time dynamic systems are presented. These controllers do not use any PI/PID schemes at all. Instead, a different and fresh control structure is defined, with new unknown parameters that are estimated online, leading to defining the control signal. In that chapter, an adaptive version of a differential Riccati equation (DRE) is utilized for online updating of the main controller gains of the proposed MFAC algorithm for continuous-time dynamic system, as well. This helps to reduce the number of tuning parameters of the controller.

Further in Chapter 6, the VRFT-based tuning of MFC and MFAC is carried out. This is advantageous, where VRFT as a direct and one-shot data-driven model-free tuning approach is applied to tune automatically the parameters of data-driven model-free controllers represented here by MFC and MFAC.

In Chapter 7, the combination of MFC and MFAC and fuzzy control is carried out. The first section is a short overview of fuzzy logic and control, which can be used by the readers who are not familiarized with fuzzy logic and control in terms of offering a stand-alone material that contains both theory and examples applied to both model-free and model-based controls.

Finally, Chapter 8 is the extension of MFAC in Chapter 5 to multiagent dynamic systems, where more than one dynamic system is cooperating in order to achieve a common tracking goal. This chapter can also be used as an independent part, which is also useful as a stand-alone material to those readers who are interested in control problems specific to multiagent systems treated in the model-free context. In that chapter, stability proofs for the cooperative control and observers are provided based on the Lyapunov and LaSalle-Yoshizawa theorems. Moreover, similar to the MFAC in Chapter 5, an adaptive DRE is utilized for online tuning of the main controller gains, while that DRE is accommodating the information from the communication graph of the network among the multiagent dynamic system, as well.

For all of the chapters, the applications of the data-driven model-free controllers are provided for a crane system, via both simulations and experiments. In addition, the application of the adaptive model-free controller in autonomous flying and ground mobile robots is presented in Chapter 8 and the Appendix.

In the context of the information specified above, the style of presentation of all chapters might not be homogeneous. This is normal accounting as **not the same performance specifications have been used**. That is also the reason why a separate concluding chapter is not inserted in the structure of this book. A fair comparison of all data-driven model-free controllers means to use controllers with a similar complexity and tune them in terms of the same performance specification. A convenient way to do that is to define the same optimization problems (with the same objective functions in both SISO and MIMO control system structures) and tune the free parameters of the controllers as solutions to these optimization problems. Metaheuristic optimization algorithms can be used in this regard; however, this requires the model-based evaluation of the objective functions, which is against model-free tuning, and a rather big number of iterations, which is also against the small number of iterations specific to data-driven control. The authors are invited to apply the data-driven model-free controllers presented in this book, and the accompanying Matlab & Simulink programs are guiding them in order to carry out the relatively easy and transparent implementations, which will next help them to design, tune and apply one or more controllers to the process that is actually controlled.

The stability analysis of data-driven MFC algorithms is a key issue. Here in this book, detailed stability proofs for some of the algorithms are presented in Chapters 3, 5 and 8. These stability proofs are based on the Lyapunov stability theorem as well as the LaSalle-Yoshizawa theorem. The latter one leads to the Ultimately Uniformly Bounded (UUB) convergence for the tracking errors and the parameter estimation errors of the proposed MFC algorithms. The UUB property of the convergence confirms that all the above-mentioned errors are bounded within a small region around origin, even in the case of unknown bounded internal dynamics and unknown bounded external disturbances. This boundedness property is further shown in the simulation and practical results in different case studies provided in the book.

As specified in Section 1.1 in the context of robustness, the stability analysis requires the mathematical model of the controlled process. Moreover, as outlined in [vanH10], no process model is available in data-driven MFC algorithms to verify the robustness margins, in general. But, in the adaptive laws proposed for online parameter estimations in model-free adaptive controllers in Chapters 5 and 8, the *leakage* method is suggested to deliver the robustness against the bounded internal dynamics and external disturbances. In the provided simulation results, it is shown that the robust parameter estimation along with the UUB stability property can bring an acceptable level of convergence and robustness.

Another shortcoming related to the stability of the data-driven MFC algorithms is that the effectiveness of these algorithms is strongly affected by measurement noise. The authors highlighted the results on stability for iterative methods based on the Vinnicombe metric analyzed in [deB99] and reported in [Rad11b] in the context of generalized stability margin. This use case is made of nonparametric process models, where one concern is the quality of estimates. For further discussions on the stability analysis, the readers are referred to the corresponding literature for model-free controllers as [Fli09, Fli13, Rom18, Pol19, Joi20, Ols20, Vil20] and for model-free adaptive controllers as [Hou97, Hou99, Hou11a, Hou11b, Saf18a, Saf18b, Yu20, Saf21].

Important results on the stability of control systems tuned by noniterative data-driven model-free techniques are presented in [vanH10], where the controller tuning problem is generalized with a constraint that guarantees the control system stability. This constraint is derived in [vanH10] from stability conditions based on the small gain theorem, and it is conservative. However, as an alternative, a nonconservative a posteriori stability test is developed in [Vanh09] on the basis of similar stability conditions.

The stability issue is discussed briefly as follows in terms of classical and fresh papers. The stability analysis of control systems with MFAC is treated systematically in [Hou19]. Stability and robustness in data-driven control are investigated in [DeP20] in relation with optimality. The data-driven dissipativity subject is tackled in [Mau17]. The stability is guaranteed in [Sel21] for VRFT-based control system structures in the context of nonconvex optimization.

All data-driven model-free controllers presented in this book are **tested and validated in terms of simulation and experimental results in both SISO and MIMO control system structures**. The control system performance is good in the context of the selected challenging nonlinear system treated in this book, thus indicating the potential of these controllers to deal with industrial and nonindustrial processes. Since the numbers of simulation and experimental results and associated figures are rather large in some chapters, and that would negatively affect the understanding of the book, the authors decided to drop out a part of the simulation and experimental results, especially the experimental ones. The full transparency of the controllers implemented as Matlab & Simulink programs and schemes is ensured as these programs and schemes are offered as accompanying material to this book. However, if interested, the readers are invited to contact the authors in order to share the missing experimental results.

The material included in this book is a result of the studies and research carried out by the authors at their institutions. The authors **acknowledge** the support and excellent cooperation of their colleagues who provided useful opinions and ideas on the book topics: Prof. Stefan Preitl, Prof. Emil M. Petriu, Dr. Mircea-Bogdan Radac, Dr. Claudia-Adina Bojan-Dragos, Dr. Alexandra-Iulia Szedlak-Stinean, Ms. Elena-Lorena Hedrea, Dr. Radu-Codrut David, Dr. Adriana Albu and Dr. Florin Dragan, for the results presented in Chapters 2–7, and Dr. Muhammad Nasiruddin Mahyuddin, the PhD supervisor of the third author as most of the basis of the continuous-time materials in Chapters 5 and 8 are developed during his PhD period and professional guidance in this regard. Many ideas and research results contained in the book were presented in journal, conference papers and book chapters coauthored with them, and we are thankful for offering us the opportunity to work together as members of well-trained research and teaching groups.

Several academics contributed with their useful advice and information. Special thanks are due to Prof. Michel Fliess, who created MFC, Prof. Zhongsheng Hou, who created MFAC, and Prof. Sergio Matteo Savaresi and Dr. Simone Formentin, for their continuous and valuable cooperation. We want to thank all of our colleagues for their efforts and contributions in the field of data-driven model-free controllers that inspired us to achieve the current book.

The Executive Agency for Higher Education, Research, Development and Innovation Funding (UEFISCDI) of Romania is constantly supporting the first and second authors' research. This is duly acknowledged in the context of this book. This work was supported by grants of the Romanian Ministry of Education and Research, CNCS – UEFISCDI, project numbers PN-III-P4-ID-PCE-2020–0269 and PN-III-P1-1.1-PD-2019-0637.

At last but not least, the contribution of Mr. Marc Gutierrez and Mr. Nick Mould from the Taylor and Francis Group, in the preparation and production of this book, is highly acknowledged and appreciated.

REFERENCES

[Aan05] W. Aangenent, D. Kostić, B. De Jager and M. Steinbuch, "Data-based optimal control," in *Proceedings of 2005 American Control Conference*, Portland, OR, 2005, pp. 1460–1465.

[AlM00] A. A. Al-Mousa, *Control of Rotary Cranes Using Fuzzy Logic and Time-Delayed Position Feedback Control*, M.Sc. Thesis, Virginia Polytechnic Institute and State University, Blacksburg, VA, 2000.

[Bag18] M. Bagheri, M. Krstić and P. Naseradinmousavi, "Multivariable extremum seeking for joint-space trajectory optimization of a high-degrees-of-freedom robot," *Journal of Dynamic Systems, Measurement, and Control*, vol. 140, no. 11, p. 111017, Nov 2018.

[Baz12] A. S. Bazanella, L. Campestrini and D. Eckhard, *Data-Driven Controller Design: The H2 Approach*, Springer, Dordrecht, 2012.

[Ben93] J. K. Bennighof, S.-H. Chang and M. Subramaniam, "Minimum time pulse response based control of flexible structure," *Journal of Guidance, Control, and Dynamics*, vol. 16, no. 5, pp. 874–881, Sep.–Oct. 1993.

[Ben18a] M. Benosman, "Model-based vs data-driven adaptive control: an overview," *International Journal of Adaptive Control and Signal Processing*, vol. 32, no. 5, pp. 753–776, May 2018.

[Ben18b] A. Benrabah, D. G. Xu and Z. Q. Gao, "Active disturbance rejection control of LCL-filtered grid-connected inverter using Padé approximation," *IEEE Transactions on Industry Applications*, vol. 54, no. 6, pp. 6179–6189, Nov.–Dec. 2018.

[Bia15] F. D. Bianchi, C. Ocampo-Martinez, C. Kunusch and R. S. Sánchez-Peña, "Fault-tolerant unfalsified control for PEM fuel cell systems," *IEEE Transactions on Energy Conversion*, vol. 30, no. 1, pp. 307–315, Mar. 2015.

[Bol18] J. Bolder, S. Kleinendorst and T. Oomen, "Data-driven multivariable ILC: enhanced performance by eliminating L and Q filters," *International Journal of Adaptive Control and Signal Processing*, vol. 28, no. 12, pp. 3728–3751, Aug. 2018.

[Bon99] G. Bontempi, M. Birattari and H. Bersini, "Lazy learning for local modelling and control design," *International Journal of Control*, vol. 72, no. 7–8, pp. 643–658, Jun. 1999.

[Bon05] G. Bontempi and M. Birattari, "From linearization to lazy learning: a survey of divide-and-conquer techniques for nonlinear control," *International Journal of Computational Cognition*, vol. 3, no. 1, pp. 56–73, Mar. 2005.

[Bru19] S. L. Brunton and J. N. Kutz, *Data-Driven Science and Engineering: Machine Learning, Dynamical Systems, and Control*, Cambridge University Press, Cambridge, 2019.

[Bus18] L. Busoniu, T. de Bruin, D. Tolić, J. Kober and I. Palunko, "Reinforcement learning for control: performance, stability, and deep approximators," *Annual Reviews in Control*, vol. 46, pp. 8–28, Dec. 2018.

[But08] M. Butcher, A. Karimi and R. Longchamp, "Iterative learning control based on stochastic approximation," in *Proceedings of 17th IFAC World Congress*, Seoul, Korea, 2008, pp. 1478–1483.

[Cam02] M. C. Campi, A. Lecchini and S. M. Savaresi, "Virtual reference feedback tuning: a direct method for the design of feedback controllers," *Automatica*, vol. 38, no. 8, pp. 1337–1346, Aug. 2002.

[Cam03] M. C. Campi, A. Lecchini and S. M. Savaresi, "An application of the Virtual Reference Feedback Tuning (VRFT) method to a benchmark active system," *European Journal of Control*, vol. 9, no. 1, pp. 66–76, Jan. 2003.

[Car19] A. Carè, F. Torricelli, M. C. Campi and S. M. Savaresi, "A toolbox for Virtual Reference Feedback Tuning (VRFT)," in *Proceedings of 2019 European Control Conference*, Napoli, Italy, 2019, pp. 4252–4257.

[Che04] H.-F. Chen and H.-T. Fang, "Output tracking for nonlinear stochastic systems by iterative learning control," *IEEE Transactions on Automatic Control*, vol. 49, no. 4, pp. 583–588, Apr. 2004.

[Che18] Y.-J. Chen Y, L. K. Norford, H. W. Samuelson and A. Malkawi, "Optimal control of HVAC and window systems for natural ventilation through reinforcement learning," *Energy and Buildings*, vol. 169, pp. 195–205, Jun. 2018.

[Chi15] R.-H. Chi, Z.-S. Hou, B. Huang and S.-T. Jin, "A unified data-driven design framework of optimality-based generalized iterative learning control," *Computers & Chemical Engineering*, vol. 77, pp. 10–23, Jun. 2015.

[Chi18] R.-H. Chi, X.-H. Liu, R.-K. Zhang, Z.-S. Hou and B. Huang, "Constrained data-driven optimal iterative learning control," *Journal of Process Control*, vol. 55, pp. 10–29, Jul. 2017.

[Dai21] T. Dai and M. Sznaier, "A semi-algebraic optimization approach to data-driven control of continuous-time nonlinear systems," *IEEE Control Systems Letters*, vol. 5, no. 2, pp. 487–492, Apr. 2021.

[daS19] G. R. G. da Silva, A. S. Bazanella, C. Lorenzini and L. Campestrini, "Data-driven LQR control design," *IEEE Control Systems Letters*, vol. 3, no. 1, pp. 180–185, Jan. 2019.

[daS18] J. da Silva Moreira, G. Acioli Júnior and G. Rezende Barros, "Time and frequency domain data-driven PID iterative tuning," *IFAC-PapersOnLine*, vol. 51, no. 15, pp. 1056–1061, Jul. 2018.

[deB99] F. de Bruyne and L.C. Kammer, "Iterative Feedback Tuning with guaranteed stability," in *Proceedings of 1999 American Control Conference*, San Diego, CA, USA, vol. 5, 1999, pp. 3317–3321.

[DeP20] C. De Persis and P. Tesi, "Formulas for data-driven control: stabilization, optimality, and robustness," *IEEE Transactions on Automatic Control*, vol. 65, no. 3, pp. 909–924, Mar. 2020.

[Din14] S. X. Ding, *Data-Driven Design of Fault Diagnosis and Fault-Tolerant Control Systems*, Springer, London, 2014.

[Fav99] W. Favoreel, B. De Moor, P. van Overschee and M. Gevers, "Model-free subspace-based LQG-design," in *Proceedings of 1999 American Control Conference*, San Diego, CA 1999, vol. 5, pp. 3372–3376.

[Fli09] M. Fliess and C. Join, "Model-free control and intelligent PID Controllers: towards a possible trivialization of nonlinear control?" *IFAC Proceedings Volumes*, vol. 42, no. 10, pp. 1531–1550, Jul. 2009.

[Fli13] M. Fliess and C. Join, "Model-free control," *International Journal of Control*, vol. 86, no. 12, pp. 2228–2252, Dec. 2013.

[Fli20] M. Fliess and C. Join, "Machine learning and control engineering: the model-free case," in *Proceedings of Future Technologies Conference 2020*, Vancouver, BC, Canada, 2020, pp. 1–20.

[For14] S. Formentin, K. van Heusden and A. Karimi, "A comparison of model-based and data-driven controller tuning," *International Journal of Adaptive Control and Signal Processing*, vol. 28, no. 10, pp. 882–897, Oct. 2014.

[For19] S. Formentin, M. C. Campi, A. Caré and S. M. Savaresi, "Deterministic continuous-time Virtual Reference Feedback Tuning (VRFT) with application to PID design," *Systems & Control Letters*, vol. 127, pp. 25–34, May 2019.

[Gal17] O. Galluppi, S. Formentin, C. Novara and S. M. Savaresi, "Nonlinear stability control of autonomous vehicles: a MIMO D^2-IBC solution," *IFAC-PapersOnLine*, vol. 50, no. 1, pp. 3691–3696, Jul. 2017.

[Gal19] O. Galluppi, S. Formentin, C. Novara and S. M. Savaresi, "Multivariable D^2-IBC and application to vehicle stability control," *ASME Journal of Dynamic Systems, Measurement and Control*, vol. 141, no. 10, 101012, Oct. 2019.

[Gao06] Z. Gao, "Active disturbance rejection control: a paradigm shift in feedback control system design," in *Proceedings of 2006 American Control Conference*, Minneapolis, MN, 2006, pp. 2399–2405.

[Gev96] M. Gevers, "Identification for control," *Annual Reviews in Control*, vol. 20, no. 1, pp. 95–106, Jan. 1996.

[Gon00] R. Gonzalez-Lima, "Pulse response based control for positive definite systems," *Journal of the Brazilian Society of Mechanical Sciences*, vol. 22, no. 2, pp. 169–177, Apr. 2000.

[Gua17] M. Guay and D. Dochain, "A proportional-integral extremum-seeking controller design technique," *Automatica*, vol. 77, pp. 61–67, Mar. 2017.

[Gua18] M. Guay, I. Vandermeulen, S. Dougherty and P. J. McLellan, "Distributed extremum-seeking control over networks of dynamically coupled unstable dynamic agents," *Automatica*, vol. 93, pp. 498–509, Jul. 2018.

[Gun01] S. Gunnarsson and M. Norrlöf, "On the design of ILC algorithms using optimization," *Automatica*, vol. 37, no. 12, pp. 2011–2016, Dec. 2001.

[Hal04] K. Halmevaara and H. Hyötyniemi, *"Process Performance Optimization using Iterative Regression Tuning,"* Research Report no. 139, Control Engineering Laboratory, Helsinki University of Technology, Helsinki, 2004.

[Hal06] K. Halmevaara and H. Hyötyniemi, "Data-based parameter optimization of dynamic simulation models," in *Proceedings of 47th Conference on Simulation and Modelling*, Helsinki, 2006, pp. 68–73.

[Hed21] E.-L. Hedrea, R.-E. Precup, R.-C. Roman and E. M. Petriu, "Tensor product-based model transformation approach to tower crane systems modeling," *Asian Journal of Control*, DOI: 10.1002/asjc.2494, Mar. 2021.

[Hja94] H. Hjalmarsson, S. Gunnarsson and M. Gevers, "A convergent iterative restricted complexity control design scheme," in *Proceedings of 33rd IEEE Conference on Decision and Control*, Orlando, FL, USA, vol. 2, 1994, pp. 1735–1740.

[Hja98] H. Hjalmarsson, M. Gevers, S. Gunnarsson and O. Lequin, "Iterative feedback tuning: theory and applications," *IEEE Control Systems Magazine*, vol. 18, no. 4, pp. 26–41, Aug. 1998.

[Hja02] H. Hjalmarsson, "Iterative feedback tuning - an overview," *International Journal of Adaptive Control and Signal Processing*, vol. 16, no. 5, pp. 373–395, Jun. 2002.

[Hou97] Z.-S. Hou and W.-H. Huang, "The model-free learning adaptive control of a class of SISO nonlinear systems," in *Proceedings of 1997 American Control Conference*, Albuquerque, NM, 1997, pp. 343–344.

[Hou99] Z.-S. Hou, *Nonparametric Models and Its Adaptive Control Theory*, Science Press, Beijing, 1999.

[Hou11a] Z.-S. Hou and S. Jin, "A novel data-driven control approach for a class of discrete-time nonlinear systems," *IEEE Transactions on Control Systems Technology*, vol. 19, pp. no. 6, 1549–1558, Nov. 2011.

[Hou11b] Z.-S. Hou and S. Jin, "Data-driven model-free adaptive control for a class of MIMO nonlinear discrete-time systems," *IEEE Transactions on Neural Networks*, vol. 22, no. 12, pp. 2173–2188, Dec. 2011.

[Hou13a] Z.-S. Hou and S.-T. Jin, *Model Free Adaptive Control: Theory and Applications*, CRC Press, Boca Raton, FL, 2013.

[Hou13b] Z.-S. Hou and Z. Wang, "From model-based control to data-driven control: survey, classification and perspective," *Information Sciences*, vol. 235, pp. 3–35, Jun. 2013.

[Hou13c] Z.-S. Hou and Y.-M. Zhou, "Model based control and MFAC, which is better in simulation?" *IFAC Proceedings Volumes*, vol. 46, no. 13, pp. 82–87, Jul. 2013.

[Hou17a] Z.-S. Hou, R.-H. Chi and H.-J. Gao, "An overview of dynamic-linearization-based data-driven control and applications," *IEEE Transactions on Industrial Electronics*, vol. 64, no. 5, pp. 4076–4090, May 2017.

[Hou17b] Z.-S. Hou, S.-D. Liu and T.-T. Tian, "Lazy-learning-based data-driven model-free adaptive predictive control for a class of discrete-time nonlinear systems," *IEEE Transactions on Neural Networks and Learning Systems*, vol. 28, no. 8, pp. 1914–1928, Aug. 2017.

[Hou19] Z.-S. Hou and S.-S. Xiong, "On model-free adaptive control and its stability analysis," *IEEE Transactions on Automatic Control*, vol. 64, no. 11, pp. 4555–4569, Nov. 2019.

[Hua20] J.-W. Huang and J.-W. Gao, "How could data integrate with control? A review on data-based control strategy," *International Journal of Dynamics and Control*, vol. 8, no. 4, pp. 1189–1199, Dec. 2020.

[Int12] Inteco, *Tower Crane, User's Manual*, Inteco Ltd, Krakow, 2012.

[Jan13] P. Janseens, G. Pipeleers and J. Swevers, "A data-driven constrained norm-optimal iterative learning control framework for LTI systems," *IEEE Transactions on Control Systems Technology*, vol. 21, no. 2, pp. 546–551, Mar. 2013.

[Jia16] P. Jiang, Y.-Q. Cheng, X.-N. Wang and Z. Feng, "Unfalsified visual serving for simultaneous object recognition and pose tracking," *IEEE Transactions on Cybernetics*, vol. 46, no. 12, pp. 3032–3046, Dec. 2016.

[Joi20] C. Join, M. Fliess and F. Chaxel, "Model-free control as a service in the industrial internet of things: Packet loss and latency issues via preliminary experiments," in *Proceedings of 28th Mediterranean Conference on Control and Automation*, Saint-Raphaël, France, 2020, pp. 1–6.

[Jun21] H. Jung, K. Jeon, J.-G. Kang and S. Oh, "Iterative feedback tuning of cascade control of two-inertia system," *IEEE Control Systems Letters*, vol. 5, no. 3, pp. 785–790, Jul. 2021.

[Kad03] R. Kadali, B. Huang and A. Rossiter, "A data driven subspace approach to predictive controller design," *Control Engineering Practice*, vol. 11, no. 3, pp. 261–278, Mar. 2003.

[Kam99] L. C. Kammer, F. De Bruyne and R. R. Bitmead, "Iterative feedback tuning via minimization of the absolute error," in *Proceedings of 38th IEEE Conference on Decision and Control*, Phoenix, AZ, USA, 1999, vol. 5, pp. 4619–4624.

[Kam00] L. C. Kammer, R. R. Bitmead and P. L. Bartlett, "Direct iterative tuning via spectral analysis," *Automatica*, vol. 36, no. 9, pp. 1301–1307, Sep. 2000.

[Kam18] R. Kamalapurkar, P. Walters, J. Rosenfeld and W. Dixon, *Reinforcement Learning for Optimal Feedback Control: A Lyapunov-Based Approach*, Springer, Cham, 2018.

[Kar04] A. Karimi, L. Mišković and D. Bonvin, "Iterative correlation-based controller tuning," *International Journal of Adaptive Control and Signal Processing*, vol. 18, no. 8, pp. 645–664, Oct. 2004.

[Kar07] A. Karimi, K. van Heusden and D. Bonvin, "Non-iterative data-driven controller tuning using the correlation approach," in *Proceedings of 2007 European Control Conference*, Kos, Greece, 2007, pp. 5189–5195.

[Ker17] P. Kergus, C. Poussot-Vassal, F. Demourant and S. Formentin, "Frequency-domain data-driven control design in the Loewner framework," *IFAC-PapersOnLine*, vol. 50, no. 1, pp. 2095–2100, Jul. 2017.

[Ker19] P. Kergus, M. Olivi, C. Poussot-Vassal and F. Demourant, "From reference model selection to controller validation: application to Loewner data-driven control," *IEEE Control Systems Letters*, vol. 3, no. 4, pp. 1008–1013, Oct. 2019.

[Ker20] P. Kergus, F. Demourant and C. Poussot-Vassal, "Identification of parametric models in the frequency-domain through the subspace framework under LMI constraints," *International Journal of Control*, vol. 93, no. 8, pp. 1879–1890, Aug. 2020.

[Kha14] S. Khadraoui, H. Nounou, M. Nounou, A. Datta and S. P. Bhattacharyya, "A model-free design of reduced-order controllers and application to a DC servomotor," *Automatica*, vol. 50, no. 8, pp. 2142–2149, Aug. 2014.

[Krs00a] M. Krstić, "Performance improvement and limitations in extremum seeking control," *Systems & Control Letters*, vol. 39, no. 5, pp. 313–326, Apr. 2000.

[Krs00b] M. Krstić and H.-H. Wang, "Stability of extremum seeking feedback for general nonlinear dynamic systems," *Automatica*, vol. 36, no. 4, pp. 595–601, Apr. 2000.

[Lec01] A. Lecchini, M. C. Campi and S. M Savaresi, "Sensitivity shaping via virtual reference feedback tuning," in *Proceedings of 40th Conference on Decision and Control*, Orlando, FL, USA, 2001, pp. 750–755.

[Lju99] L. Ljung, *System Identification: Theory for the User*, 2nd Edition, Prentice Hall, Englewood Cliffs, NJ, 1999.

[Luc20] A. Lucchini, S. Formentin, M. Corno, D. Piga and S. M. Savaresi, "Torque vectoring for high-performance electric vehicles: a data-driven MPC approach," *IEEE Control Systems Letters*, vol. 4, no. 3, pp. 725–730, Jul. 2020.

[Mad19] R. Madonski, S. Shao, H. Zhang, Z. Gao, J. Yang and S. Li, "General error-based active disturbance rejection control for swift industrial implementations," *Control Engineering Practice*, vol. 84, pp. 218–229, Mar. 2019.

[Mar17] A. Marjanović, M. Krstić, Ž Đurović and B. Kovačević, "Control of thermal power plant combustion distribution using extremum seeking," *IEEE Transactions on Control Systems Technology*, vol. 25, no. 5, pp. 1670–1682, Sep. 2017.

[Mau17] T. M. Maupong, J. C. Mayo-Maldonado and P. Rapisarda, "On Lyapunov functions and data-driven dissipativity," *IFAC-PapersOnLine*, vol. 50, no. 1, pp. 7783–7788, Jul. 2017.

[McD12] A. J. McDaid, K. C. Aw, E. Haemmerle and S. Q. Xie, "Control of IPMC actuators for microfluidics with adaptive "online" iterative feedback tuning," *IEEE/ASME Transactions on Mechatronics*, vol. 17, no. 4, pp. 789–797, Aug. 2012.

[Mis11] S. Mishra, U. Topcu and M. Tomizuka, "Optimization-based constrained iterative learning control," *IEEE Transactions on Control Systems Technology*, vol. 19, no. 6, pp. 1613–1621, Nov. 2011.

[Miš07] L. Mišković, A. Karimi, D. Bonvin and M. Gevers, "Correlation-based tuning of decoupling multivariable controllers," *Automatica*, vol. 43, no. 9, pp. 1481–1494, Sep. 2007.

[Ngo18] P. D. Ngo, S. Wei, A. Holubová, J. Muzik and F. Godtliebsen, "Reinforcement-learning optimal control for type-1 diabetes," in *Proceedings of 2018 IEEE EMBS International Conference on Biomedical & Health Informatics*, Las Vegas, NV, 2018, pp. 333–336.

[Nor02] M. Norrlöf and S. Gunnarsson, "Time and frequency domain convergence properties in iterative learning control," *International Journal of Control*, vol. 75, no. 14, pp. 1114–1126, Aug. 2002.

[Nov15] C. Novara, S. Formentin, S. M. Savaresi and M. Milanese, "A data-driven approach to nonlinear braking control," in *Proceedings of 54th IEEE Conference on Decision and Control*, Osaka, Japan, 2015, pp. 1–6.

[Ols20] J. M. Olszanecki Barth, J.-P. Condomines, M. Bronz, G. Hattenberger, J.-M. Moschetta, C. Join, and M. Fliess, "Towards a unified model-free control architecture for tailsitter micro air vehicles: flight simulation analysis and experimental flights," in *Proceedings of AIAA Scitech 2020 Forum*, Orlando, FL, 2020, pp. 1–20.

[Owe05] D. H. Owens and J. Hätönen, "Iterative learning control - an optimization paradigm," *Annual Reviews in Control*, vol. 29, no. 1, pp. 57–70, Jun. 2005.

[Owe15] D. H. Owens, *Iterative Learning Control: An Optimization Paradigm*, Springer, London, 2015.

[Pol19] P. Polack, S. Delprat and B. d'Andréa-Novel, "Brake and velocity model-free control on an actual vehicle," *Control Engineering Practice*, vol. 92, 104072, Nov. 2019.

[Pre07] S. Preitl, R.-E. Precup, Z. Preitl, S. Vaivoda, S. Kilyeni and J. K. Tar, "Iterative feedback and learning control. Servo systems applications," *IFAC Proceedings Volumes*, vol. 40, no. 8, pp. 16–27, May 2007.

[Pre10] R.-E. Precup, C. Borchescu, M.-B. Radac, S. Preitl, C.-A. Dragos, E. M. Petriu and J. K. Tar, "Implementation and signal processing aspects of iterative regression tuning," in *Proceedings of 2010 IEEE International Symposium on Industrial Electronics*, Bari, Italy, 2010, pp. 1657–1662.

[Pre15] R.-E. Precup, P. Angelov, B. S. J. Costa and M. Sayed-Mouchaweh, "An overview on fault diagnosis and nature-inspired optimal control of industrial process applications," *Computers in Industry*, vol. 74, pp. 75–94, Dec. 2015.

[Pre20a] R.-E. Precup, S. Preitl, E. M. Petriu, R.-C. Roman, C.-A. Bojan-Dragos, E.-L. Hedrea and A.-I. Szedlak-Stinean, "A center manifold theory-based approach to the stability analysis of state feedback Takagi-Sugeno-Kang fuzzy control systems," *Facta Universitatis, Series: Mechanical Engineering*, vol. 18, no. 2, pp. 189–204, Jul. 2020.

[Pre20b] R.-E. Precup, R.-C. Roman, T.-A. Teban, A. Albu, E. M. Petriu and C. Pozna, "Model-free control of finger dynamics in prosthetic hand myoelectric-based control systems," *Studies in Informatics and Control*, vol. 29, no. 4, pp. 399–410, Dec. 2020.

[Qi19] X.-W. Qi, Y.-D. Luo, G.-Y. Wu, K. Boriboonsomsin and M. Barth, "Deep reinforcement learning enabled self-learning control for energy efficient driving," *Transportation Research Part C: Emerging Technologies*, vol. 99, pp. 67–81, Feb. 2019.

[Rad11a] M.-B. Radac, *Iterative Techniques for Controller Tuning*, PhD thesis, Editura Politehnica, Timisoara, 2011.

[Rad11b] M.-B. Radac, R.-E. Precup, E. M. Petriu and S. Preitl, "Application of IFT and SPSA to servo system control," *IEEE Transactions on Neural Networks*, vol. 22, no. 12, pp. 2363–2375, Dec. 2011.

[Rad15] M.-B. Radac, R.-E. Precup and E. M. Petriu, "Constrained data-driven model-free ILC-based reference input tuning algorithm," *Acta Polytechnica Hungarica*, vol. 12, no. 1, pp. 137–160, Feb. 2015.

[Rad17] M.-B. Radac, R.-E. Precup and R.-C. Roman, "Model-free control performance improvement using virtual reference feedback tuning and reinforcement Q-learning," *International Journal of Systems Science*, vol. 48, no. 5, pp. 1071–1083, Apr. 2017.

[Rad18] M.-B. Radac and R.-E. Precup, "Data-driven model-free slip control of anti-lock braking systems using reinforcement Q-learning," *Neurocomputing*, vol., pp. 317–329, Jan. 2018.

[Rom18] R.-C. Roman, R.-E. Precup and R.-C. David, "Second order intelligent proportional-integral fuzzy control of twin rotor aerodynamic systems," *Procedia Computer Science*, vol. 139, pp. 372–380, Oct. 2018.

[Rom21] R.-C. Roman, R.-E. Precup and E. M. Petriu, "Hybrid data-driven fuzzy active disturbance rejection control for tower crane systems," *European Journal of Control*, vol. 58, pp. 373–387, Mar. 2021.

[Saf97] M. G. Safonov and T.-C. Tsao, "The unfalsified control concept and learning," *IEEE Transactions on Automatic Control*, vol. 42, no. 6, pp. 843–847, Jun. 1997.

[Saf16] A. Safaei and M. N. Mahyuddin, "Lyapunov-based nonlinear controller for quadrotor position and attitude tracking with GA optimization," in *Proceedings of IEEE Industrial Electronics and Applications Conference*, Kota Kinabalu, Malaysia, 2016, pp. 342–347.

[Saf18a] A. Safaei and M. N. Mahyuddin, "Optimal model-free control for a generic MIMO nonlinear system with application to autonomous mobile robots," *International Journal of Adaptive Control and Signal Processing*, vol. 32, no. 6, pp. 792–815, Jun. 2018.

[Saf18b] A. Safaei and M. N. Mahyuddin, "Adaptive model-free control based on an ultra-local model with model-free parameter estimations for a generic SISO system," *IEEE Access*, vol. 6, pp. 4266–4275, Jun. 2018.

[Saf21] A. Safaei, "Cooperative adaptive model-free control with model-free estimation and online gain tuning," *IEEE Transactions on Cybernetics*, DOI: 10.1109/TCYB.2021.3059200, 2021.

[San15] R. S. Sánchez-Peña, P. Colmegna and F. Bianchi, "Unfalsified control based on the H$_\infty$ controller parameterization," *International Journal of Systems Science*, vol. 46, no. 15, pp. 2820–2831, Nov. 2015.

[Sat20] T. Sato, T. Kusakabe, K. Himi, N. Arakim and Y. Konishi, "Ripple-free data-driven dual-rate controller using lifting technique: application to a physical rotation system," *IEEE Transactions on Control Systems Technology*, DOI: 10.1109/TCST.2020.2988613, Apr. 2020.

[Sel21] D. Selvi, D. Piga, G. Battistelli and A. Bemporad, "Optimal direct data-driven control with stability guarantees," *European Journal of Control*, vol. 59, pp. 175–187, May 2021.

[She18] D. Shen, "Iterative learning control with incomplete information: a survey," *IEEE/CAA Journal of Automatica Sinica*, vol. 5, no. 5, pp. 885–901, Sep. 2018.

[Shi00] G. Shi and R. E. Skelton, "Markov data-based LQG control," *Journal of Dynamic Systems, Measurement, and Control*, vol. 122, no. 3, pp. 551–559, Sep. 2000.

[Spa98] J. C. Spall and J. A. Cristion, "Model-free control of nonlinear stochastic systems with discrete-time measurements," *IEEE Transactions on Automatic Control*, vol. 43, no. 9, pp. 1198–1210, Sep. 1998.

[Sun19] J. K. Sun, J. Yang, S. H. Li and W. X. Zheng, "Sampled-data-based event-triggered active disturbance rejection control for disturbed systems in networked environment," *IEEE Transactions on Cybernetics*, vol. 49, no. 2, pp. 556–566, Feb. 2019.

[Sut92] R. S. Sutton, A. G. Barto and R. J. Williams, "Reinforcement learning is direct adaptive optimal control," *IEEE Control Systems Magazine*, vol. 12, no. 2, pp. 19–22, Apr. 1992.

[Sut17] R. S. Sutton and A. G. Barto, *Reinforcement Learning: an Introduction*, 2nd Edition, MIT Press, Cambridge, MA, 2017.

[Tsa01] T.-C. Tsao and M. G. Safonov, "Unfalsified direct adaptive control of a two-link robot arm," *International Journal of Adaptive Control and Signal Processing*, vol. 15, no. 3, pp. 319–334, May 2001.

[VanH07] J. van Helvoort, B. de Jager and M. Steinbuch, "Direct data-driven recursive controller unfalsification with analytic update," *Automatica*, vol. 43, no. 12, pp. 2034–2046, Dec. 2007.

[VanH08] J. van Helvoort, B. de Jager and M. Steinbuch, "Data-driven controller unfalsification with analytic update applied to a motion system," *IEEE Transactions on Control Systems Technology*, vol. 16, no. 6, pp. 1207–1217, Nov. 2008.

[vanH09] K. van Heusden, A. Karimi and D. Bonvin, "Data-driven controller validation," *IFAC Proceedings Volumes*, vol. 42, no. 10, pp. 1050–1055, Jul. 2009.

[vanH10] K. van Heusden, *Non-Iterative Data-Driven Model Reference Control*, PhD thesis, École Polytechnique Fédérale de Lausanne, Lausanne, 2010.

[vanH11] K. van Heusden, A. Karimi and D. Bonvin, "Data-driven model reference control with asymptotically guaranteed stability," *International Journal of Adaptive Control and Signal Processing*, vol. 25, no. 4, pp. 331–351, Apr. 2011.

[VanW20] H. J. Van Waarde, J. Eising, H. L. Trentelman and M. K. Camlibel, "Data informativity: a new perspective on data-driven analysis and control," *IEEE Transactions on Automatic Control*, vol. 65, no. 11, pp. 4753–4768, Nov. 2020.

[Váz13] C. Vázquez, J. Collado and L. Friedman, "Control of a parametrically excited crane: a vector Lyapunov approach," *IEEE Transactions on Control Systems Technology*, vol. 21, no. 6, pp. 2332–2340, Nov. 2013.

[Vil20] J. Villagra, C. Join, R. Haber and M. Fliess, "Model-free control for machine tools," in *Proceedings of 21st IFAC World Congress*, Berlin, Germany, 2020, pp. 1–5.

[Wan98] I.-J. Wang and J. C. Spall, "Stochastic optimisation with inequality constraints using simultaneous perturbations and penalty functions," *International Journal of Control*, vol. 81, no. 8, pp. 1232–1238, Aug. 2008.

[Wan07] X. Wang, B. Huang, and T. Chen, "Data-driven predictive control for solid oxide fuel cells," *Journal of Process Control*, vol. 17, no. 2, pp. 103–114, Feb. 2007.

[Wan17] D. Wang, H. He and D. Liu, "Adaptive critic nonlinear robust control: a survey," *IEEE Transactions on Cybernetics*, vol. 47, no. 10, pp. 3429–3451, Oct. 2017.

[Xu03] J. X. Xu and Y. Tan, *Linear and Nonlinear Iterative Learning Control*, Springer-Verlag, Berlin, 2003.

[Yin14] S. Yin, H.-J. Gao and O Kaynak, "Data-driven control and process monitoring for industrial applications - Part I," *IEEE Transactions on Industrial Electronics*, vol. 61, no. 11, pp. 6356–6359, Nov. 2014.

[Yin15a] S. Yin, H.-J. Gao and O Kaynak, "Data-driven control and process monitoring for industrial applications - Part II," *IEEE Transactions on Industrial Electronics*, vol. 62, no. 1, pp. 583–586, Jan. 2015.

[Yin15b] S. Yin, X.-W. Li, H.-J. Gao and O Kaynak, "Data-based techniques focused on modern industry: an overview," *IEEE Transactions on Industrial Electronics*, vol. 62, no. 1, pp. 657–667, Jan. 2015.

[Yu20] W. Yu, R. Wang, X.-H. Bu and Z.-S. Hou, "Model free adaptive control for a class of nonlinear systems with fading measurements," *Journal of the Franklin Institute*, vol. 357, no. 12, pp. 7743–7760, Aug. 2020.

[Zam20] M. Zamanipour, "A novelty in Blahut-Arimoto type algorithms: optimal control over noisy communication channels," *IEEE Transactions on Vehicular Technology*, vol. 69, no. 6, pp. 6348–6358, Jun. 2020.

[Zho18] D. Zhou, A. Al-Durra, I. Matraji, A. Ravey and F. Gao, "Online energy management strategy of fuel cell hybrid electric vehicles: a fractional-order extremum seeking method," *IEEE Transactions on Industrial Electronics*, vol. 65, no. 8, pp. 6787–6799, Aug. 2018.

2 Iterative Feedback Tuning

2.1 BACKGROUND

As specified in Chapter 1, IFT [Hja94, Hja98b, Hja02] is a well-acknowledged iterative data-driven model-free technique that tunes the controller parameters iteratively
along the gradient direction of an objective function.

Definition 2.1

IFT aims to minimize **objective functions** J that are specific to linear-quadratic
Gaussian (LQG) controllers expressed as

$$J(\chi) = \frac{1}{2N} E \left\{ \sum_{k=1}^{N} \left\{ [L_y(q^{-1}) \delta y(k, \chi)]^2 + \lambda [L_u(q^{-1}) u(k, \chi)]^2 \right\} \right\}, \qquad (2.1)$$

where χ is the parameter vector of the controller, $L_y(q^{-1})$ and $L_u(q^{-1})$ are weighting
filters that penalize the output error (or the tracking error) δy and the control input (or
the control signal) u to give importance to certain frequency regions, q^{-1} is the unit
delay operator, referred to also as the backward shift operator, λ is the control signal
weighting parameter, k is the discrete-time index, and N is the number of samples
or the length of the experiment (the trial). The mathematical expectation $E\{\Xi\}$ is
taken with respect to the stochastic probability distribution of the disturbance inputs
applied to the process and thus affects the control system behavior. The disturbance
inputs are assumed to be zero-mean discrete-time stochastic processes, and it is also
assumed that sequences in different experiments are mutually independent in order
to obtain unbiased estimates of the gradient.

Definition 2.2

The expression of vector variable of the objective function in (2.1), which is also **the
parameter vector** χ of the controller, is

$$\chi = \begin{bmatrix} \chi_1 & \chi_2 & \cdots & \chi_n \end{bmatrix}^T \in \Re^n, \qquad (2.2)$$

where the superscript T indicates matrix transposition, and χ_l, $l = 1 \dots n$, are the controller tuning parameters.

DOI: 10.1201/9781003143444-2

Remark 2.1

The expression $J(\chi)$ of the objective function in (2.1) should be specified, including the weighting filters $L_y(q^{-1})$, $L_u(q^{-1})$ and the weighting parameter λ, such that the minimization of the objective function will ensure the fulfillment of the performance specifications imposed to the control system. The expression of the output error (or the tracking error) $\delta y(k,\chi)$ is

$$\delta y(k,\chi) = y(k,\chi) - y_d(k),\qquad(2.3)$$

where $y(k,\chi)$ is the controlled output, and $y_d(k)$ is the output of the reference model, i.e., the desired output of the control system to be tracked by the controlled output.

Definition 2.3

The objective of IFT is to compute the optimal parameter vector χ^* as the solution to **the optimization problem**:

$$\chi^* = \arg\min_{\chi} J(\chi),\qquad(2.4)$$

which minimizes the objective function $J(\chi)$ expressed in (2.1). The expression of the optimal parameter vector χ^* is

$$\chi^* = \begin{bmatrix} \chi_1^* & \chi_2^* & \cdots & \chi_n^* \end{bmatrix}^T \in \Re^n,\qquad(2.5)$$

where χ_l^*, $l = 1 \ldots n$, are the optimal tuning parameters of the controller.

Proposition 2.1

The major hint in solving the optimization problem in (2.4), which is actually **an optimal control problem**, is the computation of the gradient of the objective function with respect to the controller parameters. The specific feature of IFT is that an estimate of the gradient of the objective function with respect to the controller parameters can be obtained by conducting special "gradient" experiments on the control system at each iteration of the IFT and thus obtaining the data-driven model-free estimation of the gradient.

Lemma 2.1

In the IFT algorithm, the solution is approached iteratively using different gradient-based search algorithms as, for example, the Gauss-Newton scheme expressed as the following IFT algorithm, which represents **the parameter update law**:

$$\chi^{[i+1]} = \chi^{[i]} - \gamma^{[i]} \left(\mathbf{R}^{[i]}\right)^{-1} \text{est}\left[\frac{\partial J}{\partial \chi}(\chi^{[i]})\right], \qquad (2.6)$$

where i, $i \in \mathbf{N}$, is the iteration number, the superscript $[i]$ indicates the value of a certain scalar or vector or matrix at iteration i, the vector $\text{est}\left[\frac{\partial J}{\partial \chi}(\chi^{[i]})\right]$ is the estimated gradient, the parameter $\gamma^{[i]}$, $\gamma^{[i]} > 0$, is the step size of the current iteration, and $\mathbf{R}^{[i]}$ is a positive definite regular matrix, which is typically chosen to be equal to an estimate of the Hessian matrix of the objective function J or to the identity matrix.

Remark 2.2

As shown in [Hja94, Hja98b, Hja02, Rad11a], IFT is not only a sensitivity-based tuning technique, but it also holds stochastic convergence results. That is important in data-based tuning, where all experiments and measurements are affected by random effects.

Proposition 2.2

It is stated in [Ham03] that IFT is applicable if an initial stabilizing controller is first designed and tuned. However, since discussing the stability in the framework of model-free control (MFC) is complicated and preferred to be avoided, the authors of this book consider, as in [Pre07] and [Jun21], that IFT is applicable if an initial appropriately parameterized controller, which ensures a finite value of the objective function, is assumed to be known.

Remark 2.3

Several versions of IFT algorithms can be built by the choice of the objective function and the controller structure. The IFT-based parameter tuning of controllers is formulated in the discrete-time domain. The control system designer selects the controller's structure, which means that the controller should be known in analytical form. The parameterization should be carried out such that the transfer function of the controller is differentiable with respect to its parameter vector χ.

Proposition 2.3

The SISO control system structure with IFT algorithm is presented in Figure 2.1, where IFTA – the iterative feedback tuning algorithm, PS – the performance specifications applied to IFT in order to define the objective function in (2.1), RM – the

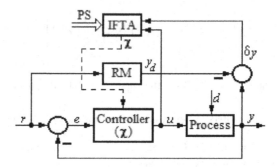

FIGURE 2.1   SISO control system structure with IFT algorithm.

reference model, r – the reference input or the set-point, and $e = r - y$ – the control error

$$e = r - y, \qquad (2.7)$$

u – the control signal or the control input and d – the disturbance input. As illustrated in Figure 2.1, the controller should be parameterized by the parameter vector χ. In addition, all signals in the control system are assumed to be differentiable with respect to χ.

Remark 2.4

The control system structure depicted in Figure 2.1 is similar to a model reference adaptive control system structure. The IFT control system structure actually carries out the parameter adaptation in terms of the parameter update law given in (2.6), which ensures **adaptation and learning** in the context of (2.4). However, the IFT control system structure does not operate permanently as an adaptive control system structure; it operates few iterations in order to improve the control system performance by the reduction of the value of the objective function. As specified in Chapter 1 and pointed out in [Pre20], the successful applications and performance proved by data-driven MFC justify to consider, in accordance with the statement given in [Fli20], that MFC and also the **iterative data-driven model-free techniques are new tools for machine learning (ML)**. The authors support this statement and highlight that it is also valid for IFT, where the parameter updated law given in (2.6) can be considered as a learning algorithm. This is especially important because IFT can also be applied in the context of modeling instead of control, and it ensures the training of several nonlinear model architectures including neural network and fuzzy logic ones provided they can be organized such that to work in feedback structures; in this ML-based modeling framework, the elements of the parameter vector χ are the parameters of the nonlinear models.

Remark 2.5

The disturbance input d is highlighted in Figure 2.1, and it can be applied in several informational parts of the process. The integral component of the controller deals with disturbance rejection.

The theoretical basics of IFT in the SISO control system structure will be discussed in the following section in the one-degree-of-freedom formulation. The MIMO control system structure will be next treated.

2.2   THEORY AND ALGORITHM IN THE SISO CASE

This section is dedicated to explaining the theory focused on the minimization of the objective function $J(\chi)$ expressed in (2.1). Various objective function structures can be used. For example, two objective functions are used in [Ham03], which are different for each controller in the control system dedicated to a two-mass-spring system with friction. A particular form of the objective function, with $\lambda = 0$, suggested in [Hja98b], is used in [Leq03] to compare the IFT-based tuning of proportional-integral-derivative (PID) controller parameters with classical tuning rules.

Definition 2.4

Dropping out the mathematical expectation operator in (2.1), which is a common practice in the control systems' applications considered in this book, the expression of **the gradient of the objective function** is

$$\frac{\partial J}{\partial \chi}(\chi) = \frac{1}{N}\sum_{k=1}^{N}\left\{L_y(q^{-1})\delta y(k,\chi)\frac{\partial \delta y}{\partial \chi}(k,\chi) + \lambda L_y(q^{-1})u(k,\chi)\frac{\partial u}{\partial \chi}(k,\chi)\right\}. \quad (2.8)$$

The necessary condition for optimality is

$$\left.\frac{\partial J}{\partial \chi}(\chi)\right|_{\chi=\chi^*} = 0, \quad (2.9)$$

where

$$\mathbf{0} = \begin{bmatrix}0 & 0 & \dots & 0\end{bmatrix}^T \in \Re^{n\times 1}. \quad (2.10)$$

Remark 2.6

Solving equation (2.9) with respect to the stationary points χ^* is not simple because the signals δy and u depend on χ, and their gradients have to be computed with respect to χ. Since the process is supposed to contain unknown or difficult-to-model dynamics, analytical dependencies cannot be derived. The main contribution of IFT is that it represents a technique that calculates the gradient $\dfrac{\partial J}{\partial \chi}(\chi)$ directly from the control system. According to (2.8), the required information is represented by the signals δy and u and the gradients of these signals, namely $\dfrac{\partial \delta y}{\partial \chi}(\chi)$ and $\dfrac{\partial u}{\partial \chi}(\chi)$, respectively. The signals δy and u can be obtained by direct measurements in the control system, and the gradients are obtained by appropriate computations using the signals measured from the control system.

Lemma 2.2

After computing the unbiased estimates of the gradients $\dfrac{\partial \delta y}{\partial \chi}(\chi)$ and $\dfrac{\partial u}{\partial \chi}(\chi)$, i.e.,

$\text{est}\left[\dfrac{\partial \delta y}{\partial \chi}(k,\chi)\right]$ and $\text{est}\left[\dfrac{\partial u}{\partial \chi}(k,\chi)\right]$, respectively, the estimated gradient of the objective function, namely $\text{est}\left[\dfrac{\partial J}{\partial \chi}(\chi)\right]$, is calculated:

$$\text{est}\left[\frac{\partial J}{\partial \chi}(\chi)\right] = \frac{1}{N}\sum_{k=1}^{N}\left\{L_y(q^{-1})\delta y(k,\chi)\text{est}\left[\frac{\partial \delta y}{\partial \chi}(k,\chi)\right] + \right.$$

$$\left. \lambda L_y(q^{-1})u(k,\chi)\text{est}\left[\frac{\partial u}{\partial \chi}(k,\chi)\right]\right\}. \tag{2.11}$$

Considering the one-degree-of-freedom control system structure given in Figure 2.1, the control error, the control signal and the controlled output depend on the actual controller tuning, and this is indicated by adding the argument χ.

Definition 2.5

The SISO process is supposed to be modeled by the discrete transfer function $P(z^{-1})$, and the controller is characterized by the discrete transfer function $C(z^{-1},\chi)$, which illustrates that it should be suitably parameterized by the parameter vector χ expressed in (2.2). The notation $S(z^{-1},\chi)$ is used for the (discrete) sensitivity function and the notation $T(z^{-1},\chi)$ for the (discrete) complementary sensitivity function. The expressions of these functions are

$$S(z^{-1},\chi) = \frac{1}{1+P(z^{-1})C(z^{-1},\chi)}, \tag{2.12}$$

$$T(z^{-1},\chi) = 1 - S(z^{-1},\chi) = \frac{P(z^{-1})C(z^{-1},\chi)}{1+P(z^{-1})C(z^{-1},\chi)}. \tag{2.13}$$

Theory 2.1

Accepting that the disturbance input d is additive on the controlled output, the following relations are obtained on the basis of Figure 2.1:

$$y(k,\chi) = T(q^{-1},\chi)r(k) + S(q^{-1},\chi)d(k), \tag{2.14}$$

$$e(k,\chi) = S(q^{-1},\chi)(r(k) - d(k)), \tag{2.15}$$

$$u(k,\chi) = C(q^{-1},\chi)S(q^{-1},\chi)(r(k) - d(k)). \tag{2.16}$$

The analytical expressions of the gradients $\dfrac{\partial \delta y}{\partial \chi}(\chi)$ and $\dfrac{\partial u}{\partial \chi}(\chi)$ are next calculated.

Remark 2.7

Since $y_d(k)$ in (2.3) does not depend on χ,

$$\frac{\partial y_d}{\partial \chi}(k,\chi) = \mathbf{0}, \tag{2.17}$$

and therefore, computing the gradient of (2.3) with respect to χ using (2.17) leads to

$$\frac{\partial \delta y}{\partial \chi}(k,\chi) = \frac{\partial y}{\partial \chi}(k,\chi). \tag{2.18}$$

Lemma 2.3

In addition, the computation of the gradient of (2.14) with respect to χ yields [Rad11a]

$$\frac{\partial y}{\partial \chi}(k,\chi) = \frac{\partial T}{\partial \chi}\left(q^{-1},\chi\right)r(k) + \frac{\partial S}{\partial \chi}\left(q^{-1},\chi\right)d(k). \tag{2.19}$$

Using the expressions of $S\left(q^{-1},\chi\right)$ and $T\left(q^{-1},\chi\right)$ in (2.12) and (2.13) with $q = z$, computing their gradients with respect to χ and replacing them in (2.19) give the following expression of $\dfrac{\partial y}{\partial \chi}(k,\chi)$, which is $\dfrac{\partial \delta y}{\partial \chi}(k,\chi)$ according to (2.18):

$$\frac{\partial \delta y}{\partial \chi}(k,\chi) = \frac{\partial C}{\partial \chi}\left(q^{-1},\chi\right)\frac{T\left(q^{-1},\chi\right)}{C\left(q^{-1},\chi\right)}S\left(q^{-1},\chi\right)(r(k) - d(k)). \tag{2.20}$$

The right-hand term of (2.14) is next substituted in (2.20) resulting in

$$\frac{\partial \delta y}{\partial \chi}(k,\chi) = \frac{\partial C}{\partial \chi}(q^{-1},\chi)\frac{T(q^{-1},\chi)}{C(q^{-1},\chi)}e(k,\chi). \tag{2.21}$$

Lemma 2.4

The gradient $\dfrac{\partial u}{\partial \chi}(\chi)$ is first computed in terms of taking the gradient of (2.16) with respect to χ:

$$\frac{\partial u}{\partial \chi}(k,\chi) = \left[\frac{\partial C}{\partial \chi}(q^{-1},\chi)S(q^{-1},\chi) + C(q^{-1},\chi)\frac{\partial S}{\partial \chi}(q^{-1},\chi)\right](r(k) - d(k)). \tag{2.22}$$

Using the expression of $S(q^{-1},\chi)$ in (2.10) with $q = z$, computing its gradient with respect to χ and replacing it in (2.22) give

$$\frac{\partial u}{\partial \chi}(k,\chi) = \frac{\partial C}{\partial \chi}(q^{-1},\chi)S^2(q^{-1},\chi)(r(k) - y(k)). \tag{2.23}$$

The right-hand term of (2.15) is next substituted in (2.23) resulting in

$$\frac{\partial u}{\partial \chi}(k,\chi) = \frac{\partial C}{\partial \chi}(q^{-1},\chi)S(q^{-1},\chi)e(k). \tag{2.24}$$

Definition 2.6

The gradient of the controller transfer function, $\dfrac{\partial C}{\partial \chi}(q^{-1},\chi)$, which appears in (2.21) and (2.24), can be expressed analytically using the expression of discrete transfer function $C(q^{-1},\chi)$, which is known. The gradient $\dfrac{\partial C}{\partial \chi}(q^{-1},\chi)$ is a column matrix expressed as

$$\frac{\partial C}{\partial \chi}(q^{-1},\chi) = \left[\frac{\partial C}{\partial \chi_1}(q^{-1},\chi) \quad \frac{\partial C}{\partial \chi_2}(q^{-1},\chi) \quad \cdots \quad \frac{\partial C}{\partial \chi_n}(q^{-1},\chi)\right]^T, \tag{2.25}$$

where all derivatives are assumed to exist.

Remark 2.8

Since the process model is accepted to be unknown in the context of model-free tuning, the gradients of δy and u cannot be calculated using the analytical relations (2.21) and (2.24). However, these two relations give the idea of IFT on how to conduct the two experiments on the real-world control system in order to enable the computation of the next parameter vector $\chi^{[i+1]}$.

Theory 2.2

The first experiment is also known as **the normal experiment**. Using the notation $\{j\}$, $j \in \{1,2\}$, for the subscript that is inserted to certain variables, the reference input $r_{\{1\}} = r$ is applied to the control system in terms of the control system structure given in Figure 2.1. The control signal $u_{\{1\}}$, the controlled output $y_{\{1\}}$ and the control error $e_{\{1\}}$ are measured when conducting this experiment. The input-output relationships specific to the normal experiment, specifying the measured variables and also the control system inputs, are

$$u_{\{1\}}(k,\chi) = C\left(q^{-1},\chi\right)S\left(q^{-1},\chi\right)(r_{\{1\}}(k) - d_{\{1\}}(k)), \qquad (2.26)$$

$$y_{\{1\}}(k,\chi) = T\left(q^{-1},\chi\right)r_{\{1\}}(k) + S\left(q^{-1},\chi\right)d_{\{1\}}(k), \qquad (2.27)$$

$$e_{\{1\}}(k,\chi) = S\left(q^{-1},\chi\right)(r_{\{1\}}(k) - d_{\{1\}}(k)). \qquad (2.28)$$

Theory 2.3

The second experiment is referred to as **the gradient experiment**. The reference input applied to the control system in the second experiment is the control error of the first experiment:

$$r_{\{2\}}(k,\chi) = e_{\{1\}}(k,\chi) = r_{\{1\}}(k) - y_{\{1\}}(k,\chi), \qquad (2.29)$$

and the control signal $u_{\{2\}}$ and the controlled output $y_{\{2\}}$ are measured during this experiment. The input-output relationships specific to the gradient experiment are

$$u_{\{2\}}(k,\chi) = C\left(q^{-1},\chi\right)S\left(q^{-1},\chi\right)(r_{\{2\}}(k,\chi) - d_{\{2\}}(k)), \qquad (2.30)$$

$$y_{\{2\}}(k,\chi) = T\left(q^{-1},\chi\right)r_{\{2\}}(k,\chi) + S\left(q^{-1},\chi\right)d_{\{2\}}(k). \qquad (2.31)$$

Remark 2.9

As outlined in [Rad11a], the disturbance inputs in the two experiments, namely $d_{\{1\}}(k)$ and $d_{\{2\}}(k)$, are assumed to be of zero mean and mutually uncorrelated. Under this assumption, the estimated gradient of the objective function is unbiased. Therefore, the computations are next done using

$$d_{\{1\}}(k) = d_{\{2\}}(k) = 0. \qquad (2.32)$$

Otherwise, the dynamic regime in which the objective function is assessed is modified, and the optimization cannot be performed anymore.

Lemma 2.5

The measurements obtained in the two experiments are next involved in the estimation of the gradients $\frac{\partial \delta y}{\partial \chi}(k,\chi)$ and $\frac{\partial u}{\partial \chi}(k,\chi)$. Using (2.29) and (2.31) in (2.21), the estimated expression of $\frac{\partial \delta y}{\partial \chi}(k,\chi)$ is

$$\text{est}\left[\frac{\partial \delta y}{\partial \chi}(k,\chi)\right] = \frac{1}{C(q^{-1},\chi)} \cdot \frac{\partial C}{\partial \chi}(q^{-1},\chi) \cdot y_2(k,\chi). \tag{2.33}$$

Using (2.29) and (2.30) in (2.24), the estimated expression of $\frac{\partial u}{\partial \chi}(k,\chi)$ is

$$\text{est}\left[\frac{\partial u}{\partial \chi}(k,\chi)\right] = \frac{1}{C(q^{-1},\chi)} \cdot \frac{\partial C}{\partial \chi}(q^{-1},\chi) \cdot u_2(k,\chi). \tag{2.34}$$

Finally, the gradient estimates in (2.33) and (2.34) are substituted in (2.11) to obtain the estimate of the gradient of the objective function, $\text{est}\left[\frac{\partial J}{\partial \chi}(\chi)\right]$.

Remark 2.10

The two estimated expressions of $\frac{\partial \delta y}{\partial \chi}(k,\chi)$ and $\frac{\partial u}{\partial \chi}(k,\chi)$ computed in accordance with (2.33) and (2.34) make use of the system responses $y_2(k,\chi)$ and $u_2(k,\chi)$, which can be viewed as nonparametric models of the control system. This fact is in accordance with the statement given in Chapter 1, specifying that data-driven MFC techniques do not use parametric models of the process but can involve nonparametric ones.

Remark 2.11

The choice of **the reference model** in the control system structure presented in Figure 2.1 requires a discussion in relation with the performance specifications and the expression of the objective function. The reference model should be chosen so that the actual output response of the control system and the reference model response are not very different, or, in other words, the output error (or the tracking error) is not very big. The reference model can be chosen either to increase or to decrease the control system bandwidth.

Remark 2.12

The windsurfing approach is applied if the tracking error is relatively big. This approach makes use of intermediate reference models that should be used in order to ensure a step-by-step convergence. Abrupt performance improvement cannot be achieved, and the tracking error should be reasonable at each step; therefore, these reference models should be chosen aiming their gradual performance improvement.

Remark 2.13

The reference models are usually considered as low-order benchmark-type models for which the step response incorporates performance specifications formulated in terms of empirical performance indices as overshoot, settling time and rise time. However, this is not a constraint, and other reference models can be chosen, but they should depend on the information available on the controlled process as, for example, the special case of nonminimum phase systems. Analyses in the case of second-order systems are performed in [Lan06] and [Rad11a], where diagrams are used in order to illustrate how the choice of the performance indices depends on the damping factor and the natural frequency.

Remark 2.14

The matrix $\mathbf{R}^{[i]}$ is needed in the parameter update law (2.6). This matrix is usually computed as follows as an approximate of the Hessian matrix of the objective function J:

$$\mathbf{R}^{[i]} = \frac{1}{N} \sum_{k=1}^{N} \left(\text{est}\left[\frac{\partial y}{\partial \chi}(k, \chi^{[i]}) \right] \text{est}\left[\frac{\partial y}{\partial \chi}(k, \chi^{[i]}) \right]^T + \lambda \text{est}\left[\frac{\partial u}{\partial \chi}(k, \chi^{[i]}) \right] \text{est}\left[\frac{\partial u}{\partial \chi}(k, \chi^{[i]}) \right]^T \right).$$

$$(2.35)$$

There are other possibilities to set $\mathbf{R}^{[i]}$. For example, if $\mathbf{R}^{[i]}$ is set as the identity matrix, the resulting gradient direction will be certainly negative. However, due to disturbances in the gradient experiment, this will give a biased approximation of the Gauss-Newton direction.

Another choice of $\mathbf{R}^{[i]}$ is to use a quasi-Newton method, and the popular Broyden-Fletcher-Goldfarb-Shanno (BFGS) algorithm is used in [Ham03]. An advantage of quasi-Newton methods is the good approximation of the Hessian matrix, which is obtained using the gradient of the objective function and the controller tuning parameters. The Hessian matrix is obtained iteratively in the BFGS algorithm in terms of [Rad11a]

$$\mathbf{R}^{[i+1]} = \mathbf{R}^{[i]} + \frac{\mathbf{z}^{[i]} (\mathbf{z}^{[i]})^T}{(\mathbf{z}^{[i]})^T \mathbf{s}^{[i]}} - \frac{\mathbf{R}^{[i]} \mathbf{s}^{[i]} (\mathbf{s}^{[i]})^T \mathbf{R}^{[i]}}{(\mathbf{s}^{[i]})^T \mathbf{R}^{[i]} \mathbf{s}^{[i]}},$$

$$(2.36)$$

where $\mathbf{R}^{[i]}$ depends on $\chi^{[i]}$, and

$$\mathbf{s}^{[i]} = \chi^{[i+1]} - \chi^{[i]}, \quad \mathbf{z}^i = \frac{\partial J}{\partial \chi}(\chi^{[i+1]}) - \frac{\partial J}{\partial \chi}(\chi^{[i]}).$$

$$(2.37)$$

Any positive definite matrix can be employed to initialize $\mathbf{R}^{[i]}$, i.e., to set the value of $\mathbf{R}^{[0]}$. Two features are specific to the BFGS algorithm:

- If $\mathbf{R}^{[i]}$ is symmetric, then $\mathbf{R}^{[i+1]}$ will be symmetric.
- if $\mathbf{R}^{[i]}$ is positive definite and $\left(\mathbf{z}^{[i]}\right)^{T} \mathbf{s}^{[i]} > 0$, then $\mathbf{R}^{[i+1]}$ will be used. If $\left(\mathbf{z}^{[i]}\right)^{T} \mathbf{s}^{[i]} > 0$ is not satisfied, then $\mathbf{R}^{[i+1]}$ will be set as equal to $\mathbf{R}^{[i]}$.

Remark 2.15

The choice of the step size sequence $\{\gamma^{[i]}\}_{i \in \mathbb{N}}$ is also important in the parameter update law (2.6), which is also a stochastic algorithm, as it influences the solving of the optimization problem in (2.4). This sequence should evolve in time such that to satisfy some bounds. In this regard, the conditions to ensure the convergence of the stochastic algorithm (2.6) are [Hja94, Hja02, Huu09a]

$$\sum_{i=0}^{\infty} \gamma^{[i]} = \infty, \quad \sum_{i=0}^{\infty} \left(\gamma^{[i]}\right)^{2} < \infty. \tag{2.38}$$

A useful choice of the step size sequence that ensures the divergence of the first series in (2.38) and also the convergence of the second series in (2.38) is [Rad11b]

$$\gamma^{[i]} = \frac{\gamma^{0}}{i^{\alpha}}, \quad i \in \mathbb{N}, \quad i \geq 1, \quad 0.5 < \alpha \leq 1, \tag{2.39}$$

where the initial step size $\gamma^{[0]}$, $\gamma^{[0]} > 0$, is set such that to ensure a tradeoff to numerical stability and convergence speed.

Proposition 2.4

Summarizing the aspects presented in this section, the general IFT algorithm dedicated to the data-driven model-free tuning of one-degree-of-freedom SISO controllers is organized in terms of the following steps.

Step 1. The reference model is set, and the expression of the objective function in (2.1) is formulated, and its parameters are set such that the minimization of the objective function by solving the optimization problem in (2.4) ensures the fulfillment of the performance specifications imposed to the control system. The parameters specific to the parameter update law in (2.6) are set. An initial appropriately parameterized controller, which ensures a finite value of the objective function in (2.1), is designed and tuned. The number of iterations is set.

Step 2. The normal experiment is conducted using the reference input $r_{[1]} = r$ applied to the control system in terms of the control system structure given in Figure 2.1. The control signal $u_{\{1\}}$, the controlled output $y_{\{1\}}$ and the control error $e_{\{1\}}$ are measured.

Step 3. The gradient experiment is conducted using the reference input $r_{\{2\}} = e_{\{1\}}$ applied to the control system in terms of the control system structure given in Figure 2.1. The control signal $u_{\{2\}}$ and the controlled output $y_{\{2\}}$ are measured.

Step 4. The estimated expressions of $\dfrac{\partial \delta y}{\partial \chi}(k,\chi)$ and $\dfrac{\partial u}{\partial \chi}(k,\chi)$, namely $\mathrm{est}\left[\dfrac{\partial \delta y}{\partial \chi}(k,\chi)\right]$ and $\mathrm{est}\left[\dfrac{\partial u}{\partial \chi}(k,\chi)\right]$, are computed using (2.33) and (2.34), respectively, and next substituted in (2.11) to calculate the estimated gradient of the objective function, i.e., $\mathrm{est}\left[\dfrac{\partial J}{\partial \chi}(\chi)\right]$. The values of $\delta y(k,\chi)$ and $u(k,\chi)$ are taken from the normal experiment and next used in (2.11).

Step 5. The next parameter vector $\chi^{[i+1]}$ is computed in terms of the parameter update law given in (2.6).

Step 1 is carried out once. Steps 2 to 5 are repeated for a predefined number of iterations set in step 1.

2.3 THEORY AND ALGORITHM IN THE MIMO CASE

The seminal extensions of IFT to MIMO systems are treated in [DeB97], [Hja98a], [Hja98b], [Hja99], [Gun03] and [Jan04]. All these approaches are focused on considering that the process is a linear time-invariant MIMO system, and several ways are used to express the input-output relationship of the control system.

Proposition 2.5

The MIMO control system structure with IFT algorithm is presented in Figure 2.2 as an extension of the SISO control system structure given in Figure 2.1 but using vector

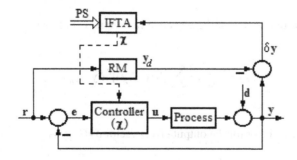

FIGURE 2.2 MIMO control system structure with IFT algorithm.

variables instead of scalar ones and accepting, as in [Jan04], the additive disturbance input vector $\mathbf{d}(k)$ on the controlled output. The disturbance input vector $\mathbf{d}(k)$
is assumed to be a zero-mean discrete-time stochastic process, and it is also assumed
in [Jan04] that sequences in different experiments are mutually independent in order
to obtain unbiased estimates of the gradient.

The main notations in Figure 2.2 have already been defined in relation with
Figure 2.1. The other notations in Figure 2.2 are $\mathbf{y}$, $\mathbf{y} \in \mathfrak{R}^{n_y}$ – the controlled output
vector, $\mathbf{r}$, $\mathbf{r} \in \mathfrak{R}^{n_y}$ – the reference input (or the set-point) vector, $\mathbf{e}$, $\mathbf{e} \in \mathfrak{R}^{n_y}$ – the control error vector

$$\mathbf{e} = \mathbf{r} - \mathbf{y}, \tag{2.40}$$

$\mathbf{d}$, $\mathbf{d} \in \mathfrak{R}^{n_y}$ – the disturbance input vector specified above, $\mathbf{u}$, $\mathbf{u} \in \mathfrak{R}^{n_u}$ – the control
signal (or the control input) vector, $\mathbf{y}_d$, $\mathbf{y}_d \in \mathfrak{R}^{n_y}$ – reference model output vector, i.e.,
the desired output vector of the control system and $\delta \mathbf{y}_d$, $\delta \mathbf{y}_d \in \mathfrak{R}^{n_y}$ – the output error
(or the tracking error) vector.

Theory 2.4

Considering that the process is modeled by the discrete transfer function matrix
$\mathbf{P}(z^{-1})$, $\mathbf{P}(z^{-1}) \in \mathfrak{R}^{n_y \times n_u}$, the controller is modeled by the discrete transfer function
matrix $\mathbf{C}(z^{-1}, \chi)$, $\mathbf{C}(z^{-1}, \chi) \in \mathfrak{R}^{n_u \times n_y}$, which should be suitably parameterized by the
parameter vector χ, $\chi \in \mathfrak{R}^n$, defined in (2.2), and also differentiable with respect
to χ, and all signals in the control system are differentiable with respect to χ, the
input-output relationship of the control system is expressed in terms of

$$\mathbf{y}(k, \chi) = \mathbf{P}(q^{-1})\mathbf{u}(k, \chi) + \mathbf{d}(k), \tag{2.41}$$

$$\mathbf{u}(k, \chi) = \mathbf{C}(q^{-1}, \chi)(\mathbf{r}(k) - \mathbf{y}(k, \chi)). \tag{2.42}$$

Definition 2.7

IFT aims the minimization of the objective function J expressed as

$$J(\chi) = \frac{1}{2N} E\left\{ \sum_{k=1}^{N} \left\{ [\delta \mathbf{y}(k, \chi)]^T \delta \mathbf{y}(k, \chi) \right\} \right\}, \tag{2.43}$$

where the tracking error (or the output error) vector $\delta \mathbf{y}(k, \chi)$ is

$$\delta \mathbf{y}(k, \chi) = \mathbf{y}(k, \chi) - \mathbf{y}_d(k). \tag{2.44}$$

Remark 2.16

The mathematical expectation $E\{\Xi\}$ in (2.43) is taken with respect to the stochastic probability distribution of the disturbance input vector. The expression of the objective function in (2.43) indicates that it is different to the expression in the SISO case, and it does not anymore depend on the control input vector, which also makes a difference between Figures 2.1 and 2.2. However, different objective functions can be used in order to express adequately the performance specifications imposed to the control system, which should be fulfilled in the MIMO case by solving the optimization problem defined in (2.43).

Remark 2.17

Similar to the SISO case, the objective of IFT is to compute the optimal parameter vector χ^* as the solution to **the optimization problem** defined in (2.4), which is another **optimal control problem**, and targets the minimization of the objective function $J(\chi)$ expressed in (2.43). In this regard, the solution is obtained iteratively using different gradient-based search algorithms as, for example, the Gauss-Newton scheme expressed as the IFT algorithm (IFTA), which represents the parameter update law given in (2.6).

Definition 2.8

The necessary condition for optimality that leads to the stationary points χ^* is expressed in (2.9). Taking the gradient of the objective function in (2.43) and imposing the condition (2.9), the stationary points are obtained as solutions to

$$\frac{\partial J}{\partial \chi}(\chi) = \frac{1}{N} E\left\{\sum_{k=1}^{N}\left\{\left[\frac{\partial \delta \mathbf{y}}{\partial \chi}(k,\chi)\right]^{T} \delta \mathbf{y}(k,\chi)\right\}\right\} = \mathbf{0}, \tag{2.45}$$

where the column matrix $\mathbf{0} \in \mathfrak{R}^{n \times 1}$ is defined in (2.10).

Details on the computation of the estimated gradient $est\left[\frac{\partial J}{\partial \chi}(\chi^{[i]})\right]$, which is involved in the parameter update law (2.6), will be given as follows in the framework of IFT.

Remark 2.18

Since $\mathbf{y}_d(k)$ in (2.44) does not depend on χ,

$$\frac{\partial \mathbf{y}_d}{\partial \chi}(k,\chi) = \mathbf{0} \in \mathfrak{R}^{n_y \times n},$$

$$\mathbf{0} = [o_{\alpha\beta}]_{\alpha=1...n_y, \beta=1...n}, \ o_{\alpha\beta} = 0, \ \alpha = 1...n_y, \ \beta = 1...n, \tag{2.46}$$

and therefore, computing the gradient with respect to χ in accordance with (2.45) using (2.46) leads to

$$\frac{\partial \delta \mathbf{y}}{\partial \chi}(k,\chi) = \frac{\partial \mathbf{y}}{\partial \chi}(k,\chi). \tag{2.47}$$

Theory 2.5

Using (2.47) in (2.45), the expression of the gradient $\dfrac{\partial J}{\partial \chi}(\chi)$ becomes

$$\frac{\partial J}{\partial \chi}(\chi) = \frac{1}{N} E \left\{ \sum_{k=1}^{N} \left\{ \left[\frac{\partial \mathbf{y}}{\partial \chi}(k,\chi) \right]^{T} \delta \mathbf{y}(k,\chi) \right\} \right\} = \mathbf{0}. \tag{2.48}$$

The following notation is introduced to highlight the gradient of the variable • with respect to the controller tuning parameter χ_l, $l = 1 \ldots n$, i.e., the partial derivative of the variable • with respect to the controller tuning parameter χ_l, $l = 1 \ldots n$ [Jan04, Rad09a]:

$$\bullet' = \frac{\partial \bullet}{\partial \chi_i}. \tag{2.49}$$

Equation (2.41) is differentiated with respect to χ_l, $l = 1 \ldots n$, leading to

$$\mathbf{y}'(k,\chi) = \mathbf{P}(q^{-1})\mathbf{u}'(k,\chi). \tag{2.50}$$

Equation (2.42) is also differentiated as a product with respect to χ_l, $l = 1 \ldots n$, leading to

$$\mathbf{u}'(k,\chi) = (\mathbf{C}(q^{-1},\chi)\mathbf{r}(k) - \mathbf{C}(q^{-1},\chi)\mathbf{y}(k,\chi))'$$

$$= \mathbf{C}'(q^{-1},\chi)\mathbf{r}(k) - \mathbf{C}'(q^{-1},\chi)\mathbf{y}(k,\chi) - \mathbf{C}(q^{-1},\chi)\mathbf{y}'(k,\chi) \tag{2.51}$$

$$= \mathbf{C}'(q^{-1},\chi)(\mathbf{r}(k) - \mathbf{y}(k,\chi)) - \mathbf{C}(q^{-1},\chi)\mathbf{y}'(k,\chi).$$

Using the notation in (2.40) for the control error vector $\mathbf{e}$, $\mathbf{e} \in \mathfrak{R}^{n_y}$, equation (2.51) becomes

$$\mathbf{u}'(k,\chi) = \mathbf{C}'(q^{-1},\chi)\mathbf{e}(k,\chi) - \mathbf{C}(q^{-1},\chi)\mathbf{y}'(k,\chi). \tag{2.52}$$

The left-hand multiplication of (2.52) with $\mathbf{P}(q^{-1})$ leads to

$$\mathbf{P}(q^{-1})\mathbf{u}'(k,\chi) = \mathbf{P}(q^{-1})\mathbf{C}'(q^{-1},\chi)\mathbf{e}(k,\chi) - \mathbf{P}(q^{-1})\mathbf{C}(q^{-1},\chi)\mathbf{y}'(k,\chi). \tag{2.53}$$

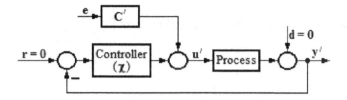

FIGURE 2.3 MIMO control system structure used in gradient experiments.

Finally, substituting the product $\mathbf{P}(q^{-1})\mathbf{u}'(k,\chi)$ from (2.50) in (2.53) results in

$$\mathbf{y}'(k,\chi) = \mathbf{P}(q^{-1})\mathbf{C}'(q^{-1},\chi)\mathbf{e}(k,\chi) - \mathbf{P}(q^{-1})\mathbf{C}(q^{-1},\chi)\mathbf{y}'(k,\chi). \qquad (2.54)$$

Equations (2.52) and (2.54) enable the organization of **the gradient experiments** in terms of the control system structure illustrated in Figure 2.3.

Remark 2.19

Equations (2.52) and (2.54) and Figure 2.3 illustrate that the control error vector $\mathbf{e}$, $\mathbf{e} \in \mathfrak{R}^{n_y}$, obtained in **the initial experiment**, referred to also as **the normal experiment**, is filtered through $\mathbf{C}'(q^{-1},\chi)$ and applied as an additional input to the control system. In addition, the reference input vector $\mathbf{r}$, $\mathbf{r} \in \mathfrak{R}^{n_y}$, is zero, i.e., $\mathbf{r} = \mathbf{0} \in \mathfrak{R}^{n_y \times 1}$.

Lemma 2.6

Since the disturbance input vector is zero in Figure 2.3, the mathematical expectation is dropped out in (2.48), which leads to the expression of the estimated expression of the gradient $\dfrac{\partial J}{\partial \chi}(\chi)$ of the objective function, i.e., $\text{est}\left[\dfrac{\partial J}{\partial \chi}(\chi)\right]$

$$\text{est}\left[\frac{\partial J}{\partial \chi}(\chi)\right] = \frac{1}{N}\sum_{k=1}^{N}\left\{\text{est}\left[\frac{\partial \mathbf{y}}{\partial \chi}(k,\chi)\right]^{T}\mathbf{y}(k,\chi)\right\}, \qquad (2.55)$$

where the controlled output vector is the column matrix

$$\mathbf{y}(k,\chi) = \left[y_1(k,\chi) \quad y_2(k,\chi) \quad \cdots \quad y_{n_y}(k,\chi)\right]^{T}, \qquad (2.56)$$

the estimated expression of the transposed gradient $\left[\dfrac{\partial \mathbf{y}}{\partial \boldsymbol{\chi}}(k,\boldsymbol{\chi})\right]^T$ of the controlled output is

$$
\text{est}\left[\frac{\partial \mathbf{y}}{\partial \boldsymbol{\chi}}(k,\boldsymbol{\chi})\right]^T =
\begin{bmatrix}
\dfrac{\partial y_1}{\partial \chi_1}(k,\boldsymbol{\chi}) & \dfrac{\partial y_2}{\partial \chi_1}(k,\boldsymbol{\chi}) & \cdots & \dfrac{\partial y_{n_y}}{\partial \chi_1}(k,\boldsymbol{\chi}) \\[2mm]
\dfrac{\partial y_1}{\partial \chi_2}(k,\boldsymbol{\chi}) & \dfrac{\partial y_2}{\partial \chi_2}(k,\boldsymbol{\chi}) & \cdots & \dfrac{\partial y_{n_y}}{\partial \chi_2}(k,\boldsymbol{\chi}) \\[2mm]
\cdots & \cdots & & \cdots \\[2mm]
\dfrac{\partial y_1}{\partial \chi_n}(k,\boldsymbol{\chi}) & \dfrac{\partial y_2}{\partial \chi_n}(k,\boldsymbol{\chi}) & \cdots & \dfrac{\partial y_{n_y}}{\partial \chi_n}(k,\boldsymbol{\chi})
\end{bmatrix}. \quad (2.57)
$$

and each row l in the matrix placed in the right-hand term of (2.57) is obtained as the output vector of the gradient experiment l conducted in terms of Figure 2.3 with respect to the controller tuning parameter χ_l, $l = 1...n$.

Proposition 2.6

The general **IFT algorithm** dedicated to the **data-driven model-free tuning of MIMO controllers** is organized in terms of the following steps, which result from the organization of the information given in this section.

Step 1. The reference model is set such that the minimization of the objective function defined in (2.43) by solving the optimization problem in (2.43) ensures the fulfillment of the performance specifications imposed to the control system. The parameters specific to the parameter update law in (2.6) are set. An initial appropriately parameterized controller, which ensures a finite value of the objective function in (2.43), is designed and tuned. The number of iterations is set.

Step 2. The normal experiment (or the initial experiment) is conducted using the reference input (or the set-point) vector $\mathbf{r}$, $\mathbf{r} \in \mathfrak{R}^{n_y}$, applied to the control system in terms of the control system structure given in Figure 2.2. The control error vector $\mathbf{e}$, $\mathbf{e} \in \mathfrak{R}^{n_y}$ is measured.

Step 3. The n gradient experiments are conducted using the zero reference input vector applied to the control system, i.e., $\mathbf{r} = \mathbf{0} \in \mathfrak{R}^{n_y \times 1}$, and the control error vector obtained in the normal experiment $\mathbf{e}$, $\mathbf{e} \in \mathfrak{R}^{n_y}$, applied to the control system in terms of the control system structure given in Figure 2.3. The controlled output vector $\mathbf{y}'$, $\mathbf{y}' \in \mathfrak{R}^{n_y}$ is measured.

Step 4. The estimated expression of the transposed gradient $\left[\dfrac{\partial \mathbf{y}}{\partial \boldsymbol{\chi}}(k,\boldsymbol{\chi})\right]^T$ of the controlled output, namely $\text{est}\left[\dfrac{\partial \mathbf{y}}{\partial \boldsymbol{\chi}}(k,\boldsymbol{\chi})\right]^T$, is obtained by placing the output vector of the gradient experiment l conducted at step 3 in the row l

of the matrix placed in the right-hand term of (2.57), where l corresponds to the experiment conducted with respect to the controller tuning parameter χ_l, $l = 1...n$. This matrix, along with the output error (or the tracking error) vector obtained in the normal experiment, is next substituted in (2.55) to calculate the estimated gradient of the objective function, i.e., est$\left[\dfrac{\partial J}{\partial \chi}(\chi)\right]$.

Step 5. The next parameter vector $\chi^{[i+1]}$ is computed in terms of the parameter update law given in (2.6).

Step 1 is carried out once. Steps 2–5 are repeated for a predefined number of iterations set in step 1.

Remark 2.20

As pointed out in [Rad09a] and applied in [Rad12] and [Rad13b], a total number of $1 + n$ experiments are needed to compute experimentally the estimates of all gradients involved in (2.45). However, this can be cost-inefficient due to the relatively high number of experiments that are conducted on the control system that does not operate in the normal operating mode. Several **approaches to reduce the number of experiments** specific to MIMO IFT are presented in [Hja98a, Hja99] and [Rad09a].

Remark 2.21

The approaches to obtain the gradients of the objective functions are kept no matter which other quadratic objective functions are used. The theoretical results presented in this section and the previous section, and the application ones to be presented in the next section, are useful for the readers in this regard.

Remark 2.22

Additional experiments are also needed if IFT is applied in **two-degree-of-freedom control system structures**. Two structures are popular in this context, (1) and (2). The first structure, (1), which carries out the simultaneous parameter tuning (of both the feedback and the feedforward controller), is proposed in [Hja94], further developed in [Hja98b] and next discussed and applied in [Hja02] and [Leq03]. The second structure, (2), which carries out separate parameter tuning (first the feedback controller parameters and next the feedforward controller parameters), is suggested in [Ham03].

Remark 2.23

The popular **state feedback control** can actually be used only when process models and the knowledge on all state variables available for feedback are assumed or appropriate observers can be designed. Alternatively, IFT offers a direct data-driven model-free approach to tune state feedback controllers, and in the context of LQG or linear-quadratic regulator (LQR) problems or generally the H_2 optimization [Baz12], which are frequently used to tune optimal state feedback control systems, IFT solves these problems by a gradient-based minimization of the objective functions using data collected from the real-world control systems. Some representative examples of IFT applied to state feedback controllers are briefly discussed as follows. Direct tuning of the model parameters and the observer design by IFT are treated in [Huu09b]. The signal processing aspects of the IFT-based state feedback control for second-order positioning systems which have an integral component are discussed in [Rad09b]. A state-space formulation of IFT is analyzed in [Huu09c], where it is proved that the solution converges to the analytical solutions for the state feedback gain matrix and the Kalman gain. An LQG formulation applied to a first-order process is reported in [Huu10]. Another LQG formulation dedicated to servo systems control with the Kalman filter state observer is suggested in [Rad11b]. An IFT-based LQR approach is applied in [Rad13a] to a servo system with actuator dead zone and control signal saturation nonlinearity. All these approaches require additional gradient experiments compared to the one-degree-of-freedom SISO IFT discussed in Section 2.2.

2.4   EXAMPLE AND APPLICATION

The tower crane system described in Section 3.1 is considered as a controlled process in order to exemplify the IFT-based SISO and MIMO data-driven MFC systems. The theoretical approaches presented in this chapter will be exemplified. Three separate SISO control systems for the three controlled outputs, namely cart position, arm angular position and payload position, will be first considered, and the MIMO control system to control all these three tower crane system outputs will be next described.

2.4.1   SISO CONTROL SYSTEMS

The optimal control problems applied to the three SISO control loops are defined as the optimization problem in (2.4), and the objective functions of each SISO control loop are expressed as follows as particular expressions of the general objective function given in (2.1), multiplied by N (because small signals are involved) and considering $L_y(q^{-1}) = 0$ and $\lambda = 0$ as only reference tracking is targeted, and the same weight is given to all frequency regions:

$$J_{(\psi)}(\chi_{(\psi)}) = \frac{1}{2}E\left\{\sum_{k=1}^{N}\left[\delta y_{(\psi)}(k,\chi_{(\psi)})\right]^2\right\}, \tag{2.58}$$

where the subscript $\psi \in \{c,a,p\}$ indicates both the controlled output and its corresponding controller, namely c – the cart position (i.e., $y_{(c)} = y_1$), a – the arm angular position (i.e., $y_{(a)} = y_2$) and p – the payload position (i.e., $y_{(p)} = y_3$). Setting the sampling period to $T_s = 0.01$ s and the time horizon to 70 s, the number of samples or the length of the experiment (the trial) is $N = 70/0.01 = 7,000$ s.

Three IFT algorithms are applied, one for each controlled output, using the general steps of the SISO IFT algorithm presented in Section 2.2. The description of the five steps is given as follows in relation with the Matlab & Simulink programs and schemes.

The performance specifications imposed to the three SISO control loops in step 1 of the three IFT algorithms, i.e., one for cart position (c), one for arm angular position (a) and one for payload position (p), are expressed in terms of the reference models with the transfer functions

$$H_{\mathrm{RM}(c)}(s) = \frac{1}{1 + 0.2s}, \tag{2.59}$$

$$H_{\mathrm{RM}(a)}(s) = \frac{1}{1 + 0.21s}, \tag{2.60}$$

$$H_{\mathrm{RM}(p)}(s) = \frac{1}{1 + 0.3s}, \tag{2.61}$$

and their corresponding discrete transfer functions implemented in the three SISO control system loops are

$$H_{\mathrm{RM}(c)}\left(z^{-1}\right) = \frac{0.0488}{1 - 0.9512z^{-1}}, \tag{2.62}$$

$$H_{\mathrm{RM}(a)}\left(z^{-1}\right) = \frac{0.0465}{1 - 0.9535z^{-1}}, \tag{2.63}$$

$$H_{\mathrm{RM}(p)}\left(z^{-1}\right) = \frac{0.0328}{1 - 0.9672z^{-1}}. \tag{2.64}$$

Three PI controllers are used to control separately the processes in the three control loops. The general expressions of the discrete transfer functions of these three PI controllers are

$$C_{(\psi)}(z^{-1}, \chi_{(\psi)}) = \frac{\chi_{(\psi),1} + \chi_{(\psi),2}z^{-1}}{1 - z^{-1}}, \tag{2.65}$$

and their parameter vectors are expressed in terms of the following particular expression of (2.2):

$$\chi_{(\psi)} = \left[\chi_{(\psi),1} \ \chi_{(\psi),2}\right]^T, \tag{2.66}$$

which points out the controller tuning parameters $\chi_{(\psi),1}$ and $\chi_{(\psi),2}$, where $\psi \in \{c, a, p\}$, i.e., two parameters for each controller because $n = 2$.

The controllers are actually nonlinear because saturation-type nonlinearities are inserted on the three controller outputs in order to match the power electronics of the actuators, which operate on the basis of the pulse width modulation (PWM) principle. As shown in Chapter 1, the three control signals (or control inputs) are within -1 and 1, ensuring the necessary connection to the tower crane system subsystem, which models the controlled process in the control system structures developed in this chapter and the next chapters and also includes three zero-order hold blocks to enable digital control. The tower crane system subsystem belongs to the Process.mdl Simulink diagram, which is included in the accompanying Matlab & Simulink programs given in Chapter 1.

The three PI controllers are initially considered as continuous-time and tuned in this regard using the extended symmetrical optimum (ESO) method [Pre96, Pre99], which works well for the tower crane systems as the behavior of the three separately controlled SISO processes is of integral type with small time constants. Therefore, a simple least-squares identification is carried out; however, this should be done on a short time horizon as the controlled process is unstable. The continuous-time PI controllers are next discretized using Tustin's method to obtain quasi-continuous digital PI controllers. In this regard, the initial parameter vectors of the PI controllers involved in step 1 of the three IFT algorithms are [Rom21]

$$\chi_{(c)}^{[0]} = \left[\chi_{(c),1}^{[0]} \ \chi_{(c),2}^{[0]} \right]^T = \left[7.9787 \ -7.7817 \right]^T, \tag{2.67}$$

$$\chi_{(a)}^{[0]} = \left[\chi_{(a),1}^{[0]} \ \chi_{(a),2}^{[0]} \right]^T = \left[4.1589 \ -4.0562 \right]^T, \tag{2.68}$$

$$\chi_{(p)}^{[0]} = \left[\chi_{(p),1}^{[0]} \ \chi_{(p),2}^{[0]} \right]^T = \left[12 \ -11.7037 \right]^T, \tag{2.69}$$

The step size sequences applied in the parameter update laws (2.6) in the three IFT algorithms are given in (2.39), with appropriate superscripts inserted. The initial step size in the parameter update law is set to $\gamma_{(c)}^{[0]} = 0.001$, and the exponent of the denominator is set to $\alpha_{(c)} = 0.75$ in the case of cart position control, $\gamma_{(a)}^{[0]} = 0.0001$ and $\alpha_{(a)} = 0.85$ in the case of arm angular position control, $\gamma_{(p)}^{[0]} = 0.01$ and $\alpha_{(p)} = 0.95$ in case of payload position control.

The parameter settings in step 1 are finished with the number of iterations, which is set to 10.

The application of the ESO method to initially tune the continuous-time PI controllers, their discretization using Tustin's method and setting the reference inputs and the reference models, which are done in step 1, are implemented in the IFT_c.m Matlab program for cart position control, the IFT_a.m Matlab program for arm angular position control and the IFT_p.m Matlab program for payload position control, which are included in the accompanying Matlab & Simulink programs. The normal experiments in step 2 at iteration 0 of three IFT algorithms are conducted by digital simulations using the CS_PI_SISO_c.mdl Simulink diagram for cart position control, the CS_PI_SISO_a.mdl Simulink diagram for arm angular position control and the CS_PI_SISO_p.mdl Simulink diagram for payload position control. These three Simulink diagrams make use of the state-space model of the tower crane system viewed as a controlled process, which is described in Chapter 1 and is implemented in the `process _ model.m` S-function.

The initial step sizes and the exponents of the denominators in the step sequences (2.39) of the parameter update laws (2.6) of the three IFT algorithms in step 1 are set in the `IFT _ c _ Part2.m` Matlab program for cart position control, the `IFT _ a _ Part2.m` Matlab program for arm angular position control and the `IFT _ p _ Part2.m` Matlab program for payload position control, which are included in the accompanying Matlab & Simulink programs. The gradient experiment and steps 3–5 at iteration 0 and steps 2–5 (including the normal experiments and the gradient ones) starting with iteration 1 in the three IFT algorithms are also implemented in these three Matlab programs. These experiments are conducted by digital simulations using the three Simulink diagrams specified in the above paragraph.

Differentiating the general form (2.65) of the discrete transfer functions of the three PI controllers, the particular expression of (2.33) involved in step 4 becomes

$$\text{est}\left[\frac{\partial \delta y}{\partial \chi_{(\psi)}}(k,\chi_{(\psi)})\right] = \frac{1}{C_{(\psi)}(q^{-1},\chi_{(\psi)})} \cdot \frac{\partial C_{(\psi)}}{\partial \chi_{(\psi)}}(q^{-1},\chi_{(\psi)}) \cdot y_{2,(\psi)}(k,\chi_{(\psi)})$$

$$= \begin{bmatrix} \dfrac{1}{C_{(\psi)}(q^{-1},\chi_{(\psi)})} \cdot \dfrac{\partial C_{(\psi)}}{\partial \chi_{(\psi),1}}(q^{-1},\chi_{(\psi)}) \cdot y_{2,(\psi)}(k,\chi_{(\psi)}) \\[3mm] \dfrac{1}{C_{(\psi)}(q^{-1},\chi_{(\psi)})} \cdot \dfrac{\partial C_{(\psi)}}{\partial \chi_{(\psi),2}}(q^{-1},\chi_{(\psi)}) \cdot y_{2,(\psi)}(k,\chi_{(\psi)}) \end{bmatrix}$$

$$= \begin{bmatrix} \dfrac{1-q^{-1}}{\chi_{(\psi),1}+\chi_{(\psi),2}z^{-1}} \cdot \dfrac{1}{1-q^{-1}} \cdot y_{2,(\psi)}(k,\chi_{(\psi)}) \\[3mm] \dfrac{1-q^{-1}}{\chi_{(\psi),1}+\chi_{(\psi),2}z^{-1}} \cdot \dfrac{q^{-1}}{1-q^{-1}} \cdot y_{2,(\psi)}(k,\chi_{(\psi)}) \end{bmatrix}$$

$$= \begin{bmatrix} \dfrac{1}{\chi_{(\psi),1}+\chi_{(\psi),2}z^{-1}} \cdot y_{2,(\psi)}(k,\chi_{(\psi)}) \\[3mm] \dfrac{q^{-1}}{\chi_{(\psi),1}+\chi_{(\psi),2}z^{-1}} \cdot y_{2,(\psi)}(k,\chi_{(\psi)}) \end{bmatrix}. \tag{2.70}$$

The estimated gradients of the objective functions, namely $\text{est}\left[\dfrac{\partial J_{(\psi)}}{\partial \chi_{(\psi)}}(\chi_{(\psi)})\right]$, are calculated in step 4 using the following particular expression of (2.11), which is obtained in the context of the particular expressions (2.58) of the objective functions:

$$\text{est}\left[\frac{\partial J_{(\psi)}}{\partial \chi_{(\psi)}}(\chi_{(\psi)})\right] = \sum_{k=1}^{N}\left\{\delta y_{(\psi)}(k,\chi_{(\psi)})\text{est}\left[\frac{\partial \delta y}{\partial \chi_{(\psi)}}(k,\chi_{(\psi)})\right]\right\}, \tag{2.71}$$

where est $\left[\dfrac{\partial \delta y}{\partial \chi_{(\psi)}}(k, \chi_{(\psi)})\right]$ is obtained from (2.70), and the output error (or the track-

ing error) $\delta y_{(\psi)}(k, \chi_{(\psi)})$ is taken from the normal experiment and employed in (2.71).

The expressions of the matrices $\mathbf{R}_{(\psi)}^{[i]}$ involved in the parameter update laws (2.6) in step 5 are set as the identity matrices in all three IFT algorithms:

$$\mathbf{R}_{(\psi)}^{[i]} = \begin{bmatrix} 1 & 0 \\ 0 & 1 \end{bmatrix}. \tag{2.72}$$

The iteration indices are also incremented in step 5 in the IFT _ c _ Part2.m Matlab program (for cart position control), the IFT _ a _ Part2.m Matlab program (for arm angular position control) and the IFT _ p _ Part2.m Matlab program (for payload position control), which are run repeatedly until the maximum number of 10 iterations is reached.

Some results obtained after running the three IFT algorithms on ten iterations to ensure the data-driven model-free tuning of the three PI controllers are briefly presented as follows. The parameter vectors of the PI controllers obtained after the ten iterations of the three IFT algorithms are as follows [Rom21]:

$$\chi_{(c)}^{[10]} = \begin{bmatrix} \chi_{(c),1}^{[10]} & \chi_{(c),2}^{[10]} \end{bmatrix}^T = [7.9695 - 7.7919]^T, \tag{2.73}$$

$$\chi_{(a)}^{[10]} = \begin{bmatrix} \chi_{(a),1}^{[10]} & \chi_{(a),2}^{[10]} \end{bmatrix}^T = [4.1577 - 4.0577]^T, \tag{2.74}$$

$$\chi_{(p)}^{[10]} = \begin{bmatrix} \chi_{(p),1}^{[10]} & \chi_{(p),2}^{[10]} \end{bmatrix}^T = [11.9781 - 11.7282]^T. \tag{2.75}$$

The evolutions of the controller parameters in (2.65) along the ten iterations of the IFT algorithms are given in Figure 2.4 for the parameters of the cart position controller, Figure 2.5 for the parameters of the arm angular position controller and Figure 2.6 for the parameters of the payload position controller. Small modifications occur for all three SISO controllers.

The evolutions of the objective functions in (2.58) along ten iterations are illustrated in Figure 2.7 for cart position control, Figure 2.8 for arm angular position control and Figure 2.9 for payload position control. A rather slow convergence is observed; however, the performance improvement of all three control systems is obtained and expressed in terms of alleviating the values of the objective functions.

The simulation results obtained for the PI controllers with the initial parameters in (2.67–2.69) and the PI controllers with the parameters in (2.73–2.75) tuned after ten iterations of the IFT algorithms are illustrated in Figure 2.10 for cart position control, Figure 2.11 for arm angular position control and Figure 2.12 for payload position control.

The experimental results are next presented. They are obtained for the PI controllers with the initial parameters in (2.67–2.69) and the PI controllers with the parameters in (2.73–2.75) tuned after ten iterations of the IFT algorithms and are given in Figure 2.13 for cart position control, Figure 2.14 for arm angular position control and Figure 2.15 for payload position control.

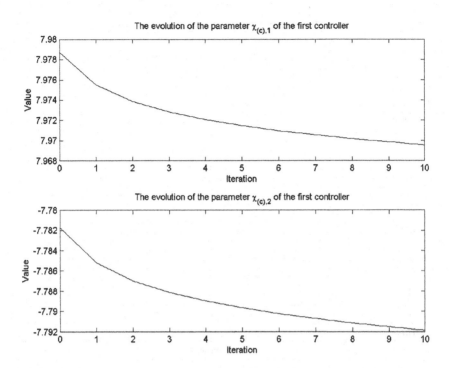

FIGURE 2.4 Parameters of cart position controller versus iteration number of IFT algorithm.

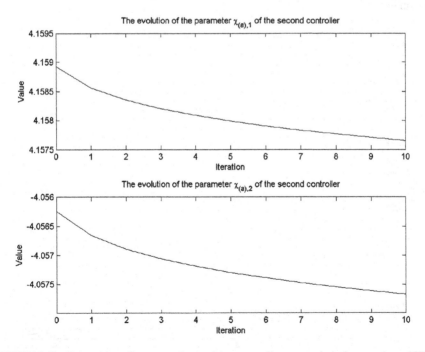

FIGURE 2.5 Parameters of arm angular position controller versus iteration number of IFT algorithm.

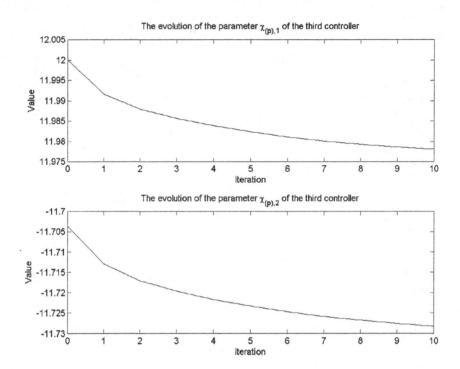

FIGURE 2.6 Parameters of payload position controller versus iteration number of IFT algorithm.

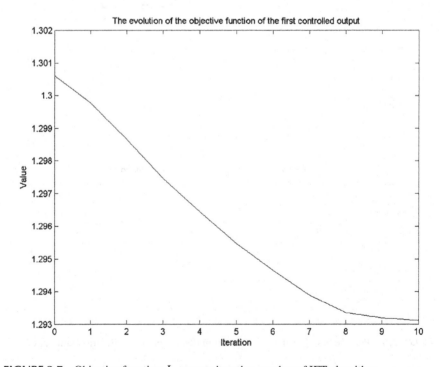

FIGURE 2.7 Objective function $J_{(c)}$ versus iteration number of IFT algorithm.

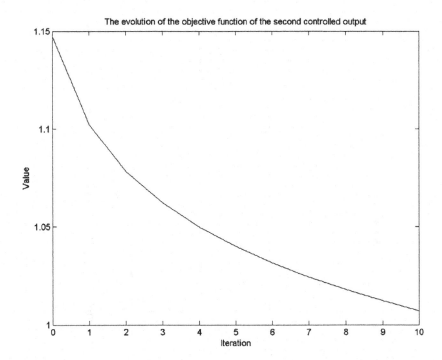

FIGURE 2.8 Objective function $J_{(a)}$ versus iteration number of IFT algorithm.

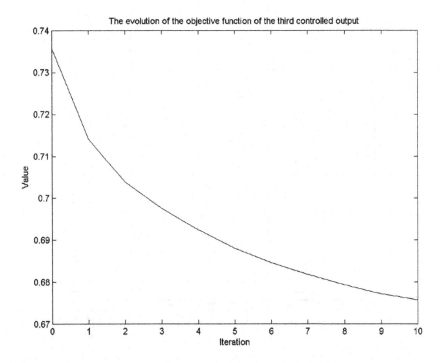

FIGURE 2.9 Objective function $J_{(p)}$ versus iteration number of IFT algorithm.

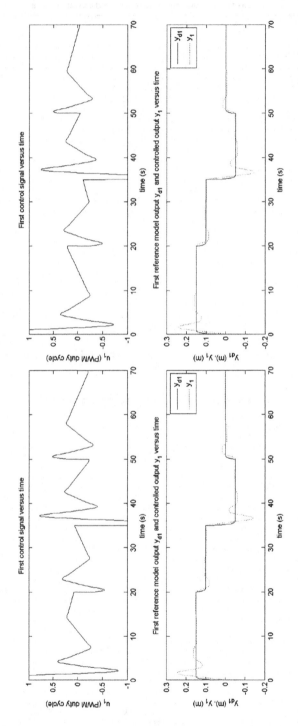

FIGURE 2.10 Simulation results of cart position control system with the initial PI controller parameters (left) and the PI controller parameters tuned after ten iterations of the IFT algorithm.

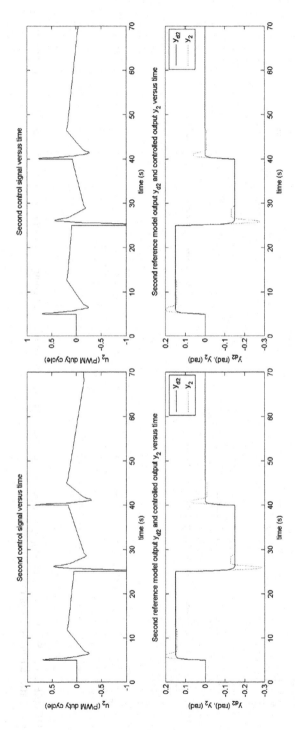

FIGURE 2.11 Simulation results of arm angular position control system with the initial **PI** controller parameters (left) and the **PI** controller parameters tuned after ten iterations of the **IFT** algorithm.

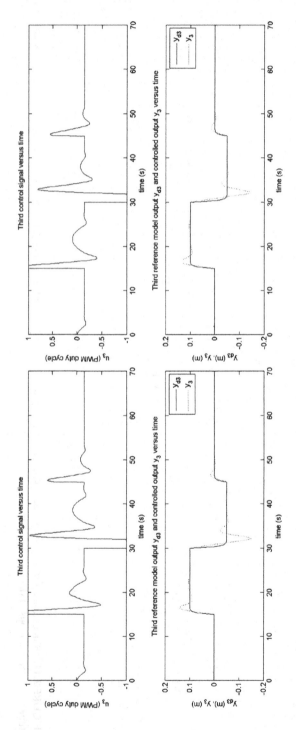

FIGURE 2.12 Simulation results of payload position control system with the initial PI controller parameters (left) and the PI controller parameters tuned after ten iterations of the IFT algorithm.

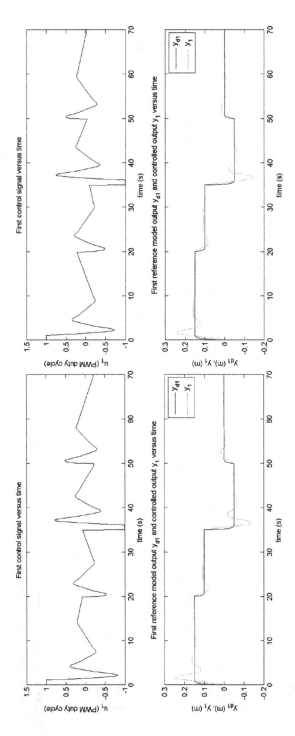

FIGURE 2.13 Experimental results of cart position control system with the initial PI controller parameters (left) and the PI controller parameters tuned after ten iterations of the IFT algorithm.

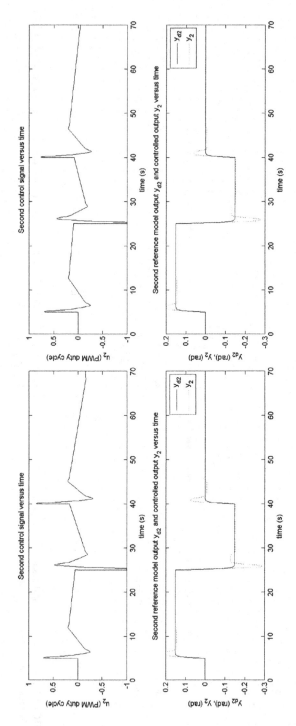

FIGURE 2.14 Experimental results of arm angular position control system with the initial PI controller parameters (left) and the PI controller parameters tuned after ten iterations of the IFT algorithm.

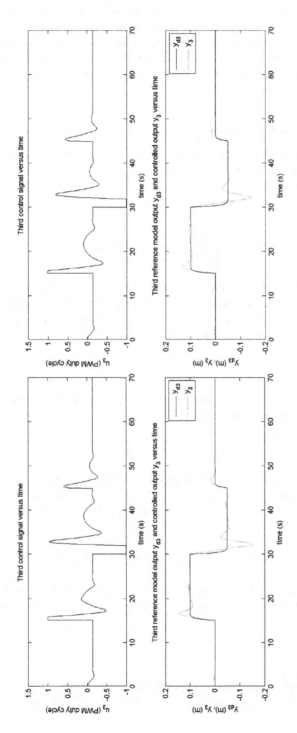

FIGURE 2.15 Experimental results of payload position control system with the initial PI controller parameters (left) and the PI controller parameters tuned after ten iterations of the IFT algorithm.

Both simulation and experimental results show the performance improvement offered by the IFT-based tuning in terms of improved tracking and improved empirical performance indices (overshoot and settling time). Additional iterations are necessary for a more significant improvement, and the readers are invited to test the control systems for different parameter settings of IFT algorithms, different numbers of iterations and other processes.

2.4.2 MIMO CONTROL SYSTEM

The optimal control problem applied to the MIMO control system is defined as the optimization problem in (2.4), and the objective function is expressed as follows as a particular expression of the objective function given in (2.43), multiplied by N (because small signals are involved):

$$J(\chi) = \frac{1}{2} E \left\{ \sum_{k=1}^{N} \left\{ [\delta \mathbf{y}(k,\chi)]^T \delta \mathbf{y}(k,\chi) \right\} \right\}, \tag{2.76}$$

where, as in the SISO case, setting the sampling period to $T_s = 0.01$ s and the time horizon to 70 s, the number of samples or the length of the experiment (the trial) is $N = 70 / 0.01 = 7,000$ s. Since $n_y = 3$ for the tower crane system (in the context of Chapter 1), the expression of the controlled output vector is

$$\mathbf{y} = \begin{bmatrix} y_1 & y_2 & y_3 \end{bmatrix}^T, \tag{2.77}$$

y_1 is the cart position, y_2 is the arm angular position and y_3 is the payload position; the expression of the reference model output vector is

$$\mathbf{y}_d = \begin{bmatrix} y_{d1} & y_{d2} & y_{d3} \end{bmatrix}^T, \tag{2.78}$$

y_{d1} is the reference model output for cart position control, y_{d2} is the reference model output for arm angular position control and y_{d3} is the reference model output for payload position control; and the expression of the output error (or the tracking error) vector is

$$\delta \mathbf{y} = \mathbf{y} - \mathbf{y}_d = \begin{bmatrix} \delta y_1 & \delta y_2 & \delta y_3 \end{bmatrix}^T = \begin{bmatrix} y - y_{d1} & y - y_{d2} & y - y_{d3} \end{bmatrix}^T, \tag{2.79}$$

δy_1 is the cart position tracking error, δy_2 is the arm angular position tracking error and δy_3 is the payload position tracking error. Using the reference input (or the set-point) vector

$$\mathbf{r} = \begin{bmatrix} r_1 & r_2 & r_3 \end{bmatrix}^T, \tag{2.80}$$

where r_1 is the reference input (or the set-point) for cart position control, r_2 is the reference input (or the set-point) for arm angular position control and r_3 is the reference input (or the set-point) for payload position control, the particular expression of the control error vector is

$$\mathbf{e} = \mathbf{r} - \mathbf{y} = \begin{bmatrix} e_1 & e_2 & e_3 \end{bmatrix}^T = \begin{bmatrix} r_1 - y_1 & r_2 - y_2 & r_3 - y_3 \end{bmatrix}^T, \qquad (2.81)$$

where e_1 is the control error for cart position control, e_2 is the control error for angular position control and e_3 is the control error for payload position control.

Since $n_u = 3$ for the tower crane system (in the context of Chapter 1), the expression of the control signal (or the control input) vector is

$$\mathbf{u} = \begin{bmatrix} u_1 & u_2 & u_3 \end{bmatrix}^T, \qquad (2.82)$$

where u_1 is the control signal (or the control input) for cart position control, u_2 is the control signal (or the control input) for arm angular position control and u_3 is the control signal (or the control input) for payload position control. Since there are no cross-couplings of the first two control channels (for cart position control and arm angular position control), the expression of the discrete transfer function matrix of the MIMO controller is set as follows in diagonal form:

$$\mathbf{C}(z^{-1}, \boldsymbol{\chi}) = \begin{bmatrix} C_{(c)}\left(z^{-1}, \boldsymbol{\chi}_{(c)}\right) & 0 & 0 \\ 0 & C_{(a)}\left(z^{-1}, \boldsymbol{\chi}_{(a)}\right) & 0 \\ 0 & 0 & C_{(p)}\left(z^{-1}, \boldsymbol{\chi}_{(p)}\right) \end{bmatrix}$$

$$= \begin{bmatrix} \dfrac{\chi_{(c),1} + \chi_{(c),2} z^{-1}}{1 - z^{-1}} & 0 & 0 \\ 0 & \dfrac{\chi_{(a),1} + \chi_{(a),2} z^{-1}}{1 - z^{-1}} & 0 \\ 0 & 0 & \dfrac{\chi_{(p),1} + \chi_{(p),2} z^{-1}}{1 - z^{-1}} \end{bmatrix}. \qquad (2.83)$$

The discrete transfer functions in (2.65) of the three SISO PI controllers are placed in the main diagonal, and the expression of the parameter vector is

$$\boldsymbol{\chi} = \begin{bmatrix} \chi_1 & \chi_2 & \chi_3 & \chi_4 & \chi_5 & \chi_6 \end{bmatrix}^T = \begin{bmatrix} \chi_{(c),1} & \chi_{(c),2} & \chi_{(a),1} & \chi_{(a),2} & \chi_{(p),1} & \chi_{(p),2} \end{bmatrix}^T, \qquad (2.84)$$

indicating that a number of $n = 6$ parameters should be tuned by MIMO IFT. These parameters are also highlighted in (2.65).

As specified in the previous section dedicated to SISO IFT, the three SISO controllers are actually nonlinear because saturation-type nonlinearities are inserted on their outputs in order to match the power electronics of the actuators, which operate on the basis of the PWM principle. That is the reason why, as shown in Chapter 1, the three control signals (or control inputs) are within -1 and 1 that belong to the Process.mdl Simulink diagram, which is included in the accompanying Matlab & Simulink programs given in Chapter 1.

The general MIMO IFT algorithm presented in Section 2.3 is applied. The description of the five steps is highlighted as follows in relation with the accompanying Matlab & Simulink programs and schemes.

The performance specifications imposed to the MIMO control system are expressed in step 1 of the MIMO IFT algorithm in terms of three reference models, one for each output, with the transfer functions presented in (2.59–2.61) and the corresponding discrete transfer functions (2.62–2.64), which are implemented in the MIMO control system structure.

As carried out in the previous section dedicated to SISO IFT, three continuous-time PI controllers that correspond to the discrete-time ones in (2.83) are initially considered and tuned using the ESO method [Pre96, Pre99]. The continuous-time PI controllers are next discretized using Tustin's method to obtain quasi-continuous digital PI controllers. This leads to the initial parameter vector MIMO controller involved in step 1 of the MIMO IFT algorithm

$$\chi^{[0]} = [\chi_{(c),1}^{[0]} \quad \chi_{(c),2}^{[0]} \quad \chi_{(a),1}^{[0]} \quad \chi_{(a),2}^{[0]} \quad \chi_{(p),1}^{[0]} \quad \chi_{(p),2}^{[0]}]^T$$

$$= [2.9787 \quad -7.7817 \quad 4.1589 \quad -4.0562 \quad 12 \quad -11.7037]^T, \tag{2.85}$$

which actually gathers the three parameter vectors in (2.67–2.69).

The step size sequence applied in the parameter update laws (2.6) in the MIMO IFT algorithms is given in (2.39). The initial step size in the parameter update law is set to $\gamma^{[0]} = 0.01$, and the exponent of the denominator is set to $\alpha = 0.55$.

The parameter settings in step 1 are finished with the number of iterations, which is set to 10.

The application of the ESO method to initially tune the continuous-time PI controllers that correspond to the discrete-time ones in (2.83), their discretization using Tustin's method and setting the reference inputs and the reference models, which are done in step 1, are implemented in the IFT.m Matlab program, which is included in the accompanying Matlab & Simulink programs. The normal experiments in step 2 at iteration 0 of MIMO IFT algorithm are conducted by digital simulations using the CS _ PI _ MIMO.mdl Simulink diagram. This Simulink diagram makes uses of the state-space model of the tower crane system viewed as a controlled process, which is described in Chapter 1 and is implemented in the process _ model.m S-function.

Building the control system structure given in Figure 2.3, which is needed to conduct the six gradient experiments in step 3, requires the computation of the partial derivatives $\mathbf{C}'(z^{-1}, \chi)$, i.e., the derivatives of $\mathbf{C}(z^{-1}, \chi)$ in (2.83) with respect to the controller tuning parameter χ_l, $l = 1...n$, which filter the control error vector $\mathbf{e}$, $\mathbf{e} \in \Re^{n_y}$, obtained in the normal experiment. Using (2.83), and the notations defined in (2.49) and (2.84), the expressions of these partial derivatives are expressed as the following matrices:

$$\frac{\partial \mathbf{C}\left(z^{-1},\boldsymbol{\chi}\right)}{\partial \chi_{(c),1}} = \begin{bmatrix} \dfrac{1}{1-z^{-1}} & 0 & 0 \\ 0 & 0 & 0 \\ 0 & 0 & 0 \end{bmatrix} \tag{2.86}$$

in the first gradient experiment,

$$\frac{\partial \mathbf{C}\left(z^{-1},\boldsymbol{\chi}\right)}{\partial \chi_{(c),2}} = \begin{bmatrix} \dfrac{z^{-1}}{1-z^{-1}} & 0 & 0 \\ 0 & 0 & 0 \\ 0 & 0 & 0 \end{bmatrix} \tag{2.87}$$

in the second gradient experiment,

$$\frac{\partial \mathbf{C}\left(z^{-1},\boldsymbol{\chi}\right)}{\partial \chi_{(a),1}} = \begin{bmatrix} 0 & 0 & 0 \\ 0 & \dfrac{1}{1-z^{-1}} & 0 \\ 0 & 0 & 0 \end{bmatrix} \tag{2.88}$$

in the third gradient experiment,

$$\frac{\partial \mathbf{C}\left(z^{-1},\boldsymbol{\chi}\right)}{\partial \chi_{(a),2}} = \begin{bmatrix} 0 & 0 & 0 \\ 0 & \dfrac{z^{-1}}{1-z^{-1}} & 0 \\ 0 & 0 & 0 \end{bmatrix} \tag{2.89}$$

in the fourth gradient experiment,

$$\frac{\partial \mathbf{C}\left(z^{-1},\boldsymbol{\chi}\right)}{\partial \chi_{(p),1}} = \begin{bmatrix} 0 & 0 & 0 \\ 0 & 0 & 0 \\ 0 & 0 & \dfrac{1}{1-z^{-1}} \end{bmatrix} \tag{2.90}$$

in the fifth gradient experiment and

$$\frac{\partial \mathbf{C}\left(z^{-1},\boldsymbol{\chi}\right)}{\partial \chi_{(p),2}} = \begin{bmatrix} 0 & 0 & 0 \\ 0 & 0 & 0 \\ 0 & 0 & \dfrac{z^{-1}}{1-z^{-1}} \end{bmatrix} \tag{2.91}$$

in the sixth gradient experiment. The calculations of the partial derivatives in terms
of (2.86–2.91) are implemented in the IFT _ Part2.m Matlab program.

The estimated expression of the transposed gradient of the controlled output, i.e.,
$\text{est}\left[\dfrac{\partial \mathbf{y}}{\partial \chi}(k, \chi)\right]^{T}$, is obtained in step 4 by placing the output vector of the gradient
experiment l conducted at step 3 in the row l of the matrix placed in the right-hand
term of the following particular expression of (2.57) obtained for $n_y = 3$:

$$\text{est}\left[\frac{\partial \mathbf{y}}{\partial \chi}(k, \chi)\right]^{T} = \begin{bmatrix} \dfrac{\partial y_1}{\partial \chi_1}(k, \chi) & \dfrac{\partial y_2}{\partial \chi_1}(k, \chi) & \dfrac{\partial y_3}{\partial \chi_1}(k, \chi) \\[2mm] \dfrac{\partial y_1}{\partial \chi_2}(k, \chi) & \dfrac{\partial y_2}{\partial \chi_2}(k, \chi) & \dfrac{\partial y_3}{\partial \chi_2}(k, \chi) \\[2mm] \cdots & \cdots & \cdots \\[2mm] \dfrac{\partial y_1}{\partial \chi_6}(k, \chi) & \dfrac{\partial y_2}{\partial \chi_6}(k, \chi) & \dfrac{\partial y_3}{\partial \chi_6}(k, \chi) \end{bmatrix}, \qquad (2.92)$$

where l corresponds to the experiment conducted with respect to the controller tuning parameter χ_l, $l = 1 \ldots 6$. This matrix, along with the output error (or the tracking
error) vector obtained in the normal experiment, is next substituted in the following
particular expression of (2.55), which is obtained in the context of the particular
expression (2.76) of the objective function, to calculate the estimated gradient of the
objective function, i.e., $\text{est}\left[\dfrac{\partial J}{\partial \chi}(\chi)\right]$:

$$\text{est}\left[\frac{\partial J}{\partial \chi}(\chi)\right] = \sum_{k=1}^{N}\left\{\text{est}\left[\frac{\partial \mathbf{y}}{\partial \chi}(k, \chi)\right]^{T} \delta \mathbf{y}(k, \chi)\right\}. \qquad (2.93)$$

The expressions of the matrices $\mathbf{R}^{[i]}$ involved in the parameter update law (2.6) in
step 5 are set as the identity matrices

$$\mathbf{R}^{[i]} = \text{diag}(1,1,1,1,1,1). \qquad (2.94)$$

The iteration index is also incremented in step 5 in the IFT _ Part2.m Matlab program, which is run repeatedly being called by the IFT.m Matlab program until the
maximum number of 10 iterations is reached.

Some results obtained after running the MIMO IFT algorithm on ten iterations
to ensure the data-driven model-free tuning of the MIMO controller are briefly presented as follows. The parameter vectors of the MIMO controller obtained after ten
iterations of the MIMO IFT algorithm are

$$\chi^{[10]} = [\chi_{(c),1}^{[10]} \ \chi_{(c),2}^{[10]} \ \chi_{(a),1}^{[10]} \ \chi_{(a),2}^{[10]} \ \chi_{(p),1}^{[10]} \ \chi_{(p),2}^{[10]}]^{T}$$

$$= [7.7431 \ -7.6207 \ 4.1589 \ -4.0563 \ 12.0001 \ -11.7036]^{T}. \quad (2.95)$$

The evolutions of the controller parameters in (2.83) along the ten iterations of the MIMO IFT algorithm are given in Figures 2.16–2.18 as follows for the parameters of the MIMO controller that consists of the three SISO controllers: Figure 2.16 is for the parameters of the cart position controller, Figure 2.17 is for the parameters of the arm angular position controller and Figure 2.18 is for the parameters of the payload position controller. Very small modifications occur for the arm angular position controller and the payload position controller and big ones for the cart position controller.

The evolution of the objective function in (2.76) along the ten iterations of the MIMO IFT algorithm is illustrated in Figure 2.19. A rather slow convergence is observed, similar to the SISO IFT algorithms, but the performance improvement of the MIMO control system is obtained and expressed in terms of alleviating the values of the objective function. In addition, the evolutions of the controller parameters and the objective function are different from the SISO IFT algorithms because a different objective function is employed.

The simulation results obtained for the MIMO controller with the initial parameters in (2.85) and the MMIMO controller with the parameters in (2.95) tuned after ten iterations of the MIMO IFT algorithm are illustrated in Figure 2.20 for cart

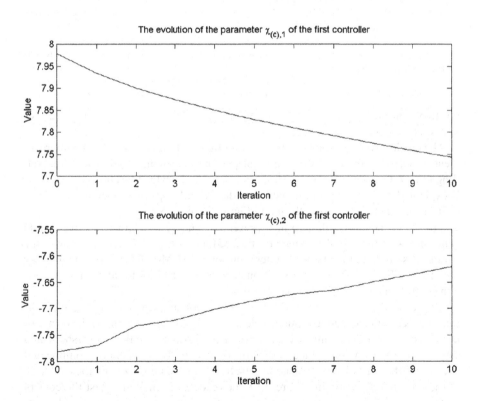

FIGURE 2.16 Parameters of cart position controller versus iteration number of MIMO IFT algorithm.

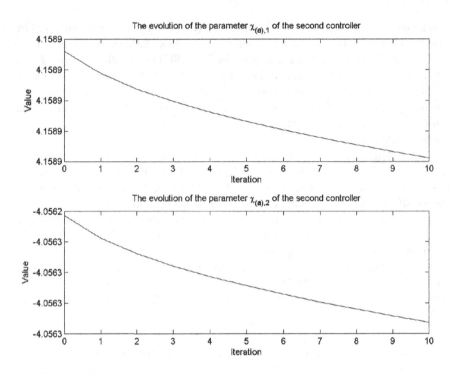

FIGURE 2.17 Parameters of arm angular position controller versus iteration number of MIMO IFT algorithm.

position control, Figure 2.21 for arm angular position control and Figure 2.22 for payload position control.

The results given in Figures 2.20–2.22 highlight that minor performance improvement is obtained as far as the empirical performance indices (overshoot and settling time) are concerned for cart position control and payload position control. This is explained because the minimization of the overall objective function defined in (2.76) is pursued and is also reflected in Figures 2.16–2.18.

The experimental results are next presented. They are obtained for the MIMO controller with the initial parameters in (2.85) and the MIMO controller with the parameters in (2.95) tuned after ten iterations of the MIMO IFT algorithm, and they are given in Figure 2.23 for cart position control, Figure 2.24 for arm angular position control and Figure 2.25 for payload position control.

Similar to the case of SITO IFT, both simulation and experimental results outline the performance improvement offered by the IFT-based tuning in terms of improved tracking and improved empirical performance indices (overshoot and settling time). However, since the optimization problem is more complicated because of the control system structure itself, which is emphasized by the evolution of the objective function in Figure 2.19, the readers are invited to test the control systems for different parameter settings of IFT algorithms, different numbers of iterations and other processes, in order to reach the expected control system performance improvement.

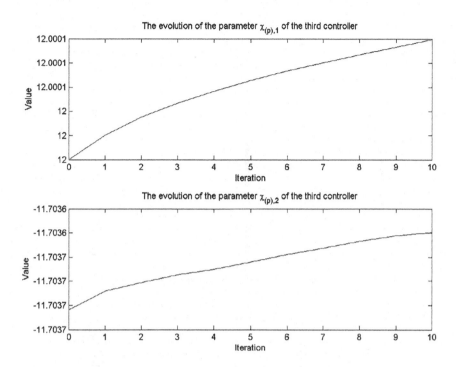

FIGURE 2.18 Parameters of payload position controller versus iteration number of MIMO IFT algorithm.

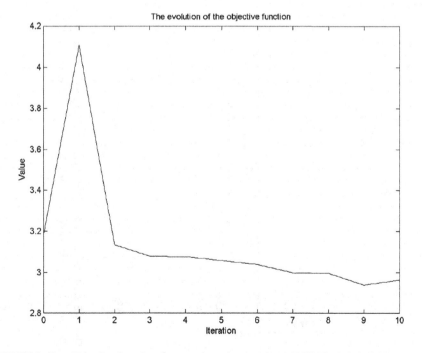

FIGURE 2.19 Objective function J versus iteration number of MIMO IFT algorithm.

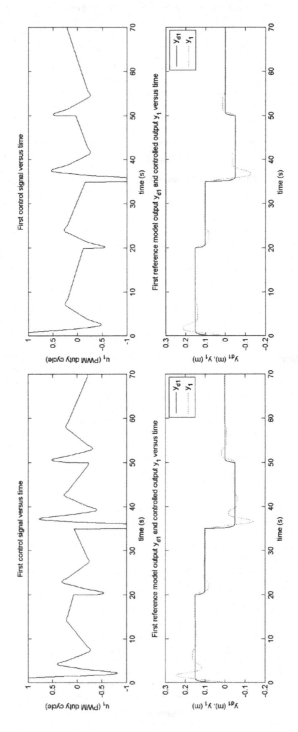

FIGURE 2.20 Simulation results of cart position control obtained by the MIMO control system with the initial MIMO controller parameters (left) and the MIMO controller parameters tuned after ten iterations of the IFT algorithm.

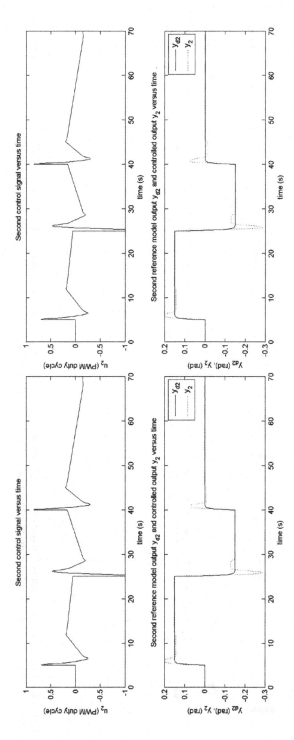

FIGURE 2.21 Simulation results of arm angular position control obtained by the MIMO control system with the initial MIMO controller parameters (left) and the MIMO controller parameters tuned after ten iterations of the IFT algorithm.

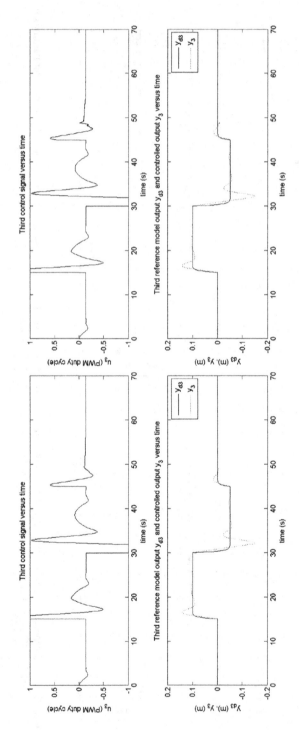

FIGURE 2.22 Simulation results of payload position control obtained by the MIMO control system with the initial MIMO controller parameters (left) and the MIMO controller parameters tuned after ten iterations of the IFT algorithm.

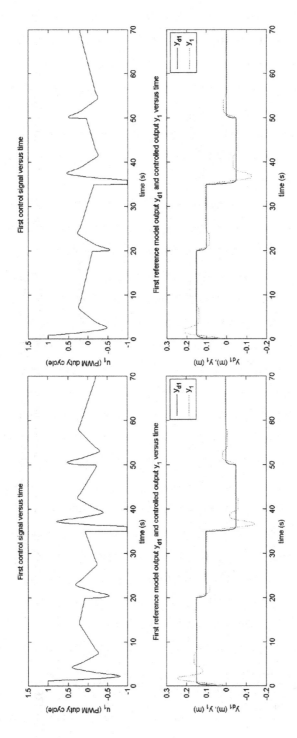

FIGURE 2.23 Experimental results of cart position control obtained by the MIMO control system with the initial MIMO controller parameters (left) and the MIMO controller parameters tuned after ten iterations of the IFT algorithm.

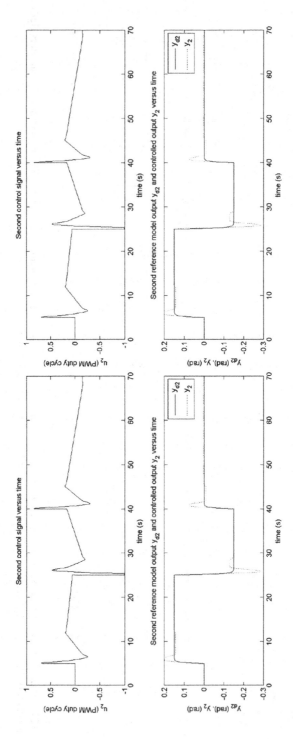

FIGURE 2.24 Experimental results of arm angular position control obtained by the MIMO control system with the initial MIMO controller parameters (left) and the MIMO controller parameters tuned after ten iterations of the IFT algorithm.

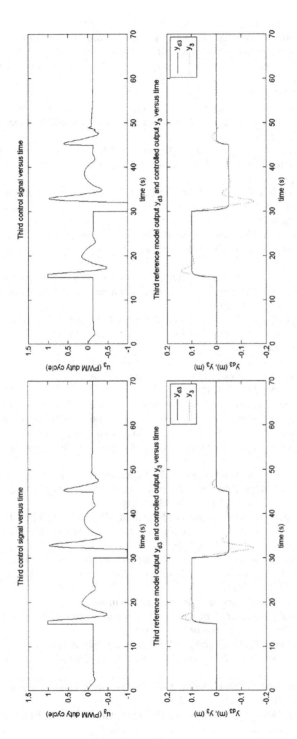

FIGURE 2.25 Experimental results of payload position control obtained by the MIMO control system with the initial MIMO controller parameters (left) and the MIMO controller parameters tuned after ten iterations of the IFT algorithm.

REFERENCES

[Baz12] A. S. Bazanella, L. Campestrini and D. Eckhard, *Data-Driven Controller Design: The H2 Approach*, Springer, Dordrecht, 2012.

[DeB97] F. De Bruyne, "Iterative feedback tuning for MIMO systems," in *Proceedings of 2nd International Symposium on Intelligent Automation and Control*, Anchorage, AK, pp. 179.1–179.8, 1997.

[Fli20] M. Fliess and C. Join, "Machine learning and control engineering: the model-free case," in *Proceedings of Future Technologies Conference 2020*, Vancouver, BC, Canada, pp. 1–20, 2020.

[Gun03] S. Gunnarsson, V. Collignon and O. Rousseaux, "Tuning of a decoupling controller for a 2×2 system using iterative feedback tuning," *Control Engineering Practice*, vol. 11, no. 9, pp. 1035–1041, Sep. 2003.

[Ham03] K. Hamamoto, T. Fukuda and T. Sugie, "Iterative feedback tuning of controllers for a two-mass-spring system with friction," *Control Engineering Practice*, vol. 11, no. 9, pp. 1061–1068, Sep. 2003.

[Hja94] H. Hjalmarsson, S. Gunnarsson and M. Gevers, "A convergent iterative restricted complexity control design scheme," in *Proceedings of 33rd IEEE Conference on Decision and Control*, Orlando, FL, vol. 2, pp. 1735–1740, 1994.

[Hja98a] H. Hjalmarsson and T. Birkeland, "Iterative feedback tuning of linear time-invariant MIMO systems," in *Proceedings of 37th IEEE Conference on Decision and Control*, Tampa, FL, pp. 3893–3898, 1998.

[Hja98b] H. Hjalmarsson, M. Gevers, S. Gunnarsson and O. Lequin, "Iterative feedback tuning: theory and applications," *IEEE Control Systems Magazine*, vol. 18, no. 4, pp. 26–41, Aug. 1998.

[Hja99] H. Hjalmarsson, "Efficient tuning of linear multivariable controllers using iterative feedback tuning," *International Journal of Adaptive Control and Signal Processing*, vol. 13, no. 7, pp. 553–572, Nov. 1999.

[Hja02] H. Hjalmarsson, "Iterative feedback tuning - an overview," *International Journal of Adaptive Control and Signal Processing*, vol. 16, no. 5, pp. 373–395, Jun. 2002.

[Huu09a] J. K. Huusom, N. K. Poulsen and S. B. Jørgensen, "Improving convergence of iterative feedback tuning," *Journal of Process Control*, vol. 19, no. 4, pp. 570–578, Apr. 2009.

[Huu09b] J. K. Huusom, N. K. Poulsen and S. B. Jørgensen, "Iterative feedback tuning of state space control loops with observers given model uncertainty," *Computer Aided Chemical Engineering*, vol. 27, pp. 1359–1364, Aug. 2009.

[Huu09c] J. K. Huusom, N. K. Poulsen and S. B. Jørgensen, "Data driven tuning of state space controllers with observers," in *Proceedings of 2009 European Control Conference*, Budapest, Hungary, pp. 1961–1966, 2009.

[Huu10] J. K. Huusom, N. K. Poulsen and S. B. Jørgensen, "Iterative feedback tuning of uncertain state space systems," *Brazilian Journal of Chemical Engineering*, vol. 27, no. 3, pp. 461–472, Sep. 2010.

[Jan04] H. Jansson and H. Hjalmarsson, "Gradient approximations in iterative feedback tuning for multivariable processes," *International Journal of Adaptive Control and Signal Processing*, vol. 18, no. 8, pp. 665–681, Oct. 2004.

[Jun21] H. Jung, K. Jeon, J.-G. Kang and S. Oh, "Iterative feedback tuning of cascade control of two-inertia system," *IEEE Control Systems Letters*, vol. 5, no. 3, pp. 785–790, Jul. 2021.

[Lan06] I. D. Landau, and G. Zito, *Digital Control Systems: Design, Identification and Implementation*, Springer-Verlag, London, 2006.

[Leq03] O. Lequin, M. Gevers, M. Mossberg, E. Bosmans and L. Triest, "Iterative feedback tuning of PID parameters: comparison with classical tuning rules," *Control Engineering Practice*, vol. 11, no. 9, pp. 1023–1033, Sep. 2003.

[Pre96] S. Preitl and R.-E. Precup, "On the algorithmic design of a class of control systems based on providing the symmetry of open-loop Bode plots," *Scientific Bulletin of UPT, Transactions on Automatic Control and Computer Science*, vol. 41 (55), no. 2, pp. 47–55, Dec. 1996.

[Pre99] St. Preitl and R.-E. Precup, "An extension of tuning relations after symmetrical optimum method for PI and PID Controllers," *Automatica*, vol. 35, no. 10, pp. 1731–1736, Oct. 1999.

[Pre07] S. Preitl, R.-E. Precup, Z. Preitl, S. Vaivoda, S. Kilyeni and J. K. Tar, "Iterative feedback and learning control. Servo systems applications," *IFAC Proceedings Volumes*, vol. 40, no. 8, pp. 16–27, May 2007.

[Pre20] R.-E. Precup, R.-C. Roman, T.-A. Teban, A. Albu, E. M. Petriu and C. Pozna, "Model-free control of finger dynamics in prosthetic hand myoelectric-based control systems," *Studies in Informatics and Control*, vol. 29, no. 4, pp. 399–410, Dec. 2020.

[Rad09a] M.-B. Radac, R.-E. Precup, S. Preitl and C.-A. Dragos, "Iterative feedback tuning in MIMO systems. Signal processing and application," in *Proceedings of 5th International Symposium on Applied Computational Intelligence and Informatics*, Timisoara, Romania, pp. 77–82, 2009.

[Rad09b] M.-B. Radac, R.-E. Precup, S. Preitl, E. M. Petriu, C.-A. Dragos, A. S. Paul and S. Kilyeni, "Signal processing aspects in state feedback control based on iterative feedback tuning," in *Proceedings of 2nd International Conference on Human System Interaction*, Catania, Italy, pp. 40–45, 2009.

[Rad11a] M.-B. Radac, *Iterative Techniques for Controller Tuning*, PhD thesis, Editura Politehnica, Timisoara, 2011.

[Rad11b] M.-B. Radac, R.-E. Precup, E. M. Petriu and S. Preitl, "Application of IFT and SPSA to servo system control," *IEEE Transactions on Neural Networks*, vol. 22, no. 12, pp. 2363–2375, Dec. 2011.

[Rad12] M.-B. Radac, B.-A. Bigher, R.-E. Precup, E. M. Petriu, C.-A. Dragos, S. Preitl and A.-I. Stinean, "Data-based tuning of PI controllers for vertical three-tank systems," in *Proceedings of 9th International Conference on Informatics in Control, Automation and Robotics*, Rome, Italy, vol. 1, pp. 31–39, 2012.

[Rad13a] M.-B. Radac, R.-E. Precup, E. M. Petriu and S. Preitl, "Experiment-based performance improvement of state feedback control systems for single input processes," *Acta Polytechnica Hungarica*, vol. 10, no. 1, pp. 5–24, Feb. 2013.

[Rad13b] M.-B. Radac, R.-C. Roman, R.-E. Precup, E. M. Petriu, C.-A. Dragos and S. Preitl, "Data-based tuning of linear controllers for MIMO twin rotor systems," in *Proceedings of IEEE Region 8 EuroCon 2013 Conference*, Zagreb, Croatia, pp. 1915–1920, 2013.

[Rom21] R.-C. Roman, R.-E. Precup, E.-L. Hedrea, S. Preitl, I. A. Zamfirache, C.-A. Bojan-Dragos and E. M. Petriu, "Iterative feedback tuning algorithm for tower crane systems," in *Proceedings of International Conference on Information Technology and Quantitative Management (ITQM 2020&2021)*, Chengdu, China, pp. 1–8, 2021.

3 Intelligent PID Controllers

3.1 INTRODUCTION

The model-free control (MFC) is a data-driven technique that uses the input/output data of the control process in the controller design; it uses the online linear approximation of the process and an estimator to update the linear approximation [Fli13, Rom16b, Rom18c].

The control algorithms tuned by MFC, also known as MFC algorithms, are well known in the state of the-art as "intelligent" controllers or MFC controllers to highlight the controllers that are implemented based on the MFC technique, usually designed based on Proportional (P), Proportional-Integral (PI) or Proportional-Integral-Derivative (PID) type controllers. The MFC controllers that contain P, PI or PID elements are also known as intelligent P, PI and PID controllers, with the abbreviations iP, iPI and iPID, respectively. The MFC technique is also known in the literature as model-free tuning [Fli13, Rom18c]. Within the MFC-based tuning, the parameters of the controllers and nonparametric mathematical models of the processes can be used in the form of process responses to various input signals, i.e., control signals [Fli13, Rom16b, Rom18c].

The MFC algorithm uses an ultra-local model $\mathbf{y}^v = \mathbf{F} + \alpha\mathbf{u}$ to replace the unknown mathematical model of the process, where the controlled output (i.e., the process output) vector $\mathbf{y}^v(t)$ is the v order derivative of $\mathbf{y}$ with $v \geq 1$ ($v = 1$ in the case of first-order MFC algorithm, and $v = 2$ in the case of the second-order MFC algorithm), $\mathbf{u}$ is the control input vector or the control signal vector, the vector $\mathbf{F}$ plays a disturbance role that is continuously updated, and α is a constant matrix.

The iPID controllers are PID controllers equipped with an online parameter estimation method for estimating the unknown nonlinear or linear dynamics of the controlled system. In this regard, their structure is generally similar to PID controllers, while estimated parameters of the system are included.

3.2 THEORY OF INTELLIGENT PID CONTROLLERS FOR CONTINUOUS-TIME DYNAMIC SYSTEMS

3.2.1 STRUCTURE OF iPID CONTROLLERS FOR CONTINUOUS-TIME DYNAMIC SYSTEMS

Definition 3.1

Here, the dynamics of a generic system is defined as follows [Fli13]:

$$\dot{\mathbf{y}} = \mathbf{F} + \mathbf{B}\,\mathbf{u}, \tag{3.1}$$

DOI: 10.1201/9781003143444-3

where $\mathbf{y} \in \mathfrak{R}^n$ is the system output (or the controlled output) vector, $\mathbf{u} \in \mathfrak{R}^n$ is the system input (the control signal or the control input) vector, $\mathbf{F} \in \mathfrak{R}^n$ represents the entire nonlinear dynamics of the system, $\mathbf{B} \in \mathfrak{R}^{n \times n}$ is the input matrix of the system, and n is the number of system inputs and outputs, i.e., supposed to be equal. The structure defined in (3.1) is known as *ultra-local* model for a generic dynamic system. In general case, $\mathbf{F}$ and $\mathbf{B}$ are assumed unknown.

Theorem 3.1

For the ultra-local model defined in *Definition 3.1*, the iP controller is defined as

$$\mathbf{u} = \hat{\mathbf{B}}^{-1}(\dot{\mathbf{y}}_d - \hat{\mathbf{F}} + \mathbf{v}), \tag{3.2}$$

where $\mathbf{v} \in \mathfrak{R}^n$ is the regular PID term of the controller (see *Remark 3.1*), and $\mathbf{y}_d \in \mathfrak{R}^n$ is the reference input vector (or the set-point vector or the desired set-points for system outputs). The term $\mathbf{v}$ is proposed based on the control error $\mathbf{e} \in \mathfrak{R}^n$, i.e., the output tracking error vector of the system, which is defined as follows:

$$\mathbf{e} = \mathbf{y}_d - \mathbf{y}. \tag{3.3}$$

Proof. Here, the proof is presented for the generic case of a PID controller for $\mathbf{v}$, but it can also be generalized for the other definitions pointed out as follows. Let us define the following Lyapunov function candidate:

$$V = \frac{1}{2}\mathbf{e}^T(\mathbf{I}_n + \mathbf{k}_d)\mathbf{e} + \xi^T\mathbf{k}_i\xi, \tag{3.4}$$

where $\xi = \int \mathbf{e}\, dt$ and $\mathbf{I}_n \in \mathfrak{R}^{n \times n}$ is an identity matrix. Then, the first time derivative of this function is

$$\dot{V} = \mathbf{e}^T\dot{\mathbf{e}} + \mathbf{e}^T\mathbf{k}_d\dot{\mathbf{e}} + \mathbf{e}^T\mathbf{k}_i\xi. \tag{3.5}$$

Replacing $\dot{\mathbf{e}}$ in the first term on the right-hand side of (3.5), the relations (3.3) and (3.1) lead to

$$\dot{V} = \mathbf{e}^T(\dot{\mathbf{y}}_d - \mathbf{F} - \mathbf{B}\,\mathbf{u}) + \mathbf{e}^T\mathbf{k}_d\dot{\mathbf{e}} + \mathbf{e}^T\mathbf{k}_i\xi. \tag{3.6}$$

Now, utilizing

$$\dot{\mathbf{y}}_d - \mathbf{F} + \mathbf{k}_p\mathbf{e} + \mathbf{k}_i\xi + \mathbf{k}_d\dot{\mathbf{e}}, \tag{3.7}$$

we reach at

$$\dot{V} = -\mathbf{e}^T\mathbf{k}_p\mathbf{e} \le 0. \tag{3.8}$$

Based on the *Lyapunov* stability theorem [Kha02], since $V > 0$ and $\dot{V} \leq 0$, $\mathbf{e}$ will be asymptotically stable and converge to zero. Moreover, by replacing $\mathbf{F}$ and $\mathbf{B}$ in (3.6), with their corresponding estimated vector variables and parameters $\hat{\mathbf{F}}$ and $\hat{\mathbf{B}}$, the controller given in (3.2) will be achieved. The usage of the estimated variables is permissible according to the *separation principle* [Ata99] if these estimated variables are computed utilizing stable observers (or state estimators). This completes the proof.

Remark 3.1

The term $\mathbf{v}$ in (3.2) is defined as follows for a regular P controller:

$$\mathbf{v} = \mathbf{k}_p \mathbf{e}, \tag{3.9}$$

for a regular PI controller is

$$\mathbf{v} = \mathbf{k}_p \mathbf{e} + \mathbf{k}_i \int \mathbf{e} \, dt, \tag{3.10}$$

for a regular PD controller is

$$\mathbf{v} = \mathbf{k}_p \mathbf{e} + \mathbf{k}_d \dot{\mathbf{e}}, \tag{3.11}$$

and finally for a regular PID controller is considered as

$$\mathbf{v} = \mathbf{k}_p + \mathbf{k}_i \int \mathbf{e} \, dt + \mathbf{k}_d \dot{\mathbf{e}}. \tag{3.12}$$

Here, $\mathbf{k}_p \in \mathfrak{R}^{n \times n}$ is the proportional controller gain matrix, $\mathbf{k}_i \in \mathfrak{R}^{n \times n}$ is the integral controller gain matrix and $\mathbf{k}_d \in \mathfrak{R}^{n \times n}$ is the derivative controller gain matrix.

Remark 3.2

As it is observed in (3.2), the estimated parameters and vectors of the dynamic system, i.e., $\hat{\mathbf{F}}$ and $\hat{\mathbf{B}}$, are incorporated into the iP controller. This is the basic feature of all model-free controllers. The values of $\hat{\mathbf{F}}$ and $\hat{\mathbf{B}}$ should be estimated online, in order to implement the proposed iPID controller in (3.2). The different approaches to this online parameter estimation problem are presented in Sections 3.2.2–3.2.4.

Remark 3.3

In (3.2), the time derivative of the desired system output vector, i.e., $\dot{\mathbf{y}}_d$, is required in the controller. In order to provide correctness and robustness into the differentiation process and consequently into the controlled system, the sliding mode differentiator [Lev03] is suggested to compute $\dot{y}_d$ as follows:

$$\dot{y}_d = \eta, \tag{3.13}$$

where for $\eta = [\eta_1 \quad \eta_2 \quad \cdots \quad \eta_n]^T$, the following relationships hold [Lev03]:

$$\dot{w}_i = \eta_i,$$

$$\eta_i = -k_1 \sqrt{|w_i - y_{di}|} \, \text{sgn}(w_i - y_{di}) + \tau_i, \tag{3.14}$$

$$\dot{\tau}_i = -k_2 \, \text{sgn}(w_i - y_{di}),$$

for $i = 1...n$, where sgn(.) is the sign function.

3.2.2 ITERATED TIME INTEGRALS FOR ONLINE PARAMETER ESTIMATIONS

Definition 3.2

For a single input-single output (SISO) dynamic system, assuming a known constant value for **B** in (3.1), named b, we reach at the following dynamic system:

$$\dot{y} = f + b \, u, \tag{3.15}$$

where f is an unknown parameter of the system.

Lemma 3.1

For the system defined in *Definition 3.2*, the iPID controller in *Theorem 3.1* is presented as follows [Fli13]:

$$u = \frac{\dot{y}_d - \hat{f} + v}{b}, \tag{3.16}$$

where $\hat{f}$ is the estimated value of f.

Lemma 3.2

The parameter f in *Definition 3.2* can be estimated by computing the averaged time integral of the system dynamics over a fixed time window of the input-output data of past operation of system. In this regard, the following estimator is suggested in [Fli13]:

$$\hat{f} = \frac{-6}{T^3} \int_{t-T}^{t} [(T - 2\tau)y + b\tau(T - \tau)u]d\tau, \tag{3.17}$$

where $T > 0$ is the length of the fixed time window for integration.

Proof (this proof is following the discussion presented in [Fli13]). Let us rewrite the dynamic system in (3.15) in terms of

$$\hat{f} = \dot{y} - b\,u, \tag{3.18}$$

where $\hat{f}$ is assumed to be constant between two consecutive time steps of the system operation. Then, the Laplace transform of (3.18) is (s is the Laplace variable)

$$\frac{\hat{f}}{s} = s\,Y - y(0) - bU, \tag{3.19}$$

where $y(0)$ is the value of system output at time $t = 0$. In order to remove this term from (3.19), let us consider the first derivative of this equation with respect to s

$$-\frac{\hat{f}}{s^2} = s\,\dot{Y} + Y - b\dot{U}. \tag{3.20}$$

Now, multiplying both sides of (3.20) with s^{-2}, we reach at

$$-\frac{\hat{f}}{s^4} = \frac{\dot{Y}}{s} + \frac{Y}{s^2} - b\frac{\dot{U}}{s^2}. \tag{3.21}$$

Then, utilizing Laplace-inverse transform over (3.21) leads to

$$-\hat{f}\left(\int_{t-T}^{t} \frac{1}{2}\tau^2\,d\tau\right) = \left(\int_{t-T}^{t} (-\tau)y\,d\tau\right) + \left(\int_{t-T}^{t} (T-\tau)y\,d\tau\right)$$

$$-b\left(\int_{t-T}^{t} (T-\tau)(-\tau)u\,d\tau\right). \tag{3.22}$$

Finally, simplifying (3.22) results in

$$-\hat{f}\left(\frac{T^3}{6}\right) = \left(\int_{t-T}^{t} [(T-2\tau)y + b\tau(T-\tau)u]\,d\tau\right), \tag{3.23}$$

and then (3.17) will be achieved. This completes the proof.

Remark 3.4

In *Lemma 3.2*, the value of parameter b is considered to be known. Generally, it is shown that $b = 1$ is a good assumption [Fli13].

3.2.3 Adaptive Observers for Online Parameter Estimations

Lemma 3.3

Based on *LaSalle-Yoshizawa* theorem [Fis13], for a positive potential function V, if its first time derivative fulfills the following inequality:

$$\dot{V} \leq -\tau + \sigma, \tag{3.24}$$

where τ and σ are constant bounded positive values, then V is uniformly ultimately bounded. It means that V and its independent time-varying variables would converge to a small bounded space around origin, i.e., $V \rightarrow (0 \pm \delta)$, where δ is a positive constant value determining the boundaries of V and its independent time-varying variables.

Definition 3.3

For a double-integrator SISO dynamic system, the ultra-local model presented in (3.1) can be expressed as [Tha14]

$$\dot{y} = f + (b-1)u + u, \tag{3.25}$$

where f and b are unknown scalar parameters of the system. Assuming next $\mathbf{x} = [y \quad \dot{y}]^T$ leads to [Tha14]

$$\dot{\mathbf{x}} = \mathbf{A}\mathbf{x} + \mathbf{B}u + \mathbf{\Psi}\theta,$$
$$\tag{3.26}$$
$$y = \mathbf{C}\mathbf{x},$$

where

$$\mathbf{A} = \begin{bmatrix} 0 & 1 \\ 0 & 0 \end{bmatrix}, \mathbf{B} = \begin{bmatrix} 0 \\ 1 \end{bmatrix}, \mathbf{C} = [0 \quad 1], \mathbf{\Psi} = \begin{bmatrix} 0 & 0 \\ 1 & u \end{bmatrix}, \theta = \begin{bmatrix} f \\ b-1 \end{bmatrix}. \tag{3.27}$$

Moreover, it is assumed that the first time derivatives of the unknown parameter are bounded, with the notation $|\dot{\theta}| \leq \beta$, where $\beta \in \mathfrak{R}^{2 \times 1}$ is a vector with positive elements, i.e., the absolute values of all elements of θ are smaller than the corresponding elements of β.

Assumption 3.1

For the dynamic system proposed in *Definition 3.3*, it is assumed that for the column matrix $\mathbf{K} \in \mathfrak{R}^{2 \times 1}$, the matrix $(\mathbf{A} - \mathbf{K}\,\mathbf{C})$ has negative eigenvalues so that the dynamic system $\dot{\mathbf{z}} = (\mathbf{A} - \mathbf{K}\,\mathbf{C})\mathbf{z}$ is *exponentially stable* for the generic time-varying variable $\mathbf{z} \in \mathfrak{R}^{2 \times 1}$ [Kha02].

Lemma 3.4

Following the model-free controller proposed in *Theorem 3.1*, a model-free controller is expressed for the dynamic system defined in (3.26), with the control law [Tha14]

$$u = \frac{\dot{y}_d - \hat{f} + v}{\hat{b}}, \tag{3.28}$$

where $\hat{f}$ and $\hat{b}$ are the estimated values for unknown parameters of the system.

Theorem 3.2

By recalling *Definition 3.3, Assumption 3.1* and *Lemma 3.4*, the values for unknown terms in (3.26), i.e.,

$$\hat{\theta} = \begin{bmatrix} \hat{f} \\ \hat{b} - 1 \end{bmatrix}, \tag{3.29}$$

can be estimated by a linear adaptive observer in terms of [Tha14]

$$\dot{\Phi} = (\mathbf{A} - \mathbf{K}\,\mathbf{C})\Phi + \Psi,$$

$$\dot{\hat{\mathbf{x}}} = \mathbf{A}\,\hat{\mathbf{x}} + \mathbf{B}\,u + \Psi\,\hat{\theta} + (\mathbf{K} + \Phi\,\Gamma\Phi^T\,\mathbf{C}^T\,\Sigma)(y - \mathbf{C}\,\hat{\mathbf{x}}), \tag{3.30}$$

$$\dot{\hat{\theta}} = \Gamma\,\Phi^T\,\mathbf{C}^T\,\Sigma\,(y - \mathbf{C}\,\hat{\mathbf{x}}),$$

where $\Gamma \in \mathfrak{R}^{2\times2}$ is a diagonal matrix of positive elements, $\mathbf{K} \in \mathfrak{R}^{2\times1}$ is a vector of positive elements and $\Sigma \in \mathbf{R}^+$ is a positive scalar, as tuning parameters of the observer, for $\Phi \in \mathfrak{R}^{2\times2}$.

Proof (here the proof is following the proposed proof in [Zha11]). Adding and subtracting the term $\mathbf{K}\,y$ in the right-hand side of the first equation in (3.26) leads to

$$\dot{\mathbf{x}} = (\mathbf{A} - \mathbf{K}\,\mathbf{C})\mathbf{x} + \mathbf{B}\,u + \Psi\,\theta + \mathbf{K}\,y. \tag{3.31}$$

Now, let us assume

$$\mathbf{x} = \mathbf{x}_u + \mathbf{x}_\theta, \tag{3.32}$$

where $\mathbf{x}_u$ is the system state vector corresponding to the input and output of the system (i.e., u and y), and $\mathbf{x}_\theta$ is the part of the system state vector corresponding to the unknown parameter θ. In this regard, we define

$$\dot{\mathbf{x}}_u = (\mathbf{A} - \mathbf{K}\,\mathbf{C})\mathbf{x}_u + \mathbf{B}\,u + \mathbf{K}\,y,$$

$$\dot{\mathbf{x}}_\theta = (\mathbf{A} - \mathbf{K}\,\mathbf{C})\mathbf{x}_\theta + \Psi\,\theta. \tag{3.33}$$

The observer for $\mathbf{x}_u$, which produces the estimated state vector $\hat{\mathbf{x}}_u$, is defined relatively simply as follows:

$$\dot{\hat{\mathbf{x}}}_u = (\mathbf{A} - \mathbf{K}\ \mathbf{C})\hat{\mathbf{x}}_u + \mathbf{B}\ u + \mathbf{K}\ y. \tag{3.34}$$

For observing $\mathbf{x}_\theta$, with the resulting estimated state vector $\hat{\mathbf{x}}_\theta$, the state-space equation of the observer is

$$\dot{\hat{\mathbf{x}}}_\theta = (\mathbf{A} - \mathbf{K}\ \mathbf{C})\hat{\mathbf{x}}_\theta + \boldsymbol{\Psi}\ \hat{\theta} + \omega, \tag{3.35}$$

where ω is the compensating term for replacing θ and $\mathbf{x}_\theta$ with $\hat{\theta}$ and $\hat{\mathbf{x}}_\theta$, respectively, in the second equation in (3.33). Then, defining

$$\hat{\mathbf{x}}_\theta = \boldsymbol{\Phi}\ \hat{\theta}, \tag{3.36}$$

we reach at

$$\dot{\boldsymbol{\Phi}}\ \hat{\theta} + \boldsymbol{\Phi}\ \dot{\hat{\theta}} = (\mathbf{A} - \mathbf{K}\ \mathbf{C})\boldsymbol{\Phi}\ \hat{\theta} + \boldsymbol{\Psi}\ \hat{\theta} + \omega. \tag{3.37}$$

Considering next the compensation term ω as

$$\omega = \boldsymbol{\Phi}\ \dot{\hat{\theta}}, \tag{3.38}$$

the following definition for $\boldsymbol{\Phi}$ will result:

$$\dot{\boldsymbol{\Phi}} = (\mathbf{A} - \mathbf{K}\ \mathbf{C})\boldsymbol{\Phi} + \boldsymbol{\Psi}. \tag{3.39}$$

Recalling *Assumption 3.1*, the homogeneous part of the differential equation in (3.39) is exponentially stable. In addition, according to (3.27), the term $\boldsymbol{\Psi}$ only includes the control inputs of the system, which are proven to be stable and bounded per *Lemma 3.4*. Therefore, the parameter $\boldsymbol{\Phi}$ computed by the dynamic system in (3.39) will be bounded within finite boundaries. This completes the proof for the first equation of the observer given in (3.30). Moreover, merging (3.38), (3.35) and (3.34) yields

$$\dot{\hat{\mathbf{x}}} = \dot{\hat{\mathbf{x}}}_u + \dot{\hat{\mathbf{x}}}_\theta = \mathbf{A}\ \hat{\mathbf{x}} + \mathbf{B}\ u + \boldsymbol{\Psi}\ \hat{\theta} + \mathbf{K}(y - \mathbf{C}\ \hat{\mathbf{x}}) + \boldsymbol{\Phi}\ \dot{\hat{\theta}}. \tag{3.40}$$

This also constructs the structure of the second equation in (3.30), except for the definition of $\dot{\hat{\theta}}$. In rest of the proof, we should show that $\dot{\hat{\theta}}$ is bounded and stable after computed using the third equation in (3.18). Let us define the estimation errors $\tilde{\mathbf{x}} = \hat{\mathbf{x}} - \mathbf{x}$ and $\tilde{\theta} = \hat{\theta} - \theta$. Then, considering (3.40), we reach at

$$\dot{\tilde{\mathbf{x}}} = (\mathbf{A} - \mathbf{K}\ \mathbf{C})\tilde{\mathbf{x}} + \boldsymbol{\Psi}\ \tilde{\theta} + \boldsymbol{\Phi}\ \dot{\hat{\theta}} + \boldsymbol{\Phi}\ \dot{\theta}. \tag{3.41}$$

Moreover, defining

$$\eta = \tilde{x} - \Phi \, \tilde{\theta}, \tag{3.42}$$

the result will be

$$\dot{\eta} = (A - K \, C)\tilde{x} + \Psi \, \tilde{\theta} + \Phi \, \dot{\tilde{\theta}} + \Phi \, \dot{\theta} - \dot{\Phi} \, \tilde{\theta} - \Phi \, \dot{\theta}, \tag{3.43}$$

which by replacing $\dot{\Phi}$ from (3.39) leads to

$$\dot{\eta} = (A - K \, C)\tilde{x} + \Psi \, \tilde{\theta} + \Phi \, \dot{\theta} - (A - K \, C)\Phi \, \tilde{\theta} - \Psi \, \tilde{\theta}. \tag{3.44}$$

Finally, this is simplified as follows:

$$\dot{\eta} = (A - K \, C)\eta + \Phi \, \dot{\theta}. \tag{3.45}$$

Now, recalling *Assumption 3.1*, we know that the homogeneous part of the dynamic system in (3.45) is exponentially stable, and hence, it is bounded. Besides, earlier in this proof, it has been shown that Φ is bounded. Moreover, by recalling $|\dot{\theta}| \le \beta$ in *Definition 3.3*, we also have $\dot{\theta}$ bounded. Hence, the vector variable η computed using the differential equation in (3.45) will always be bounded.

At this stage of the proof, let us take a look at the third equation in (3.30). In this regard, the first time derivative of the estimation error $\tilde{\theta}$ is computed as follows:

$$\dot{\tilde{\theta}} = \Gamma \, \Phi^T \, C^T \, \Sigma \, (y - C \, \hat{x}) - \dot{\theta}. \tag{3.46}$$

Then, replacing $y = C \, x$, we have

$$\dot{\tilde{\theta}} = -\Gamma \, \Phi^T \, C^T \, \Sigma \, C \, \tilde{x} - \dot{\theta}. \tag{3.47}$$

Moreover, by replacing $\tilde{x}$ from (3.42), we reach at

$$\dot{\tilde{\theta}} = -\Gamma \, \Phi^T \, C^T \, \Sigma \, C(\eta + \Phi \, \tilde{\theta}) - \dot{\theta}, \tag{3.48}$$

and therefore,

$$\dot{\tilde{\theta}} = -\Gamma \, \Phi^T \, C^T \, \Sigma \, C \, \Phi \, \tilde{\theta} - \Gamma \, \Phi^T \, C^T \, \Sigma \, C \, \eta - \dot{\theta}. \tag{3.49}$$

The homogeneous part of the differential equation in (3.49) is exponentially stable, because the eigenvalues of the term $(\Gamma \, \Phi^T \, C^T \, \Sigma \, C \, \Phi)$ are always positive. In addition, since the variables Φ and η are shown to be bounded previously, the second term in the right-hand side of (3.49) will also be bounded. Finally, recalling $|\dot{\theta}| \le \beta$, the entire dynamic system in (3.49) is stable and hence $\tilde{\theta}$ as well as $\hat{\theta}$ will be stable. Moreover, the error $\tilde{x}$ computed in terms of (3.42) will be bounded. Consequently, the estimated state vector $\hat{\mathbf{x}}$ will also be bounded and stable. Note that the second equation in the observer given in (3.30) is achieved by replacing $\dot{\hat{\theta}}$ in (3.40) from the third equation in (3.30). This completes the proof.

Remark 3.5

According to the observer defined in (3.30), the signal Ψ should be persistently excited in order to let the first equation in (3.30) and consequently the presented observer converge. This persistent excitation requirement is presented as follows [Zha11]:

$$\lambda_1 \mathbf{I}_2 \leq \int_t^{t+T} \Psi^T \Psi d\tau \leq \lambda_2 \mathbf{I}_2, \tag{3.50}$$

where $\mathbf{I}_2 \in \mathfrak{R}^{2 \times 2}$ is a unity matrix, and λ_1 and λ_2 are two positive scalar gains.

3.2.4 ADAPTIVE MODEL-BASED PARAMETER ESTIMATORS

Definition 3.4

For the SISO dynamic system defined in *Definition 3.2*, applying the Laplace transform assuming zero initial conditions leads to

$$s\,y = f + b\,u, \tag{3.51}$$

where s is the Laplace variable. Then, by multiplying both sides of (3.51) by $\dfrac{1}{s+a}$ with a constant positive value of a, we reach at

$$\frac{s}{s+a} y = \frac{1}{s+a} f + \frac{b}{s+a} u. \tag{3.52}$$

Then, the model in (3.52) can be written in the following linear estimation model form [Ioa06]:

$$z = \varphi\,\theta, \tag{3.53}$$

where

$$z = \frac{s}{s+a} y, \quad \varphi = \left[\begin{array}{cc} \dfrac{1}{s+a} & \dfrac{1}{s+a} u \end{array} \right], \tag{3.54}$$

and

$$\theta = [f \quad b]^T. \tag{3.55}$$

Assuming that both f and b are unknown variables and parameters, we define

$$\hat{\theta} = [\hat{f} \quad \hat{b}]^T, \tag{3.56}$$

and furthermore

$$\hat{z} = \varphi \, \hat{\theta}, \tag{3.57}$$

as the estimated output of the system based on the estimated parameters. The model-free controller for this system is suggested as the one presented in *Lemma 3.4*.

Theorem 3.3

Considering the linear estimation model defined in *Definition 3.4*, a model-based adaptive online estimator is formulated as follows [Ioa06]:

$$\dot{\hat{\theta}} = \gamma \, \varphi^T \, \varepsilon, \tag{3.58}$$

where $\gamma \in \Re^{2 \times 2}$ is a positive definite learning rate matrix and

$$\varepsilon = z - \hat{z}. \tag{3.59}$$

Proof. Let us consider the following objective function (or cost function):

$$J = \frac{1}{2}\varepsilon^2. \tag{3.60}$$

Then, minimizing J toward its gradient descent, we can update $\hat{\theta}$ using

$$\dot{\hat{\theta}} = -\gamma(\nabla J)_{\hat{\theta}}, \tag{3.61}$$

where $(\nabla J)_{\hat{\theta}}$ is the gradient of J in the direction of $\hat{\theta}$ as follows:

$$(\nabla J)_{\hat{\theta}} = \left[\begin{array}{cc} \dfrac{\partial J}{\partial \hat{f}} & \dfrac{\partial J}{\partial \hat{b}} \end{array} \right]^T, \tag{3.62}$$

and it results according to

$$(\nabla J)_{\hat{\theta}} = (\nabla \varepsilon)_{\hat{\theta}} \varepsilon = -(\nabla \hat{z})_{\hat{\theta}} \varepsilon. \tag{3.63}$$

Moreover, we have

$$(\nabla \hat{z})_{\hat{\theta}} = \varphi^T. \tag{3.64}$$

Finally, substituting $(\nabla J)_{\hat{\theta}}$ from (3.63) and (3.64) into (3.61), the result will be (3.58). This completes the proof.

Remark 3.6

According to the parameter estimator given in (3.58), the signal φ should be persistently excited in order to let the estimator converge. This persistent excitation requirement is presented as follows [Ioa06]:

$$\lambda_3 \leq \int\limits_{t}^{t+T} \varphi\, \varphi^T d\tau \leq \lambda_4, \tag{3.65}$$

where λ_3 and λ_4 are two positive scalar gains.

Remark 3.7

The robustness of the proposed adaptive estimator in *Theorem 3.3* can be provided by adding a *leakage* term as follows [Ioa06]:

$$\hat{\theta} = \gamma\, \varphi^T\, \varepsilon - \rho\, \gamma\, \hat{\theta} \tag{3.66}$$

where $\rho > 0$ is the leakage gain.

3.3 THEORY OF INTELLIGENT PID CONTROLLERS FOR DISCRETE-TIME DYNAMIC SYSTEMS

3.3.1 FIRST-ORDER DISCRETE-TIME iP/iPI/iPID CONTROLLERS

In the case of the first-order discrete-time MFC algorithm, the assumed ultra-local model in Section 3.1 becomes $\dot{y} = F + \alpha\, u$, and it is discretized using the Euler discretization method.

Definition 3.5

The discrete-time first-order MFC algorithms have been developed considering the nonlinear multi input-multi output (MIMO) discrete-time model of the process [Fli13, Rom16b, Rom18c]

$$\mathbf{y}(k+1) = \mathbf{f}\big(\mathbf{y}(k),\ldots,\mathbf{y}(k-n_y),\mathbf{u}(k),\ldots,\mathbf{u}(k-n_u)\big), \tag{3.67}$$

where $\mathbf{y}(k+1) = [y_1(k+1) \ldots y_n(k+1)]^T \in \mathfrak{R}^n$ is the controlled output (process output) vector, $\mathbf{u}(k) \in \mathfrak{R}^n$ is control input (or the control signal) vector, $n \in \mathbf{Z}, n \geq 0$ is the number of inputs of the process, and it is equal to the number of the outputs of the process, n is known, $\mathbf{y}(k - n_y) \in \mathfrak{R}^n$ is the controlled output vector at the $k - n_y$ time moment, $\mathbf{u}(k - n_u) \in \mathfrak{R}^n$ is the control input vector at the $k - n_u$ time moment, and $\mathbf{f} : \mathfrak{R}^{n(n_u + n_y + 2)} \to \mathfrak{R}^n$ is a known nonlinear vector function of vector variable [Fli13, Rom16b, Rom18c].

Remark 3.8

The theoretical results presented as follows in the MIMO case are starting with the MIMO nonlinear model in (3.67) and are applicable to nonlinear processes that have the number of inputs (control inputs or control signals) equal to the number of outputs (controlled outputs).

Proposition 3.1

The mathematical model of the process is substituted with an (ultra-)local model as

$$\mathbf{y}(k+1) = \mathbf{y}(k) + \alpha \, \mathbf{u}(k) + \mathbf{F}(k), \tag{3.68}$$

where the vector $\mathbf{F}(k) \in \mathfrak{R}^n$ plays a disturbance role, resulted from the input/output data, is continually updated at each moment of time k and includes the unknown parts of the process model and possible disturbances, $\alpha = \text{diag}(\alpha_1, \ldots, \alpha_n) \in \mathfrak{R}^{n \times n}$ is a constant matrix with user-chosen parameters such that $\Delta \mathbf{y}(k+1) = \mathbf{y}(k+1) - \mathbf{y}(k)$ and $\alpha \, \mathbf{u}(k)$ have the same order of magnitude, and $k \in \mathbf{Z}, k \geq 0$ is the discrete-time argument [Fli13, Rom16b, Rom18c]. The block diagram of the control system structure with first-order MFC algorithm is presented in Figure 3.1, where $\mathbf{e}(k) \in \mathfrak{R}^n$ is the control error vector.

In the design of the first-order MFC algorithms, the ultra-local model in (3.67) is used to substitute the unknown mathematical model of the process.

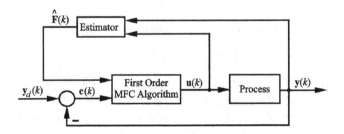

FIGURE 3.1 MIMO control system structure with first-order MFC algorithm. (Adapted from Rom16b, Rom17a, Rom17b, Rom18a, Rom18b and Rom18c.)

3.3.1.1 The First-Order Discrete-Time iP Controller
Theorem 3.4

The control law of the first-order MFC algorithm with P component is [Fli13, Rad14, Rom18c]

$$\mathbf{u}(k) = \alpha^{-1}(-\hat{\mathbf{F}}(k) + \mathbf{y}_d(k+1) - \mathbf{y}_d(k) - \mathbf{K}_1\ \mathbf{e}(k)), \tag{3.69}$$

where vector $\hat{\mathbf{F}}(k) \in \mathfrak{R}^n$ is the estimate of $\mathbf{F}(k)$, $\mathbf{y}_d(k) = [y_{d1}(k)\ \dots\ y_{dn}(k)]^T \in \mathfrak{R}^n$ is the reference input vector (or the set-point vector or the output of the reference model) and $\mathbf{K}_1 \in \mathfrak{R}^{n \times n}$ is a parameter matrix. The control law in (3.69) is also known as first-order iP controller [Fli13, Rad14, Rom18c].

Lemma 3.5

The estimate $\hat{\mathbf{F}}(k)$ is obtained in terms of rearranging (3.68) using the input/output data obtained from the controlled process at the one-step-behind and current moments of time

$$\hat{\mathbf{F}}(k) = \mathbf{y}(k) - \mathbf{y}(k-1) - \alpha\ \mathbf{u}(k-1). \tag{3.70}$$

The estimation error vector $\delta(k) \in \mathfrak{R}^n$ is defined as [Fli13, Rad14, Rom18c]

$$\delta(k) = \mathbf{F}(k) - \hat{\mathbf{F}}(k), \tag{3.71}$$

and it is considered as disturbance that will be negligible in the design of the first-order MFC iP controller. Substituting the control input vector $\mathbf{u}(k)$ from (3.69) in the local model in (3.68), the dynamics equation of the control system structure with first-order iP controller is obtained [Fli13, Rad14, Rom18c]:

$$\mathbf{y}(k+1) = \mathbf{y}(k) - \hat{\mathbf{F}}(k) + \mathbf{y}_d(k+1) - \mathbf{y}_d(k) - \mathbf{K}_1\ \mathbf{e}(k) + \mathbf{F}(k). \tag{3.72}$$

Vector $\mathbf{e}(k) \in \mathbf{R}^n$ of the control error is defined as follows:

$$\mathbf{e}(k) = \mathbf{y}_d(k) - \mathbf{y}(k) = [e_1(k)\ \dots\ e_n(k)]^T,$$
$$e_i(k) = y_{di}(k) - y_i(k),\ i = 1\dots n. \tag{3.73}$$

Substituting the expression of $\delta(k)$ from (3.71) in the control system dynamics equation (3.72) and next in the expression of the error vector in (3.73), the equation that characterizes the control error dynamics results as follows [Fli13, Rad14, Rom18c]:

$$\mathbf{e}(k+1) - (\mathbf{I} + \mathbf{K}_1)\ \mathbf{e}(k) - \delta(k) = \mathbf{0}. \tag{3.74}$$

It is assumed that the norm of $\delta(k)$ is bounded [Fli13, Rad14, Rom18c]

$$\parallel \delta(k) \parallel < \delta_{max}, \tag{3.75}$$

and the upper limit δ_{max} has a very small value. The necessary conditions that ensure reference trajectory tracking, namely the convergence to zero of the control errors, are that if the roots of the characteristic polynomial resulted from (3.74) are located inside of the unit circle, then the control system is stable [Rad14, Rom15, Rom16a, Rom16b, Pre17, Rom18c].

Proposition 3.2

Based on the results presented in this section, the following **design steps for the control system with first-order iP controller** are carried out [Fli13, Rom17b, Rom18c]:

Step 1. Choosing the value of the design parameter $\alpha = \text{diag}(\alpha_1,...,\alpha_n) \in \mathfrak{R}^{n \times n}$ such that $\Delta \mathbf{y}(k+1) = \mathbf{y}(k+1) - \mathbf{y}(k)$ and $\alpha \, \mathbf{u}(k)$ will have the same order of magnitude.

Step 2. Establishing the values of the parameters in $\mathbf{K}_1 = \text{diag}(K_{11},...,K_{1n}) \in \mathfrak{R}^{n \times n}$ in order to fulfill the performance specifications in terms of imposing the desired behavior of the control system and accounting for the stability conditions related to (3.74).

3.3.1.2 The First-Order Discrete-Time iPI Controller

Theorem 3.5

The control law of the first-order MFC algorithm with PI component is [Fli13, Rom17b, Rom18c]

$$\mathbf{u}(k) = \alpha^{-1}(-\hat{\mathbf{F}}(k) + \mathbf{y}_d(k+1) - \mathbf{y}_d(k) - \mathbf{K}_1 \, \mathbf{e}(k) - \mathbf{K}_2 \, \mathbf{e}(k-1)), \tag{3.76}$$

where the vectors $\hat{\mathbf{F}}(k) \in \mathfrak{R}^n$, $\mathbf{y}_d(k) \in \mathfrak{R}^n$ and $\mathbf{e}(k) \in \mathfrak{R}^n$ have the same significance as in Section 3.3.1.1, $\mathbf{K}_1 \in \mathfrak{R}^{n \times n}$ and $\mathbf{K}_2 \in \mathfrak{R}^{n \times n}$ are parameter matrices, and the control law in (3.76) is also known as first-order iPI controller.

Lemma 3.6

Recalling *Lemma 3.5*, the estimate $\hat{\mathbf{F}}(k)$ is obtained in terms of (3.68) using the input/output data obtained from the controlled process at the one-step-behind and current moments of time leading to (3.70), and the estimation error $\delta(k) \in \mathfrak{R}^n$ defined according to (3.71) is considered as a disturbance that will be negligible in the design of the first-order MFC iPI controller. Substituting the control input vector $\mathbf{u}(k)$ from (3.76) in the local model in (3.68), the dynamics equation of the control system (in closed-loop structure) with first-order iPI controller is obtained [Fli13, Rom17b, Rom18c]

$$\mathbf{y}(k+1) = \mathbf{y}(k) - \hat{\mathbf{F}}(k) + \mathbf{y}_d(k+1) - \mathbf{y}_d(k) - \mathbf{K}_1 \, \mathbf{e}(k) - \mathbf{K}_2 \, \mathbf{e}(k-1) + \mathbf{F}(k). \tag{3.77}$$

As in Section 3.3.1.1, substituting the expression of $\delta(k)$ from (3.71) in the control system dynamics equation (3.77) and next in the control error vector equation (3.73), the equation that characterizes the control error dynamics of the control system with iPI controller results as follows [Fli13, Rom17b, Rom18c]:

$$\mathbf{e}(k+1) - (\mathbf{I} + \mathbf{K}_1)\,\mathbf{e}(k) - \mathbf{K}_2\,\mathbf{e}(k-1) - \delta(k) = \mathbf{0}. \tag{3.78}$$

It is assumed that the norm of $\delta(k)$ is bounded as in (3.75), and the upper limit $\delta_{\max}$ has a very small value. Thus, first-order MFC iPI guarantees the control system stability if the roots of the characteristic polynomial of (3.78) are inside the unit circle [Fli13, Rom17b, Rom18c].

Proposition 3.3

Based on the results presented in this section, the following **design steps for the control system structure with first-order iPI controller** are carried out [Fli13, Rom17b, Rom18c]:

> *Step 1.* Choosing the value of the design parameter $\alpha = \operatorname{diag}(\alpha_1, \ldots, \alpha_n) \in \mathfrak{R}^{n \times n}$ such that $\Delta \mathbf{y}(k+1) = \mathbf{y}(k+1) - \mathbf{y}(k)$ and $\alpha\,\mathbf{u}(k)$ will have the same order of magnitude.
> *Step 2.* Establishing the values of the parameters in $\mathbf{K}_1 = \operatorname{diag}(K_{11}, \ldots, K_{1n}) \in \mathfrak{R}^{n \times n}$ and $\mathbf{K}_2 = \operatorname{diag}(K_{21}, \ldots, K_{2n}) \in \mathfrak{R}^{n \times n}$ in order to fulfill the performance specifications in terms of imposing the desired behavior of the control system and accounting for the stability conditions related to (3.78).

3.3.1.3   The First-Order Discrete-Time iPID Controller
Theorem 3.6

The control law of the first-order MFC algorithm with PID component is [Fli13, Rom17b, Rom18c]:

$$\mathbf{u}(k) = \alpha^{-1}(-\hat{\mathbf{F}}(k) + \mathbf{y}_d(k+1) - \mathbf{y}_d(k) - \mathbf{K}_1\,\mathbf{e}(k) - \mathbf{K}_2\,\mathbf{e}(k-1) - \mathbf{K}_3\,\mathbf{e}(k-2)), \tag{3.79}$$

where the vectors have the same significance as in Sections 3.3.1.1 and 3.3.1.2, $\mathbf{K}_1 \in \mathfrak{R}^{n \times n}$, $\mathbf{K}_2 \in \mathfrak{R}^{n \times n}$ and $\mathbf{K}_3 \in \mathfrak{R}^{n \times n}$ are parameter matrices, and the control law in (3.79) is also known as first-order iPID controller.

Lemma 3.7

Similar to *Lemmas 3.5* and *3.6*, the estimate $\hat{\mathbf{F}}(k)$ is obtained using (3.68), the input/output data are obtained from the controlled process at the one-step-behind and current moments of time in terms of (3.70), and the estimation error $\delta(k) \in \Re^n$ is defined according to (3.71) and considered as disturbance that will be negligible in the design of the first-order MFC iPID controller. Substituting the control input $\mathbf{u}(k)$ from (3.79) in the local model in (3.68), the dynamics equation of the control system (in closed-loop structure) with first-order iPID controller is obtained as follows [Fli13, Rom17b, Rom18c]:

$$\mathbf{y}(k+1) = \mathbf{y}(k) - \hat{\mathbf{F}}(k) + \mathbf{y}_d(k+1) - \mathbf{y}_d(k) - \mathbf{K}_1 \ \mathbf{e}(k)$$

$$- \mathbf{K}_2 \ \mathbf{e}(k-1) - \mathbf{K}_3 \ \mathbf{e}(k-2) + \mathbf{F}(k). \tag{3.80}$$

As in Sections 3.3.3.1 and 3.3.3.2, the control error vector $\mathbf{e}(k) \in \Re^n$ is defined in (3.73), and substituting the expression of $\delta(k)$ from (3.71) in the control system dynamics equation (3.80) and next in (3.73), the equation that characterizes the control error dynamics of the control system with first-order iPID controller results [Fli13, Rom17b, Rom18c]:

$$\mathbf{e}(k+1) - (\mathbf{I} + \mathbf{K}_1) \ \mathbf{e}(k) - \mathbf{K}_2 \ \mathbf{e}(k-1) - \mathbf{K}_3 \ \mathbf{e}(k-2) - \delta(k) = \mathbf{0}. \tag{3.81}$$

It is assumed that the norm of $\delta(k)$ is bounded according to (3.75) and the upper limit δ_{max} has a very small value. Thus, the first-order iPID controller guarantees the control system stability if all roots of the characteristic polynomial of (3.81) are inside the unit circle [Fli13, Rom17b, Rom18c].

Proposition 3.4

Based on the results presented in this section, the following design steps for the control structure with first-order iPID controller are carried out [Fli13, Rom17b, Rom18c]:

Step 1. Choosing the value of the design parameter $\alpha = \mathrm{diag}(\alpha_1, \ldots, \alpha_n) \in \Re^{n \times n}$ such that $\Delta \mathbf{y}(k+1) = \mathbf{y}(k+1) - \mathbf{y}(k)$ and $\alpha \ \mathbf{u}(k)$ will have the same order of magnitude.

Step 2. Establishing the values of the parameters in $\mathbf{K}_1 = \mathrm{diag}(K_{11}, \ldots, K_{1n}) \in \Re^{n \times n}$, $\mathbf{K}_2 = \mathrm{diag}(K_{21}, \ldots, K_{2n}) \in \mathbf{R}^{n \times n}$ and $\mathbf{K}_3 = \mathrm{diag}(K_{31}, \ldots, K_{3n}) \in \mathbf{R}^{n \times n}$ in order to fulfill the performance specifications in terms of imposing the desired behavior of the control system and accounting for the stability conditions related to (3.81).

3.3.2 SECOND-ORDER DISCRETE-TIME IP/IPI/IPID CONTROLLERS

In the case of the second-order discrete-time MFC algorithm, the ultra-local model of the process assumed in Section 3.3.1 becomes $\ddot{\mathbf{y}} = \mathbf{F} + \alpha \ \mathbf{u}$, and it is discretized using the Euler discretization method.

Remark 3.9

The discrete-time second-order MFC algorithms have been developed considering the same nonlinear (MIMO) model of the discrete-time process in (3.67) [Fli13, Rom16b, Rom18c], where $\mathbf{y}(k+1) \in \Re^n$, $\mathbf{u}(k) \in \Re^n$, $n \in \mathbf{Z}$, $n \geq 0$, $\mathbf{y}(k - n_y) \in \Re^n$, $\mathbf{u}(k - n_u) \in \Re^n$ and $\mathbf{f} : \Re^{n(n_u + n_y + 2)} \to \Re^n$ have the same significance as in Section 3.3.1 [Fli13, Rom16b, Rom18c].

Remark 3.10

The theoretical results are presented as follows in the MIMO case starting with the MIMO nonlinear model in (3.67), and they are applicable to nonlinear processes that have the number of inputs (control inputs) equal to the number of outputs (controlled outputs).

Proposition 3.5

The mathematical model of the process is substituted with the local model

$$\mathbf{y}(k) = 2\mathbf{y}(k-1) - \mathbf{y}(k-2) + \alpha\,\mathbf{u}(k-1) + \mathbf{F}(k-1), \tag{3.82}$$

where $\mathbf{F}(k) \in \Re^n$ and $\alpha = \mathrm{diag}(\alpha_1,\ldots,\alpha_n) \in \Re^{n \times n}$ have the same significance given in Section 3.3.1, $\Delta\mathbf{y}(k+1) = \mathbf{y}(k+1) - \mathbf{y}(k)$ and $\alpha\,\mathbf{u}(k)$ have the same order of magnitude, and $k \in \mathbf{Z}, k \geq 0$ is the discrete-time argument [Fli13, Rom16b, Rom18a, Rom18b, Rom18c]. The block diagram of the control system structure with second-order MFC algorithm is illustrated in Figure 3.2, which is identical to Figure 3.1; however, a different controller is involved.

In the design of the second-order MFC algorithms, the ultra-local model in (3.67) is used to substitute the unknown mathematical model of the process.

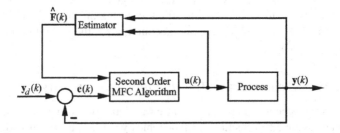

FIGURE 3.2   MIMO control system structure with second-order MFC algorithm. (Adapted from Rom16b, Rom17a, Rom17b, Rom18a, Rom18b and Rom18c.)

3.3.2.1 The Second-Order Discrete-Time iP Controller
Theorem 3.7

The control law of the second-order MFC algorithm with P component is [Fli13, Rom18a, Rom18b, Rom18c]

$$\mathbf{u}(k) = \alpha^{-1}(-\hat{\mathbf{F}}(k) + \mathbf{y}_d(k+1) - 2\mathbf{y}_d(k) + \mathbf{y}_d(k-1) - \mathbf{K}_1\ \mathbf{e}(k)), \qquad (3.83)$$

where the vectors and matrices are specified in Section 3.3.1. The control law in (3.83) is also known as iP controller.

Lemma 3.8

The estimate $\hat{\mathbf{F}}(k)$ is obtained as follows in terms of processing (3.82) using the input/output data obtained from the controlled process at the one-step-behind and current moments of time [Fli13, Rom16b, Rom18a, Rom18b, Rom18c]:

$$\hat{\mathbf{F}}(k) = \mathbf{y}(k-2) - 2\mathbf{y}(k-1) + \mathbf{y}(k) - \alpha\ \mathbf{u}(k-1), \qquad (3.84)$$

where the estimation error vector $\delta(k) \in \Re^n$ is defined in (3.71). Substituting the control input $\mathbf{u}(k)$ from (3.83) in the local model in (3.82), the dynamics equation of the control system (in closed-loop structure) with second-order iP controller is obtained [Rom18a, Rom18b, Rom18c]:

$$\mathbf{y}(k) = 2\mathbf{y}(k-1) - \mathbf{y}(k-2) - \hat{\mathbf{F}}(k-1) + \mathbf{y}_d(k) - 2\mathbf{y}_d(k-1) + \mathbf{y}_d(k-2)$$
$$- \mathbf{K}_1\ \mathbf{e}(k-1) + \mathbf{F}(k-1), \qquad (3.85)$$

with the control error vector $\mathbf{e}(k) \in \Re^n$ defined in (3.73).

Substituting the expression of $\delta(k)$ according to (3.71) in the control system structure dynamics equation (3.85) and next in (3.73), the equation that characterizes the control error dynamics is as follows [Rom18a, Rom18b, Rom18c]:

$$\mathbf{e}(k) + (-2\mathbf{I} - \mathbf{K}_1)\mathbf{e}(k-1) + \mathbf{e}(k-2) + \delta(k-1) = \mathbf{0}. \qquad (3.86)$$

As in Section 3.3.1, it is assumed that the norm of $\delta(k)$ is bounded according to (3.75), and the upper limit δ_{max} has a very small value. Thus, second-order MFC iP controller guarantees the control system stability if all roots of the characteristic polynomial of (3.86) are inside the unit circle.

Proposition 3.6

Based on the results presented in this section, the following design steps for the control system structure with second-order iP controller are carried out [Fli13, Rom16b, Rom18a, Rom18b, Rom18c]:

Step 1. Choosing the value of the design parameter $\alpha = \text{diag}(\alpha_1,\dots,\alpha_n) \in \mathfrak{R}^{n \times n}$ such that $\mathbf{y}(k) - 2\mathbf{y}(k-1) - \mathbf{y}(k-2)$ and $\alpha\,\mathbf{u}(k)$ will have the same order of magnitude.

Step 2. Establishing the values of the parameters in $\mathbf{K}_1 = \text{diag}(K_{11},\dots,K_{1n}) \in \mathfrak{R}^{n \times n}$ in order to fulfill the performance specifications in terms of imposing the desired behavior of the control system and accounting for the stability conditions related to (3.86).

3.3.2.2   The Second-Order Discrete-Time iPI Controller
Theorem 3.8

The control law of the second-order MFC algorithm with PI component is [Fli13, Rad14, Rom18a, Rom18b, Rom18c]

$$\mathbf{u}(k) = \alpha^{-1}(-\hat{\mathbf{F}}(k) + \mathbf{y}_d(k+1) - 2\mathbf{y}_d(k) + \mathbf{y}_d(k-1) - \mathbf{K}_1\,\mathbf{e}(k) - \mathbf{K}_2\,\mathbf{e}(k-1)), \qquad (3.87)$$

and it is also known as iPI controller.

Lemma 3.9

The estimate $\hat{\mathbf{F}}(k)$ is obtained as follows in terms of processing (3.82) using the input/output data obtained from the controlled process at the one-step-behind and current moments of time leading to (3.84). Substituting the control input $\mathbf{u}(k)$ from (3.87) in the local model in (3.82), the dynamics equation of the control system (in closed-loop structure) with second-order iPI controller is obtained [Rom18a, Rom18b, Rom18c]

$$\mathbf{y}(k) = 2\mathbf{y}(k-1) - \mathbf{y}(k-2) - \hat{\mathbf{F}}(k-1) + \mathbf{y}_d(k) - 2\mathbf{y}_d(k-1) + \mathbf{y}_d(k-2)$$
$$- \mathbf{K}_1\,\mathbf{e}(k-1) - \mathbf{K}_2\,\mathbf{e}(k-2) + \mathbf{F}(k-1). \qquad (3.88)$$

Substituting the expression of $\delta(k)$ according to (3.71) in the control system structure dynamics equation (3.88) and next in (3.73), the equation that characterizes the control error dynamics is [Rom18a, Rom18b, Rom18c]

$$\mathbf{e}(k) + (-2\mathbf{I} - \mathbf{K}_1)\mathbf{e}(k-1) + (\mathbf{I} - \mathbf{K}_2)\mathbf{e}(k-2) + \delta(k-1) = \mathbf{0}. \qquad (3.89)$$

As in Section 3.3.1, it is assumed that the norm of $\delta(k)$ is bounded according to (3.75), and the upper limit δ_{max} has a very small value. Thus, second-order MFC iPI controller guarantees the control system stability if all roots of the characteristic polynomial of (3.89) are inside the unit circle.

Proposition 3.7

Based on the results presented in this section, the following **design steps for the control system structure with second-order iPI controller** are carried out [Fli13, Rom16b, Rom18a, Rom18b, Rom18c]:

Step 1. Choosing the value of the design parameter $\alpha = \mathrm{diag}(\alpha_1,\ldots,\alpha_n) \in \mathfrak{R}^{n \times n}$ such that $\mathbf{y}(k) - 2\mathbf{y}(k-1) - \mathbf{y}(k-2)$ and $\alpha\,\mathbf{u}(k)$ will have the same order of magnitude.

Step 2. Establishing the values of the parameters in $\mathbf{K}_1 = \mathrm{diag}(K_{11},\ldots,K_{1n}) \in \mathfrak{R}^{n \times n}$ and $\mathbf{K}_2 = \mathrm{diag}(K_{21},\ldots,K_{2n}) \in \mathfrak{R}^{n \times n}$ in order to fulfill the performance specifications in terms of imposing the desired behavior of the control system and accounting for the stability conditions related to (3.89).

3.3.2.3 The Second-Order Discrete-Time iPID Controller
Theorem 3.9

The control law of the second-order MFC algorithm with PID component is [Fli13, Rad14, Rom18a, Rom18b, Rom18c]

$$\mathbf{u}(k) = \alpha^{-1}(-\hat{\mathbf{F}}(k) + \mathbf{y}_d(k+1) - 2\mathbf{y}_d(k) + \mathbf{y}_d(k-1) - \mathbf{K}_1\,\mathbf{e}(k)$$

$$- \mathbf{K}_2\,\mathbf{e}(k-1) - \mathbf{K}_3\,\mathbf{e}(k-2)), \tag{3.90}$$

and it is also known as second-order iPID controller.

Lemma 3.10

The estimate $\hat{\mathbf{F}}(k)$ is obtained as follows in terms of processing (3.82) using the input/output data obtained from the controlled process at the one-step-behind and current moments of time leading to (3.84). Substituting the control input $\mathbf{u}(k)$ from (3.90) in the local model in (3.82), the dynamics equation of the control system (in closed-loop structure) with second-order iPID controller is obtained [Rom18a, Rom18b, Rom18c]:

$$\mathbf{y}(k) = 2\mathbf{y}(k-1) - \mathbf{y}(k-2) - \hat{\mathbf{F}}(k-1) + \mathbf{y}_d(k) - 2\mathbf{y}_d(k-1) + \mathbf{y}_d(k-2)$$

$$- \mathbf{K}_1\,\mathbf{e}(k-1) - \mathbf{K}_2\,\mathbf{e}(k-2) - \mathbf{K}_3\,\mathbf{e}(k-3) + \mathbf{F}(k-1). \tag{3.91}$$

Substituting the expression of $\delta(k)$ according to (3.71) in the control system structure dynamics equation (3.91) and next in (3.73), the equation that characterizes the control error dynamics is [Rom18a, Rom18b, Rom18c]

$$\mathbf{e}(k) + (-2\mathbf{I} - \mathbf{K}_1)\mathbf{e}(k-1) + (\mathbf{I} - \mathbf{K}_2)\mathbf{e}(k-2) - \mathbf{K}_3\mathbf{e}(k-3) + \delta(k-1) = 0. \tag{3.92}$$

As in Section 3.3.1, it is assumed that the norm of $\delta(k)$ is bounded according to (3.75) and the upper limit δ_{max} has a very small value. Thus, second-order MFC iPI controller guarantees the control system stability if all roots of the characteristic polynomial of (3.92) are inside the unit circle.

Proposition 3.8

Based on the results presented in this subchapter, the following *design steps for the control system structure with second-order iPID controller* are carried out [Fli13, Rom16b, Rom18a, Rom18b, Rom18c]:

> *Step 1.* Choosing the value of the design parameter $\alpha = \mathrm{diag}(\alpha_1,\ldots,\alpha_n) \in \mathfrak{R}^{n\times n}$ such that $y(k) - 2y(k-1) - y(k-2)$ and $\alpha\, \mathbf{u}(k)$ will have the same order of magnitude.
>
> *Step 2.* Establishing the values of the parameters in $\mathbf{K}_1 = \mathrm{diag}(K_{11},\ldots,K_{1n}) \in \mathfrak{R}^{n\times n}$, $\mathbf{K}_2 = \mathrm{diag}(K_{21},\ldots,K_{2n}) \in \mathfrak{R}^{n\times n}$ and $\mathbf{K}_3 = \mathrm{diag}(K_{31},\ldots,K_{3n}) \in \mathfrak{R}^{n\times n}$ in order to fulfill the performance specifications in terms of imposing the desired behavior of the control system and accounting for the stability conditions related to (3.92).

3.4 EXAMPLE AND APPLICATION

The tower crane system described in Chapter 1 is considered as a controlled process in order to exemplify the first-order and the second-order discrete-time iP/iPi/iPID controllers in SISO and MIMO data-driven MFC system structures. The theoretical approaches presented in this chapter will be exemplified as follows. Three separate SISO control systems for the three controlled outputs, namely cart position, arm angular position and payload position, will be first considered, and the MIMO control system to control all these three tower crane system outputs will be next described.

3.4.1 SISO CONTROL SYSTEMS

The objective function used to assess the performance of each SISO control loop of the control system structures with the MFC algorithms presented in Section 3.3 is

$$J_{(\psi)}\left(\chi_{(\psi)}\right) = \frac{1}{2}E\left\{\sum_{k=1}^{N}\left[e_{(\psi)}(k,\chi_{(\psi)})\right]^2\right\}, \tag{3.93}$$

where the control error is $e_{(\psi)} = y_{(\psi)}^* - y_{(\psi)}$ and the subscript $\psi \in \{c,a,p\}$ indicates both the controlled output and its corresponding controller, namely c – the cart position (i.e., $y_{(c)} = y_1$), a – the arm angular position (i.e., $y_{(a)} = y_2$) and p – the payload position (i.e., $y_{(p)} = y_3$). The mathematical expectation $E\{\Xi\}$ is taken with respect to the stochastic probability distribution of the disturbance inputs applied to the process and thus affects the control system behavior. The sampling period is $T_s = 0.01$ s and the time horizon is 70 s, leading to a number of $N = 70/0.01 = 7{,}000$ s samples.

The MFC algorithms in the continuous-time and discrete-time versions are applied, one for each controlled output of the process, respecting the steps of the SISO MFC algorithm in previous sections. The description of the steps is given in relation with the Matlab & Simulink programs and schemes.

The reference input vector in all simulations and experiments is obtained by first applying the following step signals:

$$y^*_{(c)}(k) = 0.15 \text{ if } k \in [0,2,000], 0.1 \text{ if } k \in (2,000,3,500], -0.05 \text{ if } k \in (3,500,5,000],$$

$$0 \text{ if } k \in (5,000,7,000] \tag{3.94}$$

for cart position control,

$$y^*_{(a)}(k) = 0 \text{ if } k \in [0,500], 0.15 \text{ if } k \in (500,2,500], -0.15 \text{ if } k \in (2,500,4,000],$$

$$0 \text{ if } k \in (4,000,7,000] \tag{3.95}$$

for arm angular position control and

$$y^*_{(p)}(k) = 0 \text{ if } k \in [0,1,500], 0.1 \text{ if } k \in (1,500,3,000], -0.05 \text{ if } k \in (3,000,4,500],$$

$$0 \text{ if } k \in (4,500,7,000] \tag{3.96}$$

for payload position control. The reference models are obtained by applying the above signals to the filters with the transfer functions

$$H_{(c)}(s) = \frac{1}{1+0.2s} \tag{3.97}$$

for cart position control,

$$H_{(a)}(s) = \frac{1}{1+0.2s} \tag{3.98}$$

for arm angular position control and

$$H_{(p)}(s) = \frac{1}{1+0.2s} \tag{3.99}$$

for payload position control, with the corresponding discrete transfer functions:

$$H_{(c)}(z^{-1}) = \frac{0.0488}{1-0.9512z^{-1}}, \tag{3.100}$$

$$H_{(a)}(z^{-1}) = \frac{0.0465}{1-0.9535z^{-1}}, \tag{3.101}$$

$$H_{(p)}(z^{-1}) = \frac{0.0328}{1-0.9672z^{-1}}. \tag{3.102}$$

3.4.1.1 The First-Order Discrete-Time iP Controllers

Three first-order discrete-time iP SISO controllers are applied separately in three SISO control system structures, where the parameter vectors of the first-order discrete-time iP controllers are

$$\chi_{(\psi)} = K_1, \tag{3.103}$$

with $\psi \in \{c, a, p\}$, i.e., one parameter for each controller.

The controllers are nonlinear since saturation-type nonlinearities are inserted on the three controller outputs in order to match the power electronics of the actuators, which operate on the basis of the pulse width modulation (PWM) principle. As shown in Chapter 1, the three control signals (or control inputs) are within −1 and 1, ensuring the necessary connection to the tower crane system subsystem, which models the controlled process in the control system structures developed in this chapter and the next chapters and also includes three zero-order hold (ZOH) blocks to enable digital control. The tower crane system subsystem belongs to the Process. mdl Simulink diagram, which is included in the accompanying Matlab & Simulink programs given in Chapter 1.

The parameter $\alpha = 0.0015$ is chosen according to step 1, and then in step 2, the parameter K_1 is established to ensure the condition (3.74), leading to

$$\chi_{(c)} = -0.7, \tag{3.104}$$

$$\chi_{(a)} = -0.6, \tag{3.105}$$

$$\chi_{(p)} = -0.14. \tag{3.106}$$

The details of steps 1 and 2 are implemented in the MFC _ iPID _ 1st _ order _ c.m Matlab program for cart position control, the MFC _ iPID _ 1st _ order _ a.m Matlab program for arm angular position control and the MFC _ iPID _ 1st _ order _ p.m Matlab program for payload position control, which are included in the accompanying Matlab & Simulink programs. Here, the user must pick up value 1 that corresponds to first-order discrete-time iP controller. The simulations of the control systems with the three first-order discrete-time iP MFC algorithms are conducted using the CS _ iPID _ SISO _ c.mdl Simulink diagram for cart position control, the CS _ iPID _ SISO _ a.mdl Simulink diagram for arm angular position control and the CS _ iPID _ SISO _ p.mdl Simulink diagram for payload position control. These three Simulink diagrams make use of the state-space model of the tower crane system viewed as a controlled process, which is described in Chapter 1 and is implemented in the process _ model.m S-function.

The simulation results obtained for the control systems with first-order discrete-time iP controllers with the parameters in (3.104–3.106) and the user-chosen parameter α are illustrated in Figure 3.3 for cart position control, Figure 3.4 for arm angular position control and Figure 3.5 for payload position control.

The experimental results obtained for the control systems with first-order discrete-time iP controllers with the parameters in (3.104–3.106) and the user-chosen

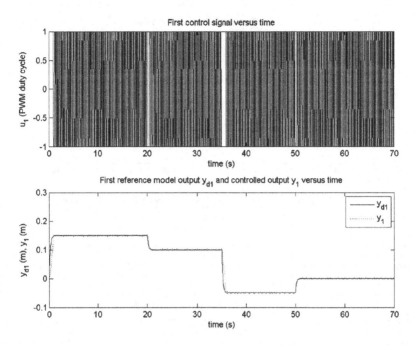

FIGURE 3.3 Simulation results of cart position control system with first-order discrete-time iP controller.

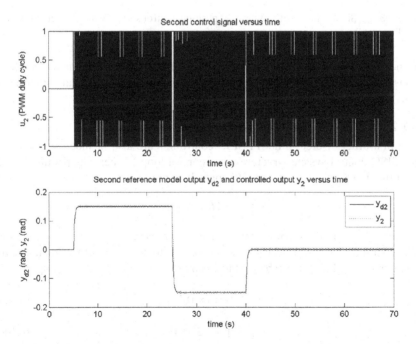

FIGURE 3.4 Simulation results of arm angular position control system with first-order discrete-time iP controller.

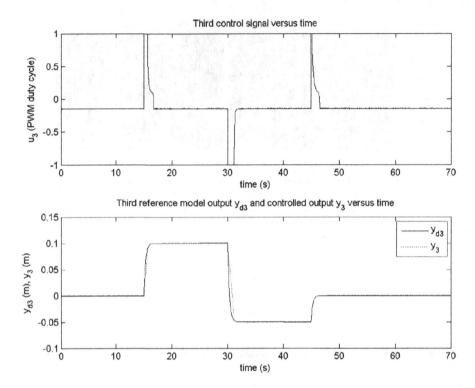

FIGURE 3.5 Simulation results of payload position control system with first-order discrete-time iP controller.

parameter α are similar to the simulation results because the mathematical model of TCS is very accurate. That is the reason why the experimental results are not inserted in this section.

3.4.1.2 The First-Order Discrete-Time iPI Controllers

Three first-order discrete-time iPI SISO controllers are also included separately in three SISO control system structures (i.e., control loops), where the parameter vectors of the first-order discrete-time iPI controllers are

$$\chi_{(\psi)} = [K_1 \ K_2]^T, \tag{3.107}$$

with $\psi \in \{c, a, p\}$, i.e., two parameters for each controller.

The parameter $\alpha = 0.0015$ is chosen according to step 1, and then in step 2, the parameters K_1 and K_2 are established to ensure the condition (3.78), leading to

$$\chi_{(c)} = [-0.75 \ 0.33]^T, \tag{3.108}$$

$$\chi_{(a)} = [-0.72 \ 0.4]^T, \tag{3.109}$$

$$\chi_{(p)} = [-0.85 \ 0.53]^T. \tag{3.110}$$

The details of steps 1 and 2 are implemented in the same files, i.e., the MFC _ iPID _ 1st _ order _ c.m Matlab program for cart position control, the MFC _ iPID _ 1st _ order _ a.m Matlab program for arm angular position control and the MFC _ iPID _ 1st _ order _ p.m Matlab program for payload position control, which are included in the accompanying Matlab & Simulink programs. Here, the user must pick up value 2 that corresponds to first-order discrete-time iPI controller. The digital simulations of the three control systems with first-order discrete-time iPI MFC algorithms are conducted using the same Simulink diagrams, i.e., the CS _ iPID _ SISO _ c.mdl for cart position control, the CS _ iPID _ SISO _ a.mdl for arm angular position control and the CS _ iPID _ SISO _ p.mdl for payload position control. These three Simulink diagrams make use of the state-space model of the tower crane system viewed as a controlled process, which is described in Chapter 1 and is implemented in the process _ model.m S-function.

The simulation results obtained for the control systems with first-order discrete-time iPI controllers with the parameters in (3.108–3.110) and the user-chosen parameter α are illustrated in Figure 3.6 for cart position control, Figure 3.7 for arm angular position control and Figure 3.8 for payload position control.

The experimental results obtained for the control systems with first-order discrete-time iPI controllers with the parameters in (3.108–3.110) and the user-chosen

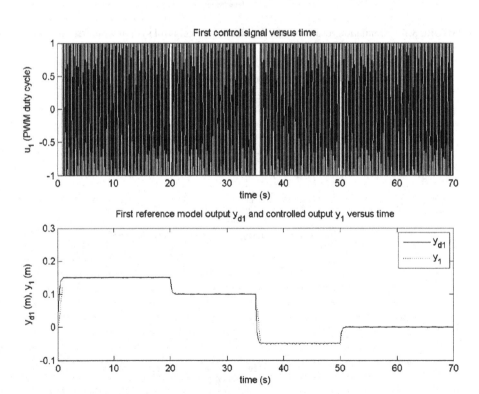

FIGURE 3.6 Simulation results of cart position control system with first-order discrete-time iPI controller.

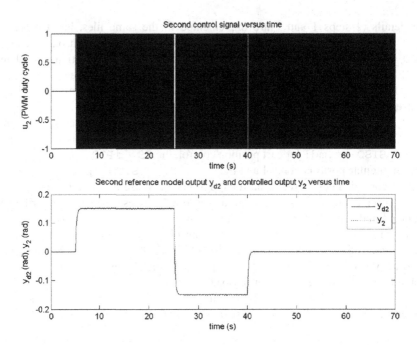

FIGURE 3.7   Simulation results of arm angular position control system with first-order discrete-time iPI controller.

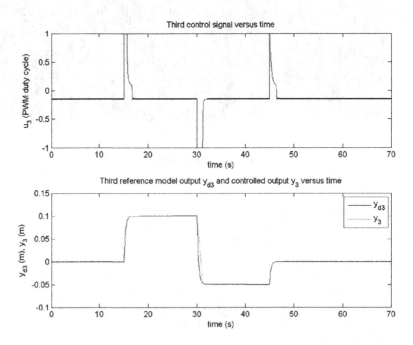

FIGURE 3.8   Simulation results of payload position control system with first-order discrete-time iPI controller.

parameter α are similar to the simulation results because the mathematical model of TCS is very accurate. This is the reason why the experimental results are not inserted in this section.

3.4.1.3 The First-Order Discrete-Time iPID Controllers

Three first-order discrete-time iPID SISO controllers are implemented separately in three control loops, where the vector parameters of the first-order discrete-time iPID controllers are

$$\chi_{(\psi)} = [K_1 \ K_2 \ K_3]^T, \tag{3.111}$$

with $\psi \in \{c, a, p\}$, i.e., three parameters for each controller.

The parameter $\alpha = 0.0015$ is chosen according to step 1, and then in step 2, the parameters K_1, K_2 and K_3 are established to ensure the condition (3.81), resulting in

$$\chi_{(c)} = [-0.77 \ 0.14 \ 0.41]^T, \tag{3.112}$$

$$\chi_{(a)} = [-0.75 \ 0.4 \ -0.1]^T, \tag{3.113}$$

$$\chi_{(p)} = [-0.1 \ -0.1 \ -0.3]^T. \tag{3.114}$$

The details of steps 1 and 2 are implemented in the same files, i.e., the MFC _ iPID _ 1st _ order _ c.m Matlab program for cart position control, the MFC _ iPID _ 1st _ order _ a.m Matlab program for arm angular position control and the MFC _ iPID _ 1st _ order _ p.m Matlab program for payload position control, which are included in the accompanying Matlab & Simulink programs. Here, the user must pick up value 3 that corresponds to first-order discrete-time iPID controller. The simulations of the control systems with the three first-order discrete-time iPID MFC algorithms are conducted using the same Simulink diagrams, i.e., the CS _ iPID _ SISO _ c.mdl for cart position control, the CS _ iPID _ SISO _ a.mdl for arm angular position control and the CS _ iPID _ SISO _ p.mdl for payload position control. These three Simulink diagrams make use of the state-space model of the tower crane system viewed as a controlled process, which is described in Chapter 1 and is implemented in the process _ model.m S-function.

The simulation results obtained for the control systems with first-order discrete-time iPID controllers with the parameters in (3.112–3.114) and the user-chosen parameter α are illustrated in Figure 3.9 for cart position control, Figure 3.10 for arm angular position control and Figure 3.11 for payload position control.

The experimental results obtained for the control systems with first-order discrete-time iPID controllers with the parameters in (3.112–3.114) and the user-chosen parameter α are similar to the simulation results because the mathematical model of TCS is very accurate. That is the reason why the experimental results are not inserted in this section.

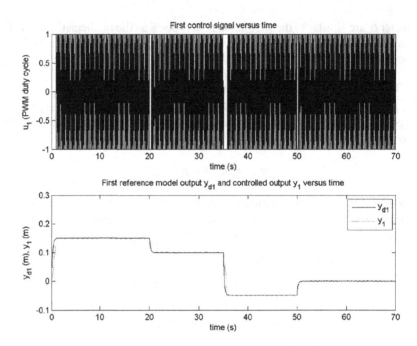

FIGURE 3.9 Simulation results of cart position control system with first-order discrete-time iPID controller.

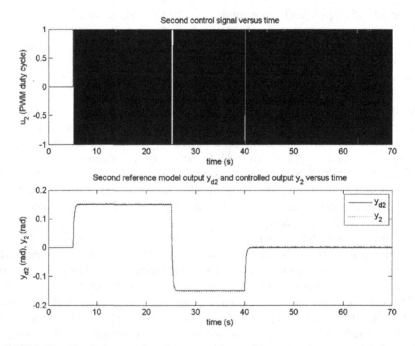

FIGURE 3.10 Simulation results of arm angular position control system with first-order discrete-time iPID controller.

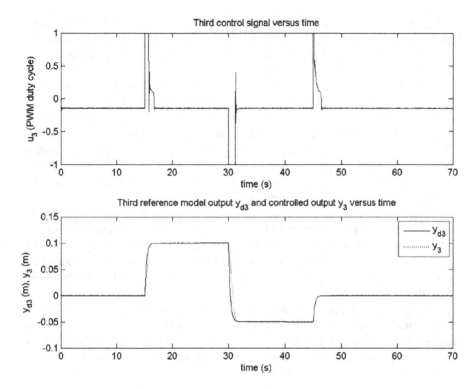

FIGURE 3.11 Simulation results of payload position control system with first-order discrete-time iPID controller.

3.4.1.4 The Second-Order Discrete-Time iP Controllers

Three second-order discrete-time iP SISO controllers are used separately in three SISO control system structures, where the parameter vectors of the second-order discrete-time iP controllers are

$$\chi_{(\psi)} = K_1, \tag{3.115}$$

with $\psi \in \{c,a,p\}$, i.e., one parameter for each controller.

The parameter $\alpha = 0.0015$ is chosen according to step 1, and then in step 2, the parameter K_1 is established to ensure the condition (3.86), resulting in

$$\chi_{(c)} = -0.75, \tag{3.116}$$

$$\chi_{(a)} = -0.89, \tag{3.117}$$

$$\chi_{(p)} = -0.8. \tag{3.118}$$

The details of steps 1 and 2 are implemented in the same files, i.e., the MFC _ iPID _ 2nd _ order _ c.m Matlab program for cart position control, the

MFC _ iPID _ 2nd _ order _ a.m Matlab program for arm angular position control and the MFC _ iPID _ 2nd _ order _ p.m Matlab program for payload position control, which are included in the accompanying Matlab & Simulink programs. Here, the user must pick up value 1 that corresponds to second-order discrete-time iP controller. The digital simulations of the control systems with the three second-order discrete-time iP MFC algorithms are conducted using the same Simulink diagrams, i.e., the CS _ iPID _ 2nd _ order _ SISO _ c.mdl for cart position control, the CS _ iPID _ 2nd _ order _ SISO _ a.mdl for arm angular position control and the CS _ iPID _ 2nd _ order _ SISO _ p.mdl for payload position control. These three Simulink diagrams make use of the state-space model of the tower crane system viewed as a controlled process, which is described in Chapter 1 and is implemented in the process _ model.m S-function.

The simulation results of the control systems with second-order discrete-time iP controllers with the parameters in (3.116–3.118) and the user-chosen parameter α are obtained in terms of running the Matlab files specified in the above paragraph. These simulation results are similar to the experimental results because the mathematical model of TCS is very accurate. This is the reason why the simulation results are not inserted in this section.

The experimental results obtained for the control systems with second-order discrete-time iP controllers with the parameters in (3.116–3.118) and the user-chosen parameter α are illustrated in Figure 3.12 for cart position control, Figure 3.13 for arm angular position control and Figure 3.14 for payload position control.

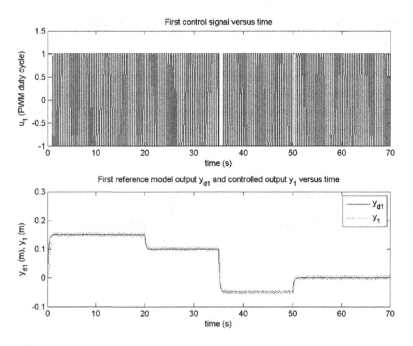

FIGURE 3.12 Experimental results of cart position control system with second-order discrete-time iP controller.

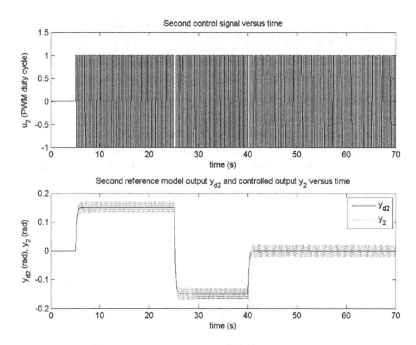

FIGURE 3.13 Experimental results of arm angular position control system with second-order discrete-time iP controller.

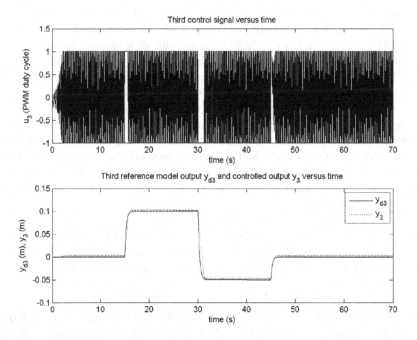

FIGURE 3.14 Experimental results of payload position control system with second-order discrete-time iP controller.

3.4.1.5   The Second-Order Discrete-Time iPI Controllers

Three second-order discrete-time iPI SISO controllers are used separately in three SISO control systems (i.e., one for each controlled output), where the parameter vectors of the second-order discrete-time iPI controllers are

$$\chi_{(\psi)} = [K_1 \ K_2]^T, \tag{3.119}$$

with $\psi \in \{c, a, p\}$, i.e., two parameters for each controller.

The parameter $\alpha = 0.0015$ is chosen according to step 1, and then in step 2, the parameters K_1 and K_2 are established to ensure the condition (3.89), leading to

$$\chi_{(c)} = [-0.75 \ 0.7]^T, \tag{3.120}$$

$$\chi_{(a)} = [-0.9 \ 0.85]^T, \tag{3.121}$$

$$\chi_{(p)} = [-0.85 \ 0.5]^T. \tag{3.122}$$

The details of steps 1 and 2 are implemented in the same files, i.e., the MFC _ iPID _ 2nd _ order _ c.m Matlab program for cart position control, the MFC _ iPID _ 2nd _ order _ a.m Matlab program for arm angular position control and the MFC _ iPID _ 2nd _ order _ p.m Matlab program for payload position control, which are included in the accompanying Matlab & Simulink programs. Here, the user must pick up value 2 that corresponds to second-order discrete-time iPI controller. The simulations of the control systems with the three second-order discrete-time iPI MFC algorithms are conducted using the same Simulink diagrams, i.e., the CS _ iPID _ 2nd _ order _ SISO _ c.mdl for cart position control, the CS _ iPID _ 2nd _ order _ SISO _ a.mdl for arm angular position control and the CS _ iPID _ 2nd _ order _ SISO _ p.mdl for payload position control. These three Simulink diagrams make use of the state-space model of the tower crane system viewed as a controlled process, which is described in Chapter 1 and is implemented in the process _ model.m S-function.

The simulation results of the control systems with second-order discrete-time iPI controllers with the parameters in (3.120–3.122) and the user-chosen parameter α are obtained in terms of running the Matlab files specified in the above paragraph. These simulation results are similar to the experimental results because the mathematical model of TCS is very accurate. This is the reason why the simulation results are not inserted in this section.

The experimental results obtained for the control systems with second-order discrete-time iPI controllers with the parameters in (3.120–3.122) and the user-chosen parameter α are illustrated in Figure 3.15 for cart position control, Figure 3.16 for arm angular position control and Figure 3.17 for payload position control.

3.4.1.6   The Second-Order Discrete-Time iPID Controllers

Three second-order discrete-time iPID SISO controllers are applied separately in three control loops (i.e., one for each controlled output), where the parameter vectors of the second-order discrete-time iPID controllers are

$$\chi_{(\psi)} = [K_1 \ K_2 \ K_3]^T, \tag{3.123}$$

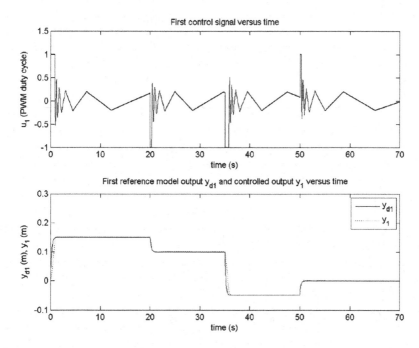

FIGURE 3.15 Experimental results of cart position control system with second-order discrete-time iPI controller.

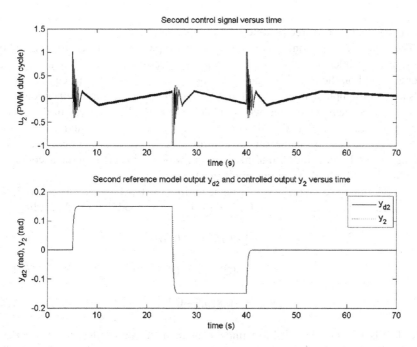

FIGURE 3.16 Experimental results of arm angular position control system with second-order discrete-time iPI controller.

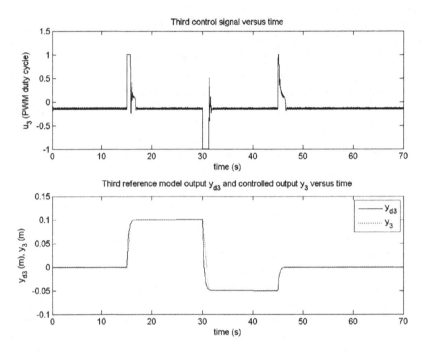

FIGURE 3.17 Experimental results of payload position control system with second-order discrete-time iPI controller.

with $\psi \in \{c, a, p\}$, i.e., three parameters for each controller.

The controllers are nonlinear since saturation-type nonlinearities are inserted on the three controller outputs to match the power electronics of the actuators, which operate based on the PWM principle. As shown in Chapter 1, the three control signals (or control inputs) are within −1 and 1, ensuring the necessary connection to the tower crane system subsystem, which models the controlled process in the control system structures developed in this chapter and the next chapters and also includes three ZOH blocks to enable digital control. The tower crane system subsystem belongs to the Process.mdl Simulink diagram, which is included in the accompanying Matlab & Simulink programs given in Chapter 1.

The parameter $\alpha = 0.0015$ is chosen according to step 1, and then in step 2, the parameters K_1, K_2 and K_3 are established to ensure the condition (3.92), leading to

$$\chi_{(c)} = [-0.75 \ \ 0.7 - 0.01]^T, \tag{3.124}$$

$$\chi_{(a)} = [-0.9 \ \ 0.85 - 0.01]^T, \tag{3.125}$$

$$\chi_{(p)} = [-0.8 \ \ 0.5 - 0.01]^T. \tag{3.126}$$

The details of steps 1 and 2 are implemented in the same files, i.e., the MFC _ iPID _ 2nd _ order _ c.m Matlab program for cart position control, the

MFC _ iPID _ 2nd _ order _ a.m Matlab program for arm angular position control and the MFC _ iPID _ 2nd _ order _ p.m Matlab program for payload position control, which are included in the accompanying Matlab & Simulink programs. Here, the user must pick up value 3 that corresponds to second-order discrete-time iPID controller. The simulations of the control systems with the three second-order discrete-time iPID MFC algorithms are conducted by using the same Simulink diagrams, i.e., the CS _ iPID _ 2nd _ order _ SISO _ c.mdl for cart position control, the CS _ iPID _ 2nd _ order _ SISO _ a.mdl for arm angular position control and the CS _ iPID _ 2nd _ order _ SISO _ p.mdl for payload position control. These three Simulink diagrams make use of the state-space model of the tower crane system viewed as a controlled process, which is described in Chapter 1 and is implemented in the process _ model.m S-function.

The simulation results of the control systems with second-order discrete-time iPID controllers with the parameters in (3.124–3.126) and the user-chosen parameter α are obtained in terms of running the Matlab files specified in the above paragraph. These simulation results are similar to the experimental results because the mathematical model of TCS is very accurate. This is the reason why the simulation results are not inserted in this section.

The experimental results obtained for the control systems with second-order discrete-time iPID controllers with the parameters in (3.124–3.126) and the user-chosen parameter α are illustrated in Figure 3.18 for cart position control, Figure 3.19 for arm angular position control and Figure 3.20 for payload position control.

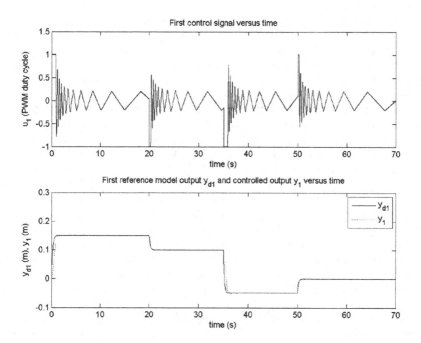

FIGURE 3.18 Experimental results of cart position control system with second-order discrete-time iPID controller.

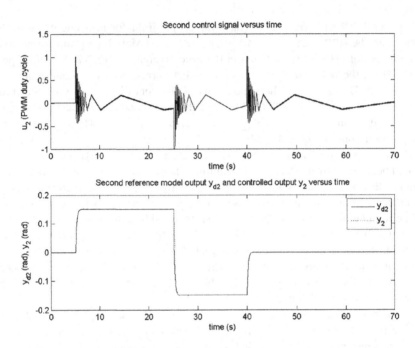

FIGURE 3.19 Experimental results of arm angular position control system with second-order discrete-time iPID controller.

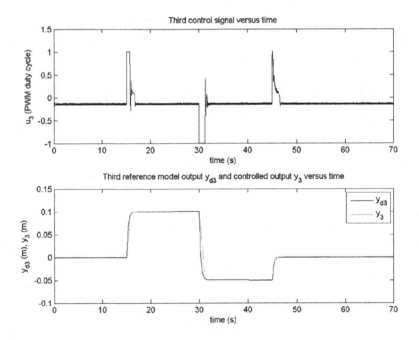

FIGURE 3.20 Experimental results of payload position control system with second-order discrete-time iPID controller.

3.4.2 MIMO CONTROL SYSTEMS

The objective function employed to measure the performance of the MIMO control system structure with MFC algorithms presented in Section 3.3 is

$$J(\chi) = \frac{1}{2} E \left\{ \sum_{k=1}^{N} \left[\mathbf{e}^T(k, \chi) \mathbf{e}(k, \chi) \right] \right\}, \qquad (3.127)$$

where the vector expression of the control error is

$$\mathbf{e} = \mathbf{y}_d - \mathbf{y} = [e_1 \ e_2 \ e_3]^T = [y_{d1} - y_1 \ y_{d2} - y_2 \ y_{d3} - y_3]^T, \qquad (3.128)$$

the scalar control errors e_1, e_2 and e_3 are related to cart position, arm angular position and payload position, respectively, the controlled output y_1, y_2 and y_3 are the cart position, the arm angular position and the payload position, respectively, the reference inputs (or the set-points, i.e., the outputs of the SISO reference models) are y_{d1}, y_{d2} and y_{d3} related to cart position, arm angular position and payload position, respectively, the mathematical expectation $E\{\Xi\}$, the sampling period, the time horizon and the number of samples are the same as in the SISO case. The reference input vector consists of the same step signals as those given in (3.94) for cart position, (3.95) for arm angular position and (3.96) for payload position, and the reference models are filtered using the filters given in (3.97) for cart position, (3.98) for arm angular position and (3.99) for payload position, with the corresponding discrete transfer functions in (3.100) for cart position, (3.101) for arm angular position and (3.102) for payload position.

The MFC algorithms in the continuous-time and discrete-time versions are applied to control all three outputs of the tower crane system process, respecting the steps of the MIMO MFC algorithms specified in Sections 3.2 and 3.3. The description of the steps is given in relation with the Matlab & Simulink programs and schemes.

3.4.2.1 The First-Order Discrete-Time iP Controllers

The first-order discrete-time iP MIMO controller is applied, where the vector parameter of the first-order discrete-time iP MIMO controller is

$$\chi = [\chi_{(c)1} \ \chi_{(a)1} \ \chi_{(p)1}]^T = [K_{11} \ K_{12} \ K_{13}]^T, \qquad (3.129)$$

which indicates three parameters of the MIMO controller, one parameter for each motion.

The controller is nonlinear since saturation-type nonlinearities are inserted on the controller outputs in order to match the power electronics of the actuators, which operate on the basis of the PWM principle. As shown in Chapter 1, the three control signals (or control inputs) are within −1 and 1, ensuring the necessary connection to the tower crane system subsystem, which models the controlled process in the control system structures developed in this chapter and the next chapters and also includes three ZOH blocks to enable digital control. The tower crane system subsystem belongs to the Process.mdl Simulink diagram, which is included in the accompanying Matlab & Simulink programs given in Chapter 1.

The parameter $\alpha = \mathrm{diag}(0.0015, 0.0015, 0.0015)$ is chosen according to step 1, and then in step 2, the parameters K_{11}, K_{12} and K_{13} are established to ensure the condition (3.74), resulting in

$$\chi = [\chi_{(c)1}\ \chi_{(a)1}\ \chi_{(p)1}]^{T} = [K_{11}\ K_{12}\ K_{13}]^{T} = [-0.7\ -0.6\ -0.14]^{T}. \quad (3.130)$$

The details of steps 1 and 2 are implemented in the MFC_iPID_1st_order.m Matlab program which is included in the accompanying Matlab & Simulink programs. Here, the user must pick up the value 1 that corresponds to the first-order discrete-time iP MIMO controller. The simulation of the control system with the first-order discrete-time iP MFC MIMO algorithm is conducted using the CS_iPID_MIMO.mdl Simulink diagram. This Simulink diagram makes use of the state-space model of the tower crane system viewed as a controlled process, which is described in Chapter 1 and is implemented in the process_model.m S-function.

The simulation results obtained for the control system with first-order discrete-time iP MIMO controller with the parameters in (3.130) and the user-chosen parameter α are illustrated in Figure 3.21 for cart position control, Figure 3.22 for arm angular position control and Figure 3.23 for payload position control.

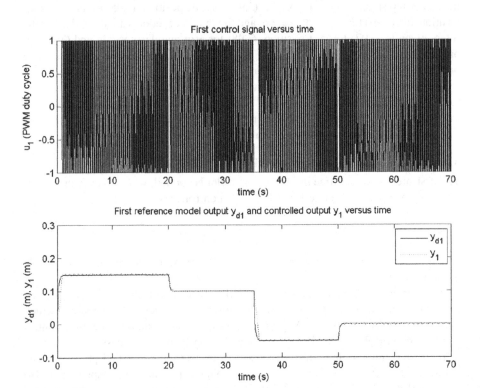

FIGURE 3.21 Simulation results of cart position control system with first-order discrete-time iP MIMO controller.

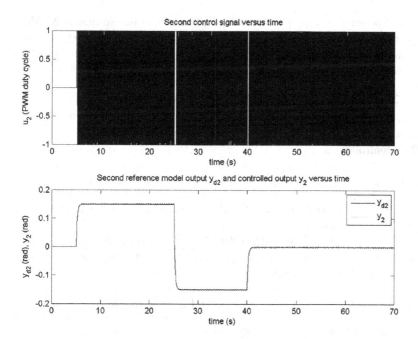

FIGURE 3.22 Simulation results of arm angular position control system with first-order discrete-time iP MIMO controller.

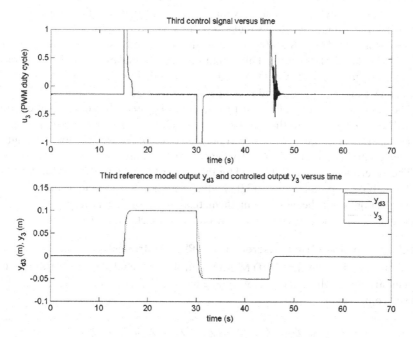

FIGURE 3.23 Simulation results of payload position control system with first-order discrete-time iP MIMO controller.

The experimental results obtained for the first-order discrete-time iP MIMO controllers with the parameters in (3.130) and the user-chosen parameter α are similar to the simulation results because the mathematical model of TCS is very accurate. This is the reason why the experimental results are not inserted in this section.

3.4.2.2 The First-Order Discrete-Time iPI Controllers

The first-order discrete-time iPI MIMO controllers are applied in a MIMO control system structure, where the vector parameter of the first-order discrete-time iPI MIMO controllers is

$$\chi = [\chi_{(c)1} \ \chi_{(c)2} \ \chi_{(a)1} \ \chi_{(a)2} \ \chi_{(p)1} \ \chi_{(p)2}]^T = [K_{11} \ K_{21} \ K_{12} \ K_{22} \ K_{13} \ K_{23}]^T, \quad (3.131)$$

which highlights six parameters of the MIMO controller, namely two parameters for each of the three motions.

The parameter $\alpha = \mathrm{diag}(0.0015, 0.0015, 0.0015)$ is chosen according to step 1, and then in step 2, the parameters K_{11}, K_{21}, K_{12}, K_{22}, K_{13} and K_{23} are established to ensure the condition (3.78), resulting in

$$\chi = [\chi_{(c)1} \ \chi_{(c)2} \ \chi_{(a)1} \ \chi_{(a)2} \ \chi_{(p)1} \ \chi_{(p)2}]^T = [K_{11} \ K_{21} \ K_{12} \ K_{22} \ K_{13} \ K_{23}]^T$$

$$= [-0.75 \ 0.33 - 0.72 \ 0.4 - 0.85 \ 0.53]^T. \tag{3.132}$$

The details of steps 1 and 2 are implemented in the MFC _ iPID _ 1st _ order.m Matlab program which is included in the accompanying Matlab & Simulink programs. Here, the user must pick up value 2 that corresponds to first-order discrete-time iPI MIMO controller. The simulations of the control system with the first-order discrete-time iPI MFC MIMO algorithm is conducted using the CS _ iPID _ MIMO.mdl Simulink diagram. This Simulink diagram makes use of the state-space model of the tower crane system viewed as a controlled process, which is described in Chapter 1 and is implemented in the process _ model.m S-function.

The simulation results obtained for the control system first-order discrete-time iPI MIMO controller with the parameters in (3.132) and the user-chosen parameter α are illustrated in Figure 3.24 for cart position control, Figure 3.25 for arm angular position control and Figure 3.26 for payload position control.

The experimental results obtained for the first-order discrete-time iPI MIMO controller with the parameters in (3.132) and the user-chosen parameter α are similar to the simulation results because the mathematical model of TCS is very accurate. That is the reason why the experimental results are not inserted in this section.

3.4.2.3 The First-Order Discrete-Time iPID Controllers

The first-order discrete-time iPID MIMO controller is included in the MIMO control system structure, where the vector parameter of the first-order discrete-time iPID MIMO controllers is

$$\chi = [\chi_{(c)1} \ \chi_{(c)2} \ \chi_{(c)3} \ \chi_{(a)1} \ \chi_{(a)2} \ \chi_{(a)3} \ \chi_{(p)1} \ \chi_{(p)2} \ \chi_{(p)3}]^T$$

$$= [K_{11} \ K_{21} \ K_{31} \ K_{12} \ K_{22} \ K_{32} \ K_{13} \ K_{23} \ K_{33}]^T, \tag{3.133}$$

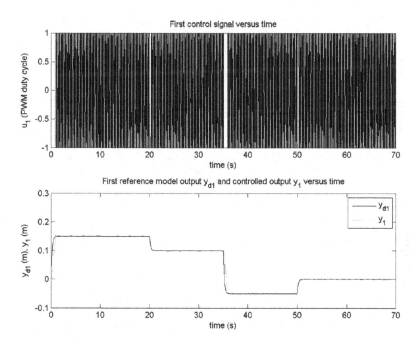

FIGURE 3.24 Simulation results of cart position control system with first-order discrete-time iPI MIMO controller.

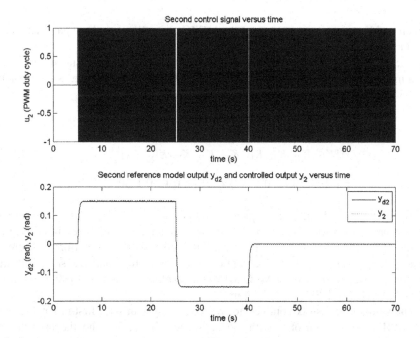

FIGURE 3.25 Simulation results of arm angular position control system with first-order discrete-time iPI MIMO controller.

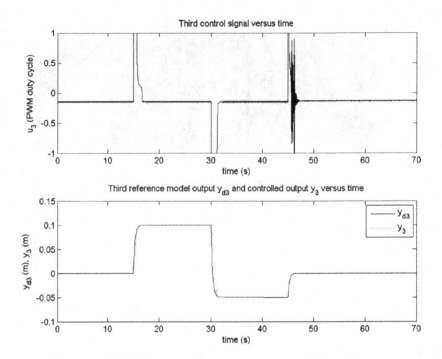

FIGURE 3.26 Simulation results of payload position control system with first-order discrete-time iPI MIMO controller.

which highlights nine parameters of the MIMO controller, i.e., three parameters for each motion.

The parameter $\alpha = \text{diag}(0.0015, 0.0015, 0.0015)$ is chosen according to step 1, and then in step 2, the parameters $K_{11}, K_{21}, K_{31}, K_{12}, K_{22}, K_{32}, K_{13}, K_{23}$ and K_{33} are established to ensure the condition (3.81), resulting in

$$\chi = [\chi_{(c)1} \ \chi_{(c)2} \ \chi_{(c)3} \ \chi_{(a)1} \ \chi_{(a)2} \ \chi_{(a)3} \ \chi_{(p)1} \ \chi_{(p)2} \ \chi_{(p)3}]^T$$

$$= [K_{11} \ K_{21} \ K_{31} \ K_{12} \ K_{22} \ K_{32} \ K_{13} \ K_{23} \ K_{33}]^T \quad (3.134)$$

$$= [-0.77 \ 0.14 \ 0.41 - 0.75 \ 0.4 - 0.1 - 0.1 - 0.1 - 0.3]^T.$$

The details of steps 1 and 2 are implemented in the MFC _ iPID _ 1st _ order.m Matlab program, which is included in the accompanying Matlab & Simulink programs. Here, the user must pick up value 3 that corresponds to first-order discrete-time iPID MIMO controller. The digital simulation of the control system with the first-order discrete-time iPID MFC MIMO algorithm is conducted using the CS _ iPID _ MIMO.mdl Simulink diagram.

The simulation results obtained for the control system with first-order discrete-time iPID MIMO controller with the parameters in (3.134) and the user-chosen parameter α are illustrated in Figure 3.27 for cart position control, Figure 3.28 for arm angular position control and Figure 3.29 for payload position control.

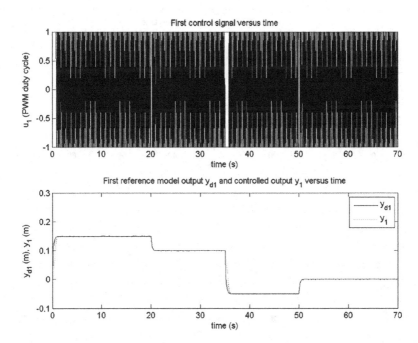

FIGURE 3.27 Simulation results of cart position control system with first-order discrete-time iPID MIMO controller.

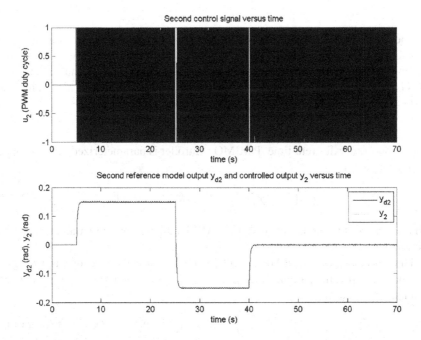

FIGURE 3.28 Simulation results of arm angular position control system with first-order discrete-time iPID MIMO controller.

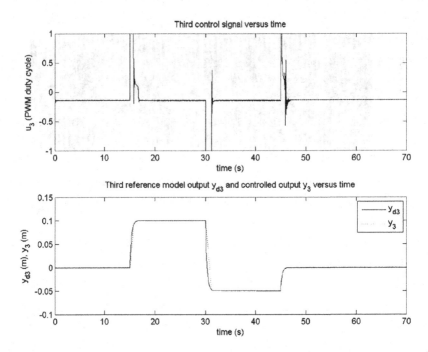

FIGURE 3.29 Simulation results of payload position control system with first-order discrete-time iPID MIMO controller.

The experimental results obtained for the control system with first-order discrete-time iPID MIMO controllers with the parameters in (3.134) and the user-chosen parameter α are similar to the simulation results because the mathematical model of TCS is very accurate. This is the reason why the experimental results are not inserted in this section.

3.4.2.4   The Second-Order Discrete-Time iP Controllers

The second-order discrete-time iP MIMO controller is parameterized in terms of the parameter vector

$$\chi = [\chi_{(c)1} \ \chi_{(a)1} \ \chi_{(p)1}]^T = [K_{11} \ K_{12} \ K_{13}]^T, \tag{3.135}$$

which illustrates three parameters of the MIMO controller, one parameter for each motion.

The parameter $\alpha = \mathrm{diag}(0.0015, 0.0015, 0.0015)$ is chosen according to step 1, and then in step 2, the parameters K_{11}, K_{12} and K_{13} are established to ensure the condition (3.86), leading to

$$\chi = [\chi_{(c)1} \ \chi_{(a)1} \ \chi_{(p)1}]^T = [K_{11} \ K_{12} \ K_{13}]^T = [-0.75 - 0.89 - 0.85]^T. \tag{3.136}$$

The details of steps 1 and 2 are implemented in the `MFC _ iPID _ 2nd _ order.m`
Matlab program which is included in the accompanying Matlab & Simulink programs. Here, the user must pick up value 1 that corresponds to second-order discrete-time iP MIMO controller. The simulation of the control system with the
second-order discrete-time iP MFC MIMO algorithm is conducted using the `CS _`
`iPID _ 2nd _ order _ MIMO.mdl` Simulink diagram.

The simulation results of the control system with second-order discrete-time iP
MIMO controller with the parameters in (3.136) and the user-chosen parameter α are
obtained in terms of running the Matlab files specified in the above paragraph. These
simulation results are similar to the experimental results because the mathematical
model of TCS is very accurate. This is the reason why the simulation results are not
inserted in this section.

The experimental results obtained for the control system with second-order
discrete-time iP MIMO controller with the parameters in (3.136) and the user-chosen
parameter α are illustrated in Figure 3.30 for cart position control, Figure 3.31 for
arm angular position control and Figure 3.32 for payload position control.

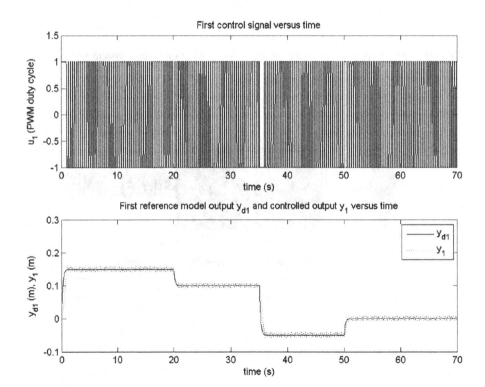

FIGURE 3.30 Experimental results of cart position control system with second-order
discrete-time iP MIMO controller.

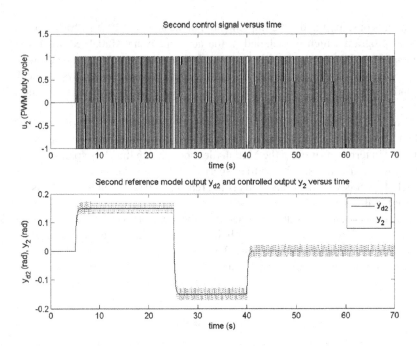

FIGURE 3.31 Experimental results of arm angular position control system with second-order discrete-time iP MIMO controller.

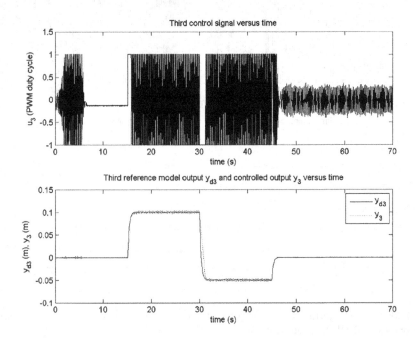

FIGURE 3.32 Experimental results of payload position control system with second-order discrete-time iP MIMO controller.

3.4.2.5   The Second-Order Discrete-Time iPI Controllers

The second-order discrete-time iPI MIMO controller employs the parameter vector

$$\chi = [\chi_{(c)1} \; \chi_{(c)2} \; \chi_{(a)1} \; \chi_{(a)2} \; \chi_{(p)1} \; \chi_{(p)2}]^T = [K_{11} \; K_{21} \; K_{12} \; K_{22} \; K_{13} \; K_{23}]^T, \quad (3.137)$$

indicating six parameters of the MIMO controller, i.e., two parameters for each of the three motions.

The parameter $\alpha = \text{diag}(0.0015, 0.0015, 0.0015)$ is chosen according to step 1, and then in step 2, the parameters $K_{11}, K_{21}, K_{12}, K_{22}, K_{13}$ and K_{23} are established to ensure the condition (3.89), which yields

$$\chi = [\chi_{(c)1} \; \chi_{(c)2} \; \chi_{(a)1} \; \chi_{(a)2} \; \chi_{(p)1} \; \chi_{(p)2}]^T = [K_{11} \; K_{21} \; K_{12} \; K_{22} \; K_{13} \; K_{23}]^T$$
$$= [-0.75 \; 0.7 - 0.9 \; 0.85 - 0.85 \; 0.5]^T. \quad (3.138)$$

The details of steps 1 and 2 are implemented in the MFC _ iPID _ 2nd _ order.m Matlab program, which is included in the accompanying Matlab & Simulink programs. Here, the user must pick up value 2 that corresponds to second-order discrete-time iPI MIMO controller. The simulation of the control system with the second-order discrete-time iPI MFC MIMO algorithm is conducted using the CS _ iPID _ 2nd _ order _ MIMO.mdl Simulink diagram.

The simulation results of the control system with second-order discrete-time iPI MIMO controller with the parameters in (3.138) and the user-chosen parameter α are obtained in terms of running the Matlab files specified in the above paragraph. These simulation results are similar to the experimental results because the mathematical model of TCS is very accurate. This is the reason why the simulation results are not inserted in this section.

The experimental results obtained for the control system with second-order discrete-time iPI MIMO controller with the parameters in (3.138) and the user-chosen parameter α are illustrated in Figure 3.33 for cart position control, Figure 3.34 for arm angular position control and Figure 3.35 for payload position control.

3.4.2.6   The Second-Order Discrete-Time iPID Controllers

The second-order discrete-time iPID MIMO controllers make use of the parameter vector

$$\chi = [\chi_{(c)1} \; \chi_{(c)2} \; \chi_{(c)3} \; \chi_{(a)1} \; \chi_{(a)2} \; \chi_{(a)3} \; \chi_{(p)1} \; \chi_{(p)2} \; \chi_{(p)3}]^T$$
$$= [K_{11} \; K_{21} \; K_{31} \; K_{12} \; K_{22} \; K_{32} \; K_{13} \; K_{23} \; K_{33}]^T, \quad (3.139)$$

which highlights nine parameters of the MIMO controller, three parameters for each motion.

This MIMO controller is also nonlinear since saturation-type nonlinearities are inserted on the controller outputs to match the power electronics of the actuators, which operate based on the PWM principle. As shown in Chapter 1, the three control signals (or control inputs) are within -1 and 1, ensuring the necessary connection

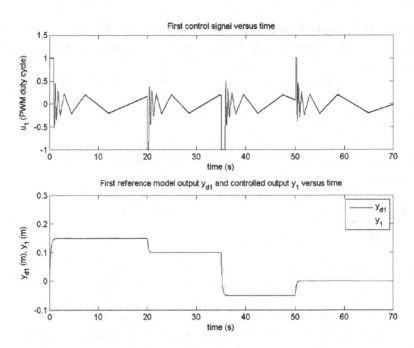

FIGURE 3.33 Experimental results of cart position control system with second-order discrete-time iPI MIMO controller.

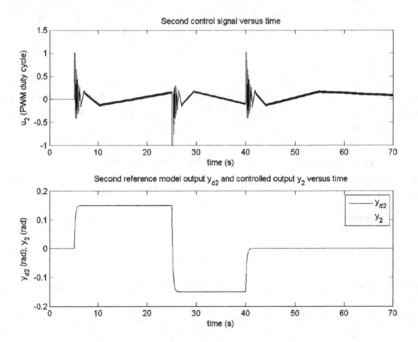

FIGURE 3.34 Experimental results of arm angular position control system with second-order discrete-time iPI MIMO controller.

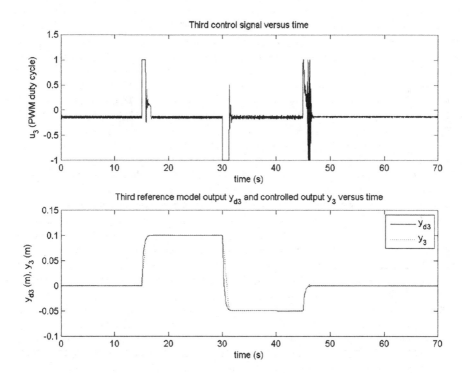

FIGURE 3.35 Experimental results of payload position control system with second-order discrete-time iPI MIMO controller.

to the tower crane system subsystem, which models the controlled process in the control system structures developed in this chapter and the next chapters and also includes three ZOH blocks to enable digital control. The tower crane system subsystem belongs to the Process.mdl Simulink diagram, which is included in the accompanying Matlab & Simulink programs given in Chapter 1.

The parameter $\alpha = \mathrm{diag}(0.0015, 0.0015, 0.0015)$ is chosen according to step 1, and then in step 2, the parameters K_{11}, K_{21}, K_{31}, K_{12}, K_{22}, K_{32}, K_{13}, K_{23} and K_{33} are established to ensure the condition (3.92), leading to

$$\chi_{(\psi)} = [\chi_{(c)1}\ \chi_{(c)2}\ \chi_{(c)3}\ \chi_{(a)1}\ \chi_{(a)2}\ \chi_{(a)3}\ \chi_{(p)1}\ \chi_{(p)2}\ \chi_{(p)3}]^T$$

$$= [K_{11}\ K_{21}\ K_{31}\ K_{12}\ K_{22}\ K_{32}\ K_{13}\ K_{23}\ K_{33}]^T \qquad (3.140)$$

$$= [-0.75\ 0.7 - 0.01 - 0.9\ 0.85 - 0.01 - 0.8\ 0.5 - 0.01]^T.$$

The details of steps 1 and 2 are implemented in the MFC _ iPID _ 2nd _ order.m Matlab program which is included in the accompanying Matlab & Simulink programs. Here, the user must pick up value 3 that corresponds to second-order discrete-time iPID MIMO controller. The simulation of the control system with the second-order discrete-time iPID MFC MIMO algorithm is conducted using the CS _ iPID _ 2nd _ order _ MIMO.mdl Simulink diagram.

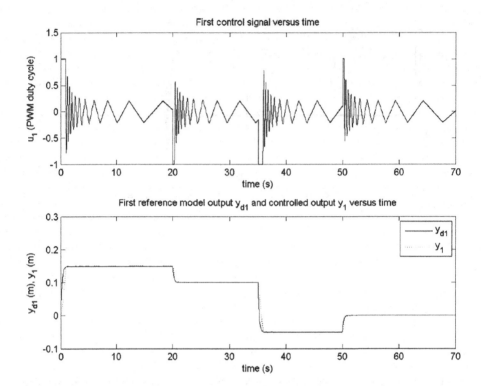

FIGURE 3.36 Experimental results of cart position control system with second-order discrete-time iPID MIMO controller.

The simulation results of the control system with second-order discrete-time iPID MIMO controller with the parameters in (3.140) and the user-chosen parameter α are obtained in terms of running the Matlab files specified in the above paragraph. These simulation results are similar to the experimental results because the mathematical model of TCS is very accurate. This is the reason why the simulation results are not inserted in this section.

The experimental results obtained for the second-order discrete-time iPID MIMO controller with the parameters in (3.140) and the user-chosen parameter α are illustrated in Figure 3.36 for cart position control, Figure 3.37 for arm angular position control and Figure 3.38 for payload position control.

3.4.2.7 Simulation and Experimental Results for the iPID Controller with the Adaptive Model-Based Parameter Estimator

Here, the simulation and the experimental results for implementing an iPID controller presented in Section 3.2.6.3 and the adaptive model-based parameter estimator presented in Section 3.2.4, applied to the second-order MIMO dynamic system of the crane, are provided. According to (3.56), both unknown parameters of the ultra-local model, i.e., f and b, are estimated online. But, since b is utilized in the denominator of the iPID controller, a saturation function is acting over the estimated values.

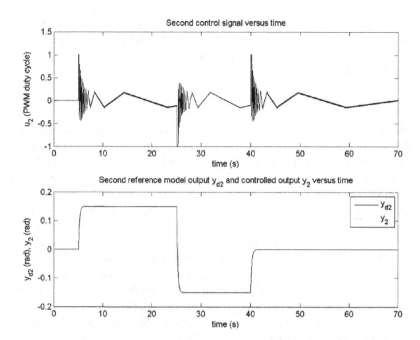

FIGURE 3.37 Experimental results of arm angular position control system with second-order discrete-time iPID MIMO controller.

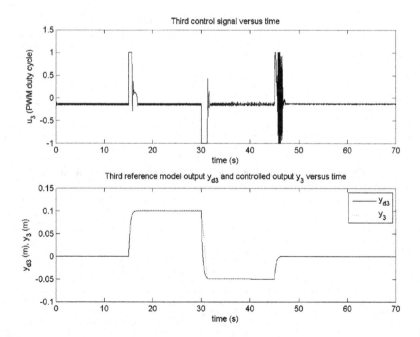

FIGURE 3.38 Experimental results of payload position control system with second-order discrete-time iPID MIMO controller.

Here, its minimum acceptable value is set to 0.001. Moreover, since the MIMO process has three dimensions, three parallel modules of the estimator are utilized for estimating the corresponding unknown parameters.

Recalling (3.66), the tuning parameters of the adaptive estimator are adjusted as

$$\gamma = \begin{bmatrix} 0.1 & 0 \\ 0 & 0.01 \end{bmatrix}, \rho = 1 \text{ and } a = 10 \text{ for each dimension of the model.}$$

The details of the controller design and tuning are given in the MFC_iPID_2nd_order_continous_time.m Matlab program. Here, the user must select the value 3 that corresponds to first-order continuous-time iPID MIMO controller. The simulation of the control system with the first-order continuous-time iPID MIMO controller is conducted using the CS_iPID_2nd_order_MIMO_AMPE_continous_time.mdl Simulink diagram. The readers are advised to use Matlab2018a or a newer version because functions that are available in relatively fresh versions of Simulink are used.

The simulation of the control system behavior is conducted in terms of running the Matlab files specified in the above paragraph. The simulation results are similar to the experimental results because the mathematical model of TCS is very accurate. This is the reason why the simulation results are not inserted in this section.

The corresponding experimental results showing the control signals and output tracking performance are given in Figures 3.39–3.41. Here, the value of $J = 1.0745$ is achieved for the cost function in (3.93). Regarding the online estimations for b, note that the corresponding estimated values are all saturated to 0.001.

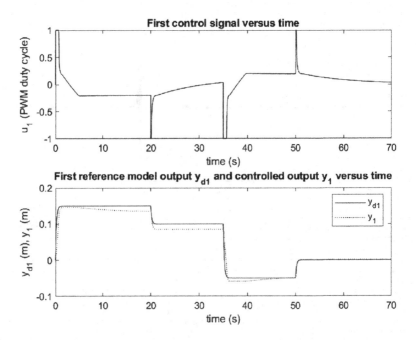

FIGURE 3.39 Experimental results: control signal and output tracking performance for the first dimension (application of iPID controller and adaptive model-based parameter estimator).

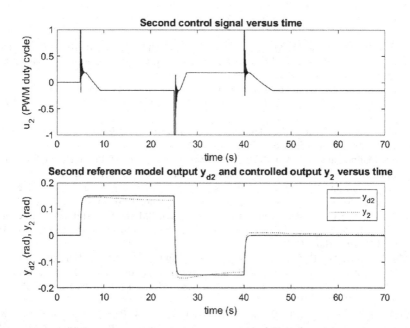

FIGURE 3.40 Experimental results: control signal and output tracking performance for the second dimension (application of iPID controller and adaptive model-based parameter estimator).

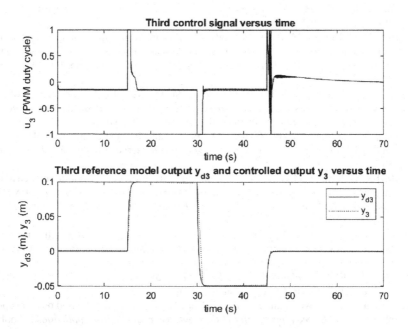

FIGURE 3.41 Experimental results: control signal and output tracking performance for the third dimension (application of iPID controller and adaptive model-based parameter estimator).

REFERENCES

[Ata99] A. N. Atassi and H. Khalil, "A separation principle for the stabilization of a class of nonlinear systems," *IEEE Transactions on Automatic Control*, vol. 44, no. 9, pp. 1672–1687, Sep. 1999.

[Fis13] N. Fischer, R. Kamalpurkar and W. E. Dixon, "LaSalle-Yoshizawa corollaries for nonsmooth systems," *IEEE Transactions on Automatic Control*, vol. 58, no. 9, pp. 2333–2338, Sep. 2013.

[Fli13] M. Fliess and C. Join, "Model-free control," *International Journal of Control*, vol. 86, no. 12, pp. 2228–2252, Dec. 2013.

[Ioa06] P. Ioannou and B. Fidan, *Adaptive Control Tutorial*, SIAM, Philadelphia, PA, 2006.

[Kha02] H. Khalil, *Nonlinear Systems*, Prentice Hall, Upper Saddle River, NJ, 2002.

[Lev03] A. Levant, "Higher-order sliding modes, differentiation and output-feedback control," *International Journal of Control*, vol. 76, no. 9–10, pp. 924–941, Sep. 2003.

[Pre17] R.-E. Precup, M.-B. Radac and R.-C. Roman, "Model-free sliding mode control of nonlinear systems: Algorithms and experiments," *Information Sciences*, vol. 381, pp. 176–192, Mar. 2017.

[Rad14] M.-B. Radac, R.-C. Roman, R.-E. Precup, and E. M. Petriu, "Data-driven model-free control of twin rotor aerodynamic systems: Algorithms and experiments," in *Proceedings of 2014 IEEE International Symposium on Intelligent Control*, Antibes, France, 2014, pp. 1889–1894.

[Rom15] R.-C. Roman, M.-B. Radac, R.-E. Precup and E.M. Petriu, "Data-driven optimal model-free control of twin rotor aerodynamic systems," in *Proceedings of 2015 IEEE International Conference on Industrial Technology*, Seville, Spain, 2015, pp. 161–166.

[Rom16a] R.-C. Roman, M.-B. Radac and R.-E. Precup, "Mixed MFC-VRFT approach for a multivariable aerodynamic system position control," in *Proceedings of 2016 IEEE International Conference on Systems, Man, and Cybernetics*, Budapest, Hungary, 2016, pp. 2615–2620.

[Rom16b] R.-C. Roman, M.-B. Radac, and R.-E. Precup, "Multi-input-multi-output system experimental validation of model-free control and virtual reference feedback tuning techniques," *IET Control Theory & Applications*, vol. 10, no. 12, pp. 1395–1403, Aug. 2016.

[Rom17a] R.-C. Roman, R.-E. Precup and M.-B. Radac, "Model-free fuzzy control of twin rotor aerodynamic systems," in *Proceedings of 25th Mediterranean Conference on Control and Automation*, Valletta, Malta, 2017, pp. 559–564.

[Rom17b] R.-C. Roman, M.-B. Radac, R.-E. Precup and E. M. Petriu, "Virtual reference feedback tuning of model-free control algorithms for servo systems," *Machines*, vol. 5, no. 4, pp. 1–15, Oct. 2017.

[Rom18a] R.-C. Roman, M.-B. Radac, C. Tureac and R.-E. Precup, "Data-driven active disturbance rejection control of pendulum cart systems," in *Proceedings of 2018 IEEE Conference on Control Technology and Applications*, Copenhagen, Denmark, 2018, pp. 933–938.

[Rom18b] R.-C. Roman, R.-E. Precup and R.-C. David, "Second order intelligent proportional-integral fuzzy control of twin rotor aerodynamic systems," *Procedia Computer Science*, vol. 139, pp. 372–380, Oct. 2018.

[Rom18c] R.-C. Roman, *Tehnici de tip model-free de acordare a parametrilor regulatoarelor automate*, PhD thesis, Editura Politehnica, Timisoara, 2018.

[Tha14] H. Thabet, M. Ayadi and F. Rotella, "Ultra-local model control based on an adaptive observer", in *Proceedings of 2014 IEEE Conference on Control Applications*, Antibes, France, 2014, pp. 122–127.

[Zha11] Q. Zhang, *Adaptive Observer for MIMO Linear Time Varying Systems*, Research Report RR-4111, INRIA, France, 2011.

4 Model-Free Sliding Mode Controllers

4.1 INTRODUCTION

Sliding mode control (SMC) is a relatively easily understandable and applicable non-linear technique with the advantage of robustness to changes in process parameters and load-type disturbances. The idea of combining data-driven and SMC techniques came about due to applications that require control system structures with superior tuning performance.

Some representative structures and applications of combinations of between data-driven control and SMC are briefly pointed out as follows because they are important by exploiting the advantages of both control techniques. In [Pre14], an iPI MFC-SMC hybrid algorithm is proposed and experimentally validated on a servosystem-type laboratory equipment. A second version of SMC combined with iPI control is suggested in [Pre17] and in [Pre21] and applied along with that given in [Pre14] to twin rotor aerodynamic systems; also MFC with SMC was applied to reverse osmosis desalination plants [Vrk18]. A comparison of several model-free control (MFC) algorithms in a quadrotor system application is performed in [Sch18]. In [Wan15], a hybrid intelligent PD-type algorithm (iPD) MFC-SMC is proposed and tested by numerical simulations on a quadrotor. In [Wan16], a hybrid MFAC-SMC algorithm is presented and tested by numerical simulations on an exoskeleton robot. In [Che12], the parameter tuning of an SMC controller by iterative learning control for a rotating process is proposed. [Fan16] proposes to tune the parameters of a SMC controller through adaptive dynamic programming; the SMC controller is validated using numerical simulation on a class of partially unknown processes with input disturbances. In [Wan03], an adaptive SMC controller dedicated to nonlinear systems with continuous time is analyzed and then validated using simulations. Model-free SMC based on linear regression estimation and optimization is theorized in [Ebr18] and tested in [Ebr20] to blood glucose control. Applications to wind turbine systems and exoskeleton systems are reported in [Li16] and [Wan17], respectively. The combination of discrete-time SML and MFAC is treated in [Cao20].

Considering the authors' results published in [Pre14], where a hybrid MFC-SMC system (MFSMCS) is proposed and experimentally validated on a servosystem-type laboratory and from the results of [Wan15], where a hybrid MFSMCS is developed and tested by numerical simulation on a quadrotor, two hybrid MFSMCS techniques are presented in [Pre17] and applied to servo system control and in [Pre21] to three-dimensional crane systems. Both techniques are based on guaranteeing stability using Lyapunov, and the theory is presented as follows focusing on these two techniques that lead to promising data-driven model-free controllers.

DOI: 10.1201/9781003143444-4

4.2   THEORY

4.2.1   THE HYBRID MODEL-FREE SLIDING MODE CONTROLLERS

Lemma 4.1

Considering the continuous-time control law of an iPI controller, an augmented control signal (or control input) $u_{\text{aug}}(t)$ of SMC type leads to the following control law specific to MFSMCS hybrid techniques [Rom18]:

$$u(t) = \frac{1}{\alpha}\left(-\hat{F}(t) + \dot{y}_d(t) - K_P e(t) - K_I \int_0^t e(\tau)d\tau\right) + u_{\text{aug}}(t), \qquad (4.1)$$

where $\hat{F}(t)$ is the estimate of $F(t)$, K_P and K_I represent the proportional and integral components of the PI controller, and this leads to the dynamic equation of control error specific to the control system structure MFSMCS algorithms:

$$\dot{e}(t) + K_P\, e(t) + K_I \int_0^t e(\tau)d\tau = e_{\text{est}}(t) + \alpha u_{\text{aug}}(t), \qquad (4.2)$$

where $e_{\text{est}}(t)$ is the estimation error of $F(t)$ and is defined as follows:

$$e_{\text{est}}(t) = \dot{y}(t) - \hat{\dot{y}}(t) = F(t) - \hat{F}(t). \qquad (4.3)$$

To estimate the first-order derivative of the controlled output $\dot{y}(t)$, the hybrid MFSMCS techniques use the low-pass (LP1) filter (a derivative with time) with the transfer function

$$H_{LP1}(s) = \frac{K_{LP1}s}{1 + T_{LP1}s}, \qquad (4.4)$$

where K_{LP1} is the gain of the filter and T_{LP1} is the time constant of the filter. The parameters of the LP1 filter must be chosen as a compromise between noise reduction and delays induced by them. The filter generates (as accurately as possible) the first-order derivative estimate of the $\dot{y}(t)$ process output, with notation $\hat{\dot{y}}(t)$, leading to the next modified form of the first-order local process model used in the design of the iPI controller equation:

$$\hat{F}(t) = \hat{\dot{y}}(t) - \alpha\, u(t). \qquad (4.5)$$

The LP1 filter with the transfer function in equation (4.4) generates the variable in equation of the iPI controller that is derived from the first order of the $y_d(t)$ reference trajectory.

The state variables $x_1(t)$ and $x_2(t)$ are defined in terms of

$$x_1(t) = \int_0^t e(\tau)d\tau, \tag{4.6}$$

$$x_2(t) = e(t),$$

and the control system structure state equations obtained by processing (4.2) are

$$\dot{x}_1(t) = x_2(t),$$

$$\dot{x}_2(t) = -K_I x_1(t) - K_P x_2(t) + \alpha\, u_{\text{aug}}(t) + e_{\text{est}}(t), \tag{4.7}$$

where $e_{\text{est}}(t)$ plays the role of disturbance input.

The switching variable $\sigma(t)$ used in the calculus of the control law $u_{\text{aug}}(t)$ specific to SMC has the expression as the equation given below:

$$\sigma(t) = x_1(t) + T\, x_2(t), \tag{4.8}$$

where $T > 0$ is a design parameter provided for determining the desired behavior of the control system structure on the sliding surface.

The Lyapunov function candidate introduced to ensure the stability of the closed-loop system modeled in (4.2) is

$$V(t) = \frac{1}{2}\sigma^2(t). \tag{4.9}$$

Using Lyapunov's stability theorem, the condition $\dot{V}(t) < 0$ is transformed into the sliding mode reaching and existence condition:

$$\sigma(t)\,\dot{\sigma}(t) < 0, \tag{4.10}$$

which is used in the design and tuning of the control law that generates $u_{\text{aug}}(t)$ as a function of $x_1(t)$ and $x_2(t)$.

The variable $\dot{\sigma}(t)$, which is the derivative of the switching variable, is obtained by differentiating (4.8)

$$\dot{\sigma}(t) = \dot{x}_1(t) + T\, \dot{x}_2(t). \tag{4.11}$$

Substituting the state equations of the control system structure from (4.7) in the expression of the derivative of the switching variable in (4.11) leads to

$$\dot{\sigma}(t) = -K_I T\, x_1(t) + (1 - K_P T)\, x_2(t) + \alpha T\, u_{\text{aug}}(t) + T\, e_{\text{est}}(t). \tag{4.12}$$

Since the estimation error $e_{\text{est}}(t)$ is unknown, in the design of the control law it is accepted that $e_{\text{est}}(t)$ is bounded and takes values in the range:

$$|e_{\text{est}}(t)| \le e_{\text{est max}}, \tag{4.13}$$

where $e_{\text{est max}}$ represents the upper limit of $|e_{\text{est}}(t)|$, and its value is known. Parameter $e_{\text{est max}}$ plays a role of design parameter.

Next two hybrid MFSMCS techniques are presented, namely MFSMCS1 and MFSMCS2, noting that the first technique is proposed in [Pre14] with a Savitzky-Golay filter and then applied in [Pre17, Wan15, Rom18] and [Pre21] in combination with iP MFC algorithms.

4.2.2   THE FIRST HYBRID MODEL-FREE SLIDING MODE CONTROLLER

To simplify the notations, the abbreviation MFSMCS1 is used for the first hybrid MFSMCS technique. This technique leads to the MFSMCS1 controller. The main details of the design of the control system structure with MFSMC1 controller are presented in this section.

Lemma 4.2

The augmented control input $u_{\text{aug}}(t)$ consists of two control signals:

$$u_{\text{aug}}(t) = u_{\text{eq}}(t) + u_{\text{cor}}(t),\qquad(4.14)$$

where $u_{\text{eq}}(t)$ is the equivalent control input, and $u_{\text{cor}}(t)$ is the correction control input. The equivalent control input $u_{\text{eq}}(t)$ is obtained from the ideal sliding model condition $\sigma(t) = 0$, which by derivation leads to

$$\dot{\sigma}(t) = 0.\qquad(4.15)$$

Substituting the expression of the augmented control input $u_{\text{aug}}(t)$ from (4.14) in the expression of the derivative of the switching variable $\dot{\sigma}(t)$ given in (4.11) with $u_{\text{aug}}(t) = u_{\text{eq}}(t)$ and solving the resulted equation considering the unknown $u_{\text{eq}}(t)$, the expression of $u_{\text{eq}}(t)$ becomes [Pre17, Pre21, Rom18]

$$u_{\text{eq}}(t) = \frac{K_I T\, x_1(t) - (1 - K_P T)\, x_2(t) - T\, e_{\text{est}}(t)}{\alpha T}.\qquad(4.16)$$

Since the value of $e_{\text{est}}(t)$ in (4.16) is unknown, its value is next substituted with $e_{\text{est max}}$, and therefore, the expression of $u_{\text{eq}}(t)$ in (4.16) is rewritten as follows [Pre17, Pre21, Rom18]:

$$u_{\text{eq}}(t) = \frac{K_I T\, x_1(t) - (1 - K_P T)\, x_2(t) - T\, e_{\text{est max}}}{\alpha T}.\qquad(4.17)$$

In order to fulfill the sliding mode reaching and existence condition given in (4.10) and to alleviate the chattering effects, a boundary layer approach is next applied. The expression of the correction control input $u_{\text{cor}}(t)$ specific to this approach is

$$u_{\text{cor}}(t) = -\frac{\eta_{\text{MFSMC}}}{\alpha\ T}\ \text{sat}(\sigma(t), e_{\text{MFSMC}}) = -\frac{\eta_{\text{MFSMC}}}{\alpha\ T} \begin{cases} -1 & \text{if } \sigma(t) < -e_{\text{MFSMC}}, \\[2mm] \dfrac{\sigma(t)}{e_{\text{MFSMC}}} & \text{if } |\sigma(t)| \le e_{\text{MFSMC}}, \\[2mm] 1 & \text{if } \sigma(t) > e_{\text{MFSMC}}, \end{cases}$$

$$(4.18)$$

where $\eta_{\text{MFSMC}} > 0$ is the convergence coefficient, and $e_{\text{MFSMC}} > 0$ is the boundary layer thickness.

Substituting the equivalent control input $u_{\text{eq}}(t)$ in (4.17) and the correction control input $u_{\text{cor}}(t)$ in (4.18) into the expression of the augmented control input $u_{\text{aug}}(t)$ in (4.14), then $u_{\text{aug}}(t)$ in the expression of the switching variable derivative $\dot{\sigma}(t)$ in (4.12), the expression of the derivative of the Lyapunov function candidate $\dot{V}(t)$ becomes

$$\dot{V}(t) = \sigma(t)\ \dot{\sigma}(t) = -\sigma(t)\ \eta_{\text{MFSMC}}\ \text{sat}(\sigma(t), \varepsilon_{\text{MFSMC}}) + \sigma(t)\ T\ [e_{\text{est}}(t) - e_{\text{est max}}]. \quad (4.19)$$

Theorem 4.1

Let the control law be defined in (4.1) for the process model $\dot{y}(t) = F(t) + \alpha\ u(t)$ with $\hat{F}(t)$ as disturbance estimator expressed in (4.5) and the state equations of the control system structure in (4.7), the switching variable $\sigma(t)$ in (4.8) and the Lyapunov function candidate in (4.9). It is assumed that the estimation error $e_{\text{est}}(t)$ is bounded according to (4.13). The control system structure with MFSMCS1 controller is stable if the following condition is fulfilled [Pre17, Rom18, Pre21]:

$$u_{\text{aug}}(t) = \frac{K_I T\ x_1(t) - (1 - K_P T)x_2(t) - Te_{\text{est max}}}{\alpha\ T} - \frac{\eta_{\text{MFSMC}}}{\alpha\ T}\ \text{sat}(\sigma(t), e_{\text{MFSMC}}). \quad (4.20)$$

Proof (this proof follows the presentation in accordance with [Pre17] and [Rom18]). Two possible cases that depend on the values of the switching variable $\sigma(t)$ to ensure the fulfillment of the reaching and existence condition (4.10) are considered. The case $\sigma(t) = 0$ is not considered because it corresponds to an ideal sliding regime, and this condition is difficult to verify in practice.

Case 1. $|\sigma(t)| \le e_{\text{MFSMC}}$ (i.e., the system state belongs to the boundary layer). Substituting the expression of the correction control input $u_{\text{cor}}(t)$ from (4.18) in (4.19), the sliding mode reaching and existence condition given in (4.10) becomes

$$\sigma(t)\ \dot{\sigma}(t) = -\frac{\sigma^2(t)\ \eta_{\text{MFSMC}}}{\varepsilon_{\text{MFSMC}}} + \sigma(t)\ T\ [e_{\text{est}}(t) - e_{\text{est max}}]. \quad (4.21)$$

To ensure the negative value of the right-hand term in (4.21), the following sufficient condition is imposed:

$$\frac{\sigma^2(t)\ \eta_{\text{MFSMC}}}{e_{\text{MFSMC}}} > |\sigma(t)|\ T\ |e_{\text{est}}(t) - e_{\text{est max}}|. \quad (4.22)$$

Since

$$| e_{\text{est}}(t) - e_{\text{est max}} | \leq 2 \, e_{\text{est max}}, \tag{4.23}$$

a sufficient condition to fulfill the inequality in (4.10) is [Pre17, Pre21, Rom18]

$$\frac{| \sigma(t) | \, \eta_{\text{MFSMC}}}{e_{\text{MFSMC}}} > 2 \, T \, e_{\text{est max}}. \tag{4.24}$$

Case 2. $|\sigma(t)| > e_{\text{MFSMC}}$ (the system state does not belong to the boundary layer). Substituting the expression of the correction control input $u_{\text{cor}}(t)$ from (4.18) in (4.19), the sliding mode reaching and existence condition expressed in (4.10) becomes

$$\sigma(t) \, \dot{\sigma}(t) = -\sigma^2(t) \, \eta_{\text{MFSMC}} + \sigma(t) \, T \, [e_{\text{est}}(t) - e_{\text{est max}}]. \tag{4.25}$$

To ensure the negative value of the right-hand term in (4.25), the following sufficient condition is imposed [Pre17, Rom18, Pre21]:

$$\sigma^2(t) \, \eta_{\text{MFSMC}} > | \sigma(t) | \, T \, | e_{\text{est}}(t) - e_{\text{est max}} |. \tag{4.26}$$

Using (4.23), the sufficient condition to fulfill the inequality in (4.10) is

$$| \sigma(t) | \, \eta_{\text{MFSMC}} > 2 \, T \, e_{\text{est max}}. \tag{4.27}$$

The dynamics of the sliding mode switching variable $\sigma(t)$ is characterized by the following expression obtained from (4.19):

$$\dot{\sigma}(t) = - \, \eta_{\text{MFSMC}} \, \text{sat}(\sigma(t), e_{\text{MFSMC}}) + \, T \, [e_{\text{est}}(t) - e_{\text{est max}}]. \tag{4.28}$$

The steady-state value σ_∞ of the sliding mode switching variable satisfies the condition $\dot{\sigma}(t) = 0$, and the boundary layer is reached if

$$\text{sat}(\sigma_\infty, e_{\text{MFSMC}}) = \frac{\sigma_\infty}{e_{\text{MFSMC}}}, \tag{4.29}$$

so the steady-state expression of (4.28) is

$$0 = -\frac{\eta_{\text{MFSMC}} \, \sigma_\infty}{e_{\text{MFSMC}}} + \, T \, \left[e_{\text{est}\infty} - e_{\text{est max}}\right], \tag{4.30}$$

where $e_{\text{est}\infty}$ is the steady-state value of the estimation error. The steady-state value σ_∞ of the switching variable results next as follows after considering (4.30) as an equation:

$$\sigma_\infty = -\frac{T e_{\text{MFSMC}} \, [e_{\text{est}\infty} - e_{\text{est max}}]}{\eta_{\text{MFSMC}}}. \tag{4.31}$$

The relations (4.28) and (4.31) indicate that high values of $\eta_{\text{MFSMC}} > 0$, low values of $e_{\text{MFSMC}} > 0$ and accurate estimates of the controlled output derivative are required to reduce the unwanted effects of the chattering phenomenon.

Substituting the equivalent control input $u_{\text{eq}}(t)$ in (4.17) and the correction control input $u_{\text{cor}}(t)$ in (4.18) into the expression of the augmented control input $u_{\text{aug}}(t)$ in (4.14) leads to the conclusion that the two cases reflected by (4.24) and (4.27) and the further processing of the dynamics of the sliding mode switching variable lead to (4.18) and to the steady-state value of the sliding mode switching variable in (4.31). Accounting for (4.20), it is guaranteed that $\dot{V}(t) < 0$. Therefore, the control system structure with the MFSMCS1 algorithm is stable, and the proof of *Theorem 4.1* is completed.

Remark 4.1

Rewriting the augmented control input $u_{\text{aug}}(t)$ using the notations in (4.6), the following final expression of the control law specific to MFSMCS1 algorithm is reached [Pre17, Rom18, Pre21]:

$$u(t) = \frac{1}{\alpha}\left(-\hat{F}(t) + \dot{y}_d(t) - \frac{e(t)}{T} - e_{\text{est max}} - \frac{\eta_{\text{MFSMC}}}{T}\ \text{sat}(\sigma(t), e_{\text{MFSMC}}) \right). \quad (4.32)$$

The control system structure with MFSMCS1 controller is presented in Figure 4.1.

Proposition 4.1

Based on the results presented in this section, the following **design and tuning steps for the control system structure with MFSMCS1 controller** are carried out [Pre17, Rom18, Pre21]:

Step 1. Choosing the value of the design parameter $\alpha > 0$ such that $\dot{y}(t)$ and $\alpha u(t)$ will have the same order of magnitude.

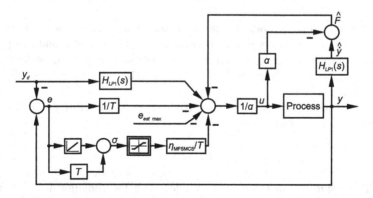

FIGURE 4.1 Control system structure with MFSMCS1 controller. (Adapted from Pre17, Rom18 and Pre21.)

Step 2. Choosing the parameters of the first-order derivative plus low-pass filter in (4.4) such that there is a trade-off to noise reduction and filter-induced delay.

Step 3. Establishing a small value for the design parameter $e_{\text{est max}}$.

Step 4. Establishing a value of the design parameter $T > 0$ so that the output of the control system performance specifications is achieved in terms of imposing the desired behavior of the control system structure in a sliding regime on the sliding manifold.

Step 5. Establishing parameters $\eta_{\text{MFSMCS}} > 0$ and $e_{\text{MFSMCS}} > 0$ using (4.26) and (4.31).

4.2.3   THE SECOND HYBRID MODEL-FREE SLIDING MODE CONTROLLER

To simplify the notations, the abbreviation MFSMCS2 is used for the second hybrid MFSMCS techniques. This technique leads to the MFSMCS2 controller. The main details on the design and tuning of the control system structure with MFSMC2 algorithm are given in this section.

Theorem 4.2

Let the control law be defined in (4.1) for the process model defined as $\dot{y} = F(t) + \alpha\, u(t)$ with $\hat{F}(t)$ as disturbance estimator computed in accordance with (4.5) and the state equations of the control system structure in (4.7), the switching variable $\sigma(t)$ in (4.8) and the Lyapunov function in (4.9). It is assumed that the estimation error $e_{\text{est}}(t)$ is bounded in terms of (4.13). The control system structure with MFSMCS2 controller is stable if [Pre17, Rom18, Pre21]

$$u_{\text{aug}}(t) = -\left(\frac{\psi + |K_I T\, x_1(t) + (K_P T - 1)x_2(t)| + T\, e_{\text{est max}}}{\alpha T} + \delta_{\text{MFSMC}} \right) \text{sgn}(\sigma(t)), \qquad (4.33)$$

where $\psi > 0$ and $\delta_{\text{MFSMC}} > 0$ are user-chosen parameters employed in the controller design.

Proof (this proof follows the presentation in accordance with [Pre17] and [Rom18]). The procedure is similar to the proof of *Theorem 4.1*, and two possible cases are considered that depend on the values of the sliding mode switching variable $\sigma(t)$ to ensure the fulfillment of the reaching and existence condition of a sliding mode regime in (4.10).

Case 1. $\sigma(t) < 0$. The condition in (4.10) leads to condition $\dot{\sigma}(t) > 0$, but a more restrictive condition is imposed:

$$\dot{\sigma}(t) > \psi, \qquad (4.34)$$

where the design parameter $\psi > 0$ guarantees that the condition $\dot{V}(t) < 0$ is met considering the limited uncertainties and the unmodeled dynamics of the system.

Substituting the expression of $\dot{\sigma}(t) > 0$ in (4.12) into (4.14), the following condition for the augmented control input $u_{\text{aug}}(t)$ results:

$$u_{\text{aug}}(t) > \frac{\psi + K_I T \, x_1(t) + (K_P T - 1)x_2(t) - T \, e_{\text{est}}(t)}{\alpha T}. \tag{4.35}$$

Substituting the inequality in (4.13) into the inequality in (4.35), a sufficient condition is obtained to fulfill the condition in (4.10) [Pre17, Rom18, Pre21]

$$u_{\text{aug}}(t) > \frac{\psi + |\, K_I T \, x_1(t) + (K_P T - 1)x_2(t)\,| + T \, e_{\text{est max}}}{\alpha T}. \tag{4.36}$$

Case 2. $\sigma(t) > 0$. The condition in (4.10) leads to $\dot{\sigma}(t) < 0$, but a more restrictive condition is imposed:

$$\dot{\sigma}(t) < -\psi \tag{4.37}$$

to guarantee, as above, that $\dot{V}(t) < 0$ if parametric disturbances occur and to ensure a short duration of the phase of reaching the sliding mode regime. Substituting the expression of $\dot{\sigma}(t)$ in (4.12) into (4.37), the following condition for the augmented control input results $u_{\text{aug}}(t)$ is obtained [Pre17, Rom18, Pre21]:

$$u_{\text{aug}}(t) < \frac{-\psi + K_I T \, x_1(t) + (K_P T - 1)x_2(t) - T \, e_{\text{est}}(t)}{\alpha T}. \tag{4.38}$$

Substituting the inequality in (4.13) into the inequality in (4.38), a sufficient condition results to satisfy the condition in (4.10):

$$u_{\text{aug}}(t) < -\frac{\psi + |\, K_I T \, x_1(t) + (K_P T - 1)x_2(t)\,| + T \, e_{\text{est max}}}{\alpha T}. \tag{4.39}$$

Concluding, two cases characterized in a unified form by (4.39) are equivalent to the expression of the augmented control input in (4.33) that guarantees the sliding mode reaching and existence condition (4.10), where $\delta_{\text{MFSMC}} > 0$ should be large enough to eliminate all the bounded uncertainties and unmodeled dynamics of the system. According to (4.33), it is guaranteed that $\dot{V}(t) < 0$. Therefore, the control system structure with MFSMCS2 controller is stable, and the proof is completed.

The dynamics of the switching variable is described by the following equation obtained from (4.12) and (4.39):

$$\dot{\sigma}(t) = - K_I T \, x_1(t) + (1 - K_P T) \, x_2(t) - (\psi + |\, K_I T \, x_1(t) + (K_P T - 1)x_2(t)\,| + T \, e_{\text{est max}}$$

$$+ \alpha T \delta_{\text{MFSMC}}) \text{sgn}(\sigma(t)) + T \, e_{\text{est}}(t). \tag{4.40}$$

The steady-state value $\dot{\sigma}_\infty$ of the switching variable satisfies the condition $\dot{\sigma}_\infty = 0$, so the steady-state expression of (4.40) is

$$K_I T\, x_1(t) + (K_P T - 1)\, x_2(t) + (\psi + |\,K_I T\, x_1(t) + (K_P T - 1)x_2(t)\,|$$

$$+ T\, e_{\text{est max}} + \alpha T \delta_{\text{MFSMC}})\,\mathrm{sgn}(\sigma(t)) = T\, e_{\text{est}\,\infty}, \qquad (4.41)$$

where $e_{\text{est}\,\infty}$ is the steady-state value of the estimation error. The condition (4.41) shows that since it is desired that $\sigma_\infty = 0$, it is recommended to choose high values for $T > 0$ and accurate estimates of the derivatives.

Remark 4.2

Substituting $u_{\text{aug}}(t)$ in (4.39) into equation (4.1) using the notations in (4.6), the following expression of the control law specific to the MFSMC2 controller is obtained [Pre17, Rom18, Pre21]:

$$u(t) = \frac{1}{\alpha}\left[-\hat{F}(t) + \dot{y}_d(t) - K_P e(t) - K_I \int_0^t e(\tau)\,d\tau \right.$$

$$\left. - \frac{1}{T}\left(\delta_{\text{MFSMC}}\alpha T + \psi + T\, e_{\text{est max}} + |\,K_I T \int_0^t e(\tau)\,d\tau + (K_P T - 1)e(t)\,| \right)\mathrm{sgn}(\sigma(t)) \right].$$

$$(4.42)$$

The control system structure with MFSMCS controller is presented in Figure 4.2.

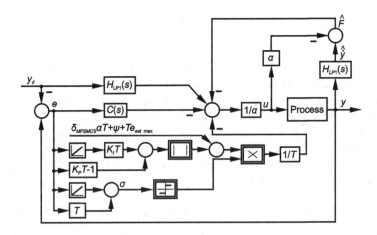

FIGURE 4.2 Control system structure with MFSMCS2 controller. (Adapted from Pre17, Rom18 and Pre21.)

Proposition 4.2

Based on the results presented in this section, the following **design steps for the control system structure with MFSMCS2 controller** are carried out [Pre17, Rom18, Pre21]:

Step 1. Choosing the value of the design parameter $\alpha > 0$ such that $\dot{y}(t)$ and $\alpha u(t)$ will have the same order of magnitude.

Step 2. Choosing the parameters of the first-order derivative plus low-pass filter in (4.4) such that there is a trade-off to noise reduction and filter-induced delay.

Step 3. Establishing a small value for the design parameter $e_{\text{est max}}$.

Step 4. Establishing small values for parameters K_P and K_I.

Step 5. Establishing a value of the design parameter $T > 0$ so that the output of the control system performance specifications is achieved in terms of imposing the desired behavior of the control system structure in a sliding regime on the sliding manifold and considering the recommendations regarding (4.41).

Step 6. Establishing small values for parameters $\psi > 0$ and $\delta_{\text{MFSMCS}} > 0$ using the experience of the control system designer.

4.3 EXAMPLE AND APPLICATION

4.3.1 SISO CONTROL SYSTEMS

The objective function used to measure the performances of the SISO control system structures with hybrid MFSMCS1 and MFSMCS2 controllers presented in Section 4.2 is

$$J_{(\psi)}(\chi_{(\psi)}) = \frac{1}{2}E\left\{\sum_{k=1}^{N}\left[e_{(\psi)}(k,\chi_{(\psi)})\right]^2\right\}, \tag{4.43}$$

where $\chi_{(\psi)}$ is the general notation for the parameter vector of each controller, $e_{(\psi)} = y_{d(\psi)} - y_{(\psi)}$ is the control error, and the subscript $\psi \in \{c,a,p\}$ indicates both the controlled output and its corresponding controller, namely c – the cart position (i.e., $y_{(c)} = y_1$), a – the arm angular position (i.e., $y_{(a)} = y_2$) and p – the payload position (i.e., $y_{(p)} = y_3$). The mathematical expectation $E\{\Xi\}$ is taken with respect to the stochastic probability distribution of the disturbance inputs applied to the process and thus affects the control system behavior. The sampling period is $T_s = 0.01$ s and the time horizon is 70 s, resulting in a number of $N = 70/0.01 = 7,000$ s samples.

The MFSMCS1 and MFSMCS2 controllers are designed and tuned, one for each controlled output of the process, with respect to the steps of the MFSMCS1 and MFSMCS2 algorithms in Section 4.2. The description of the steps is given in relation with the Matlab & Simulink programs and schemes.

The reference input vector is first obtained by applying the following inputs to the control systems:

$$y_c^*(k) = 0.15 \text{ if } k \in [0,2,000], 0.1 \text{ if } k \in (2,000,3,500], -0.05 \text{ if } k \in (3,500,5,000],$$

$$0 \text{ if } k \in (5,000,7,000] \tag{4.44}$$

for cart position control,

$$y_a^*(k) = 0 \text{ if } k \in [0,500], 0.15 \text{ if } k \in (500,2,500], -0.15 \text{ if } k \in (2,500,4,000],$$

$$0 \text{ if } k \in (4,000,7,000] \tag{4.45}$$

for arm angular position control and

$$y_p^*(k) = 0 \text{ if } k \in [0,1,500], 0.1 \text{ if } k \in (1,500,3,000], -0.05 \text{ if } k \in (3,000,4,500],$$

$$0 \text{ if } k \in (4,500,7,000] \tag{4.46}$$

for payload position control. These signals are filtered to obtain the reference inputs, i.e. the desired reference trajectory in terms of $y_{d(\psi)} = y_{(\psi)}^* H_{(\psi)}$ using the filters

$$H_{(c)}(s) = \frac{1}{1 + 0.2s} \tag{4.47}$$

for cart position control,

$$H_{(a)}(s) = \frac{1}{1 + 0.2s} \tag{4.48}$$

for arm angular position control and

$$H_{(p)}(s) = \frac{1}{1 + 0.2s} \tag{4.49}$$

for payload position control, with the corresponding discrete transfer functions

$$H_{(c)}(z^{-1}) = \frac{0.0488}{1 - 0.9512z^{-1}}, \tag{4.50}$$

$$H_{(a)}(z^{-1}) = \frac{0.0465}{1 - 0.9535z^{-1}}, \tag{4.51}$$

$$H_{(p)}(z^{-1}) = \frac{0.0328}{1 - 0.9672z^{-1}}. \tag{4.52}$$

For the sake of reducing the big volume of this chapter and enabling the relatively easy understanding of the application of the controllers, only the results of MFSMCS1 and MFSMCS2 controllers are included. However, first-order continuous-time iP, iPI and iPID and second-order continuous-time iP, iPI and iPID controllers can be easily merged with the SMC technique.

4.3.1.1 The First Hybrid Model-Free Sliding Mode Controller

Three MFSMCS1 controllers are used separately in the processes in the three control loops, where the parameter vectors of the MFSMCS1 controllers are

$$\chi_{(\psi)} = [e_{\text{MFSMCS}} \ T \ e_{\text{est max}} \ \eta_{\text{MFSMCS}}]^T, \tag{4.53}$$

where $\psi \in \{c, a, p\}$, i.e., one parameter for each controller.

The controllers are nonlinear since saturation-type nonlinearities are inserted on the three controller outputs in order to match the power electronics of the actuators, which operate on the basis of the pulse width modulation (PWM) principle. As shown in Chapter 1, the three control signals (or control inputs) are within −1 and 1, ensuring the necessary connection to the tower crane system subsystem, which models the controlled process in the control system structures developed in this chapter and the next chapters and also includes three Zero-Order Hold (ZOH) blocks to enable digital control. The tower crane system subsystem belongs to the Process.mdl Simulink diagram, which is included in the accompanying Matlab & Simulink programs given in Chapter 1.

The parameter $\alpha = 0.0015$ is chosen according to step 1 of the MFSMCS1 controller design and tuning, and then the parameters of the low-pass filter $K_{LP1} = 0.85$ and $T_{LP1} = 0.15$ s are chosen according to step 2. Finally, the remaining parameters $e_{\text{est max}}$, $T > 0$, $\eta_{\text{MFSMCS}} > 0$ and $e_{\text{MFSMCS}} > 0$ are established according to steps 3, 4 and 5 of the MFSMCS1 controller design and tuning:

$$\chi_{(c)} = [1.5 \ 15.5 \ 0.001 \ 5]^T, \tag{4.54}$$

$$\chi_{(a)} = [0.5 \ 0.5 \ 0.001 \ 7]^T, \tag{4.55}$$

$$\chi_{(p)} = [12 \ 0.105 \ 0.001 \ 4]^T. \tag{4.56}$$

The details related to the MFSMCS1 algorithm and the programs of steps 1–5 are implemented in MFSMCS1_c.m Matlab program for cart position control, the MFSMCS1 _ a.m Matlab program for arm angular position control and the MFSMCS1 _ p.m Matlab program for payload position control, which are included in the accompanying Matlab & Simulink program. The simulations of three MFSMCS1 algorithms are conducted by digital simulations using the CS _ MFC _ SMC1 _ SISO _ c.mdl Simulink diagram for cart position control, the CS _ MFC _ SMC1 _ SISO _ a.mdl Simulink diagram for arm angular position control and the CS _ MFC _ SMC1 _ SISO _ p.mdl Simulink diagram for payload position control. These three Simulink diagrams make use of the state-space model of the tower crane system viewed as a controlled process, which is described in Chapter 1 and is implemented in the process _ model.m S-function.

The simulation results obtained for the control systems with MFSMCS1 controllers with the parameters in (4.54–4.56) and the user-chosen parameters α, K_{LP1} and T_{LP1} are illustrated in Figure 4.3 for cart position control, Figure 4.4 for arm angular position control and Figure 4.5 for payload position control.

The experimental results obtained for the control systems with the second MFSMCS controllers with the parameters in (4.54–4.56) and the user-chosen parameters α, K_{LP1} and T_{LP1} are illustrated in Figure 4.6 for cart position control, Figure 4.7 for arm angular position control and Figure 4.8 for payload position control.

4.3.1.2   The Second Hybrid Model-Free Sliding Mode Controller

Three MFSMCS2 controllers are used separately in the processes in the three SISO control system structures, where the parameters of the MFSMCS2 controllers are

$$\chi_{(\psi)} = [K_P \ \ K_I \ \ \psi \ \ T \ \ e_{\text{est max}} \ \ \delta_{\text{MFSMCS}}]^T, \qquad (4.57)$$

with $\psi \in \{c, a, p\}$, i.e., one parameter for each controller.

The parameter $\alpha = 0.0015$ is chosen according to step 1 of the MFSMCS2 controller design and tuning, and then the parameters of the low-pass filter $K_{LP1} = 0.85$ and $T_{LP1} = 0.15$ s are chosen according to step 2. Finally, the parameters $e_{\text{est max}}$, K_P, $K_I, T > 0, \psi > 0$ and $\delta_{\text{MFSMCS}} > 0$ are established according to steps 3–6 leading to

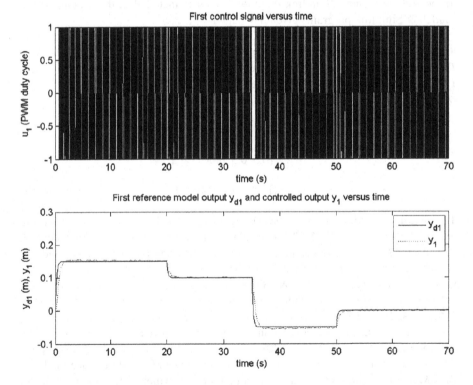

FIGURE 4.3   Simulation results of cart position control system with MFSMCS1 controller.

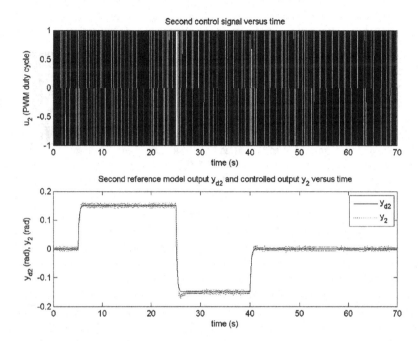

FIGURE 4.4 Simulation results of arm angular position control system with MFSMCS1 controller.

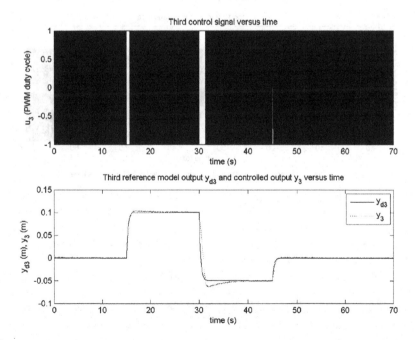

FIGURE 4.5 Simulation results of payload position control system with MFSMCS1 controller.

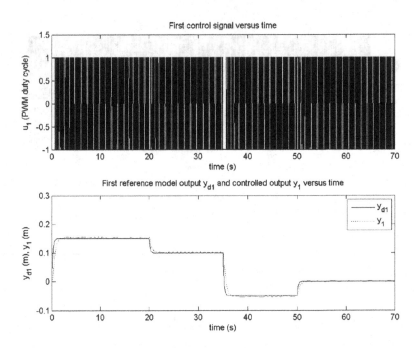

FIGURE 4.6 Experimental results of cart position control system with MFSMCS1 controller.

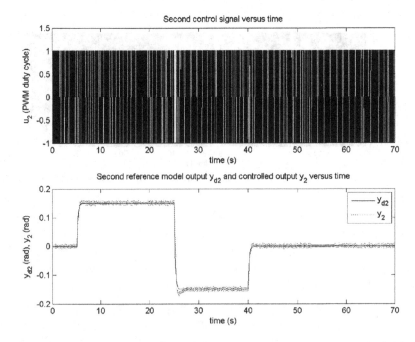

FIGURE 4.7 Experimental results of arm angular position control system with MFSMCS1 controller.

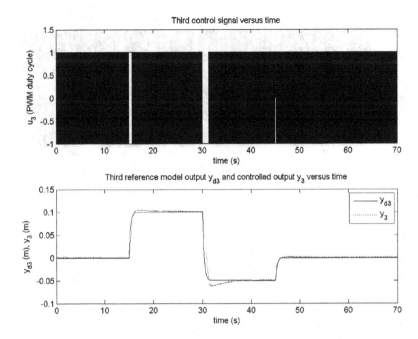

FIGURE 4.8 Experimental results of payload position control system with MFSMCS1 controller.

$$\chi_{(c)} = [3.5 \ 0.001 \ 0.005 \ 115 \ 0.001 \ 0.001]^{T}, \tag{4.58}$$

$$\chi_{(a)} = [3.5 \ 0.0001 \ 0.002 \ 68 \ 0.001 \ 0.001]^{T}, \tag{4.59}$$

$$\chi_{(p)} = [3.5 \ 0.001 \ 0.002 \ 101 \ 0.001 \ 0.001]^{T}. \tag{4.60}$$

The details related to the MFSMCS2 algorithm and the programs of steps 1–6 are implemented in MFSMCS2 _ c.m Matlab program for cart position control, the MFSMCS2 _ a.m Matlab program for arm angular position control and the MFSMCS2 _ p.m Matlab program for payload position control, which are included in the accompanying Matlab & Simulink program. The simulations of three MFSMCS2 algorithms are conducted by digital simulations using the CS _ MFC _ SMC2 _ SISO _ c.mdl Simulink diagram for cart position control, the CS _ MFC _ SMC2 _ SISO _ a.mdl Simulink diagram for arm angular position control and the CS _ MFC _ SMC2 _ SISO _ p.mdl Simulink diagram for payload position control. These three Simulink diagrams make use of the state-space model of the tower crane system viewed as a controlled process, which is described in Chapter 1 and is implemented in the process _ model.m S-function.

The simulation results obtained for the control systems with MFSMCS2 controllers with the parameters in (4.58–4.60) and the user-chosen parameters α, K_{LP1} and T_{LP1} are illustrated in Figure 4.9 for cart position control, Figure 4.10 for arm angular position control and Figure 4.11 for payload position control.

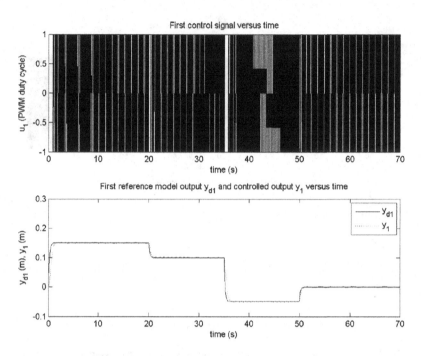

FIGURE 4.9   Simulation results of cart position control system with MFSMCS2 controller.

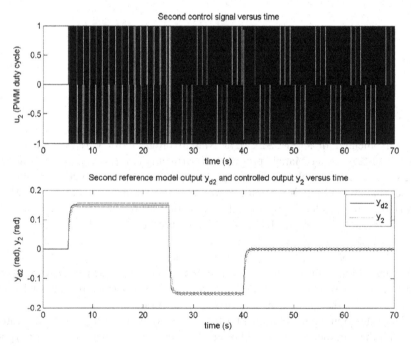

FIGURE 4.10   Simulation results of arm angular position control system with MFSMCS2 controller.

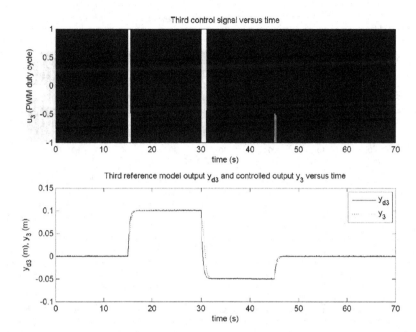

FIGURE 4.11 Simulation results of payload position control system with MFSMCS2 controller.

The experimental results obtained for the second MFSMCS controllers with the parameters in (4.58–4.60) and the user-chosen parameters α, K_{LP1} and T_{LP1} are illustrated in Figure 4.12 for cart position control, Figure 4.13 for arm angular position control and Figure 4.14 for payload position control.

4.3.2 MIMO CONTROL SYSTEMS

In this section, the validation of the first MFSMCS1 and MFSMCS2 controllers is carried out on the tower crane system equipment. The exemplification of the MIMO loops is given considering the implementation that makes use of three SISO loops that are running in parallel.

MFSMCS1 and MFSMCS2 controllers are implemented on the basis of the results presented in Section 4.2 in the SISO formulation. The control system performance is measured using the objective function

$$J(\chi) = \frac{1}{2}E\left\{\sum_{k=1}^{N}\left[\mathbf{e}(k,\chi)\right]^2\right\}, \tag{4.61}$$

where χ is the general notation for the parameter vector of MFSMCS1 and MFSMCS2 controllers, $\mathbf{e}$ is the control error vector

$$\mathbf{e} = \mathbf{y}_d - \mathbf{y} = [e_1 \ e_2 \ e_3]^T = [y_{d1} - y_1 \ y_{d2} - y_2 \ y_{d3} - y_3]^T, \tag{4.62}$$

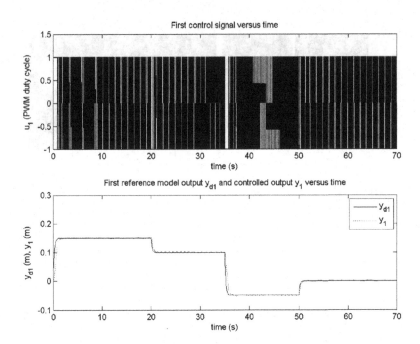

FIGURE 4.12 Experimental results of cart position control system with MFSMCS2 controller.

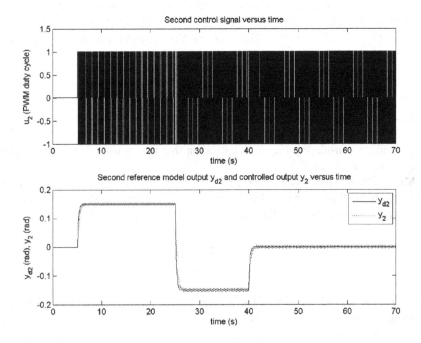

FIGURE 4.13 Experimental results of arm angular position control system with MFSMCS2 controller.

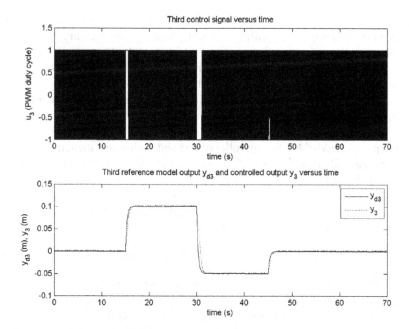

FIGURE 4.14 Experimental results of payload position control system with MFSMCS2 controller.

and the control errors e_1, e_2 and e_3 are related to cart position, arm angular position and payload position, the controlled outputs y_1, y_2 and y_3 are also related to related to cart position, arm angular position and payload position, and the references (the output of the reference model) y_{d1}, y_{d2} and y_{d3} similarly are related to cart position, arm angular position and payload position. The mathematical expectation $E\{\Xi\}$, the sampling period, the time horizon and the number of samples are the same as in the SISO case.

The reference input vector is generated by first producing the step signals according to (4.44) for cart position, (4.45) for arm angular position and (4.46) for payload position; they are next filtered using (4.47) for cart position, (4.48) for arm angular position and (4.49) for payload position, where their discrete-time transfer functions are expressed in (4.50) for cart position, (4.51) for arm angular position and (4.52) for payload position.

4.3.2.1 The First Hybrid Model-Free Sliding Mode Controller

The MIMO MFSMCS1 controller uses the parameter vector

$$\chi = \left[\chi_{(c)}^T \ \chi_{(a)}^T \ \chi_{(p)}^T \right]^T$$

$$= \left[\begin{matrix} e_{(c) \text{ MFSMCS}} \ T_{(c)} \ e_{(c) \text{ est max}} \ \eta_{(c) \text{ MFSMCS}} \\ e_{(a) \text{ MFSMCS}} \ T_{(a)} \ e_{(a) \text{ est max}} \ \eta_{(a) \text{ MFSMCS}} \ e_{(p) \text{ MFSMCS}} \ T_{(p)} \ e_{(p) \text{ est max}} \ \eta_{(p) \text{ MFSMCS}} \end{matrix} \right]^T .$$

$$(4.63)$$

The controller is nonlinear since saturation-type nonlinearities are inserted on the controller outputs in order to match the power electronics of the actuators, which operate on the basis of the PWM principle. As shown in Chapter 1, the three control signals (or control inputs) are within -1 and 1, ensuring the necessary connection to the tower crane system subsystem, which models the controlled process in the control system structures developed in this chapter and the next chapters and also includes three ZOH blocks to enable digital control. The tower crane system subsystem belongs to the Process.mdl Simulink diagram, which is included in the accompanying Matlab & Simulink programs given in Chapter 1.

The parameters $\alpha_{(c)} = 0.0015$, $\alpha_{(a)} = 0.0015$ and $\alpha_{(p)} = 0.0015$ are chosen according to step 1 of the MFSMCS1 controller design and tuning, and then, the parameters of the low-pass filters $K_{(c)LP1} = 0.85$, $T_{(c)LP1} = 0.15$ s, $K_{(a)LP1} = 0.85$, $T_{(a)LP1} = 0.15$ s, $K_{(p)LP1} = 0.85$ and $T_{(p)LP1} = 0.15$ s are chosen according to step 2. Finally, the remaining parameters $e_{(c)est max}$, $T_{(c)} > 0$, $\eta_{(c)MFSMCS} > 0$, $e_{(c)MFSMCS} > 0$, $e_{(a)est max}$, $T_{(a)} > 0$, $\eta_{(a)MFSMCS} > 0$, $e_{(a)MFSMCS} > 0$, $e_{(p)est max}$, $T_{(p)} > 0$, $\eta_{(p)MFSMCS} > 0$ and $e_{(p)MFSMCS} > 0$ are established according to steps 3–5 leading to

$$\chi = [1.5\ 15.5\ 0.001\ 5\ 0.5\ 0.5\ 0.001\ 7\ 12\ 0.105\ 0.001\ 4]^T. \tag{4.64}$$

The details related to the MIMO MFSMCS1 algorithm and the programs of steps 1–5 are implemented in MFSMCS1.m Matlab program for tower crane system control, which is included in the accompanying Matlab & Simulink programs. The simulations of the control system with the MIMO MFSMCS1 are conducted using the same Simulink diagram, i.e., the CS _ MFC _ SMC1 _ MIMO.mdl Simulink diagram for the tower crane system. The Simulink diagram makes use of the state-space model of the tower crane system viewed as a controlled process, which is described in Chapter 1 and is implemented in the process _ model.m S-function.

The simulation results obtained for the control system with MIMO MFSMCS1 controller with the parameters in (4.64) and the user-chosen parameters $\alpha_{(c)}$, $\alpha_{(a)}$, $\alpha_{(p)}$, $K_{(c)LP1}$, $T_{(c)LP1}$, $K_{(a)LP1}$, $T_{(a)LP1}$, $K_{(p)LP1}$ and $T_{(p)LP1}$ are illustrated in Figure 4.15 for cart position control, Figure 4.16 for arm angular position control and Figure 4.17 for payload position control.

The experimental results obtained for the control system with MIMO MFSMCS1 controller with the parameters in (4.64) and the user-chosen parameter $\alpha_{(c)}$, $\alpha_{(a)}$, $\alpha_{(p)}$, $K_{(c)LP1}$, $T_{(c)LP1}$, $K_{(a)LP1}$, $T_{(a)LP1}$, $K_{(p)LP1}$ and $T_{(p)LP1}$ are illustrated in Figure 4.18 for cart position control, Figure 4.19 for arm angular position control and Figure 4.20 for payload position control.

4.3.2.2   The Second Hybrid Model-Free Sliding Mode Controller

The MIMO MFSMCS2 controller operates on the basis of the parameter vector

$$\chi = \left[\chi_{(c)}^T\ \chi_{(a)}^T\ \chi_{(p)}^T \right]^T$$

$$= \begin{bmatrix} K_{(c)P}\ K_{(c)I}\ \psi_{(c)}\ T_{(c)}\ e_{(c)est max}\ \delta_{(c)MFSMCS}\ K_{(a)P}\ K_{(a)I}\ \psi_{(a)} \\ T_{(a)}\ e_{(a)est max}\ \delta_{(a)MFSMCS}\ K_{(p)P}\ K_{(p)I}\ \psi_{(p)}\ T_{(p)}\ e_{(p)est max}\ \delta_{(p)MFSMCS} \end{bmatrix}^T. \tag{4.65}$$

The parameters $\alpha_{(c)} = 0.0015$, $\alpha_{(a)} = 0.0015$ and $\alpha_{(p)} = 0.0015$ are chosen according to step 1 of the MFSMCS2 controller design and tuning, and next, the parameters of the

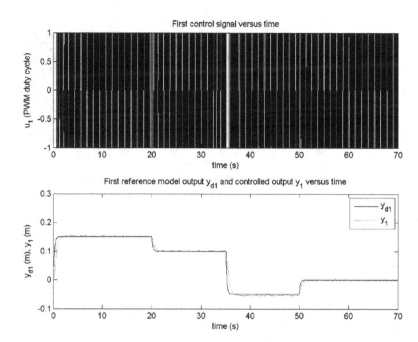

FIGURE 4.15 Simulation results of cart position control system with MIMO MFSMCS1 controller.

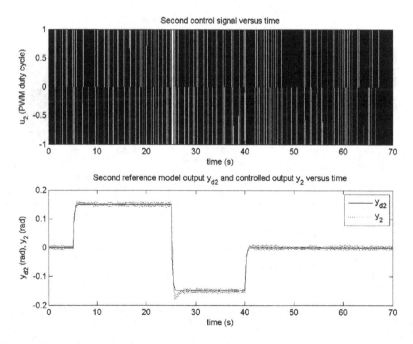

FIGURE 4.16 Simulation results of arm angular position control system with MIMO MFSMCS1 controller.

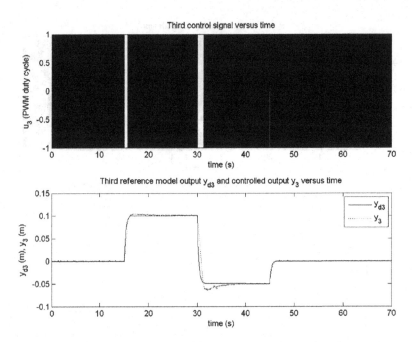

FIGURE 4.17   Simulation results of payload position control system with MIMO MFSMCS1 controller.

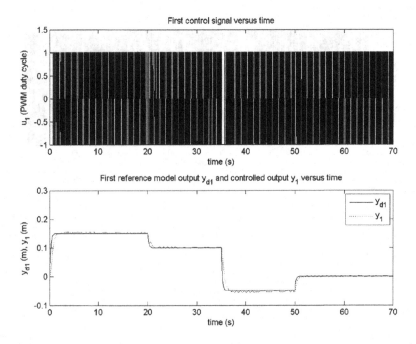

FIGURE 4.18   Experimental results of cart position control system with MIMO MFSMCS1 controller.

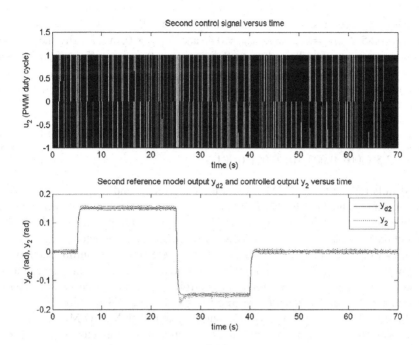

FIGURE 4.19 Experimental results of arm angular position control system with MIMO MFSMCS1 controller.

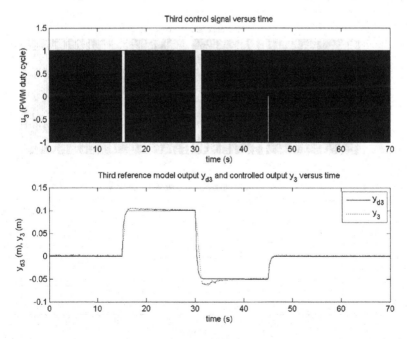

FIGURE 4.20 Experimental results of payload position control system with MIMO MFSMCS1 controller.

low-pass filters are chosen as follows according to step 2: $K_{(c)LP1} = 0.85$, $T_{(c)LP1} = 0.15$ s, $K_{(a)LP1} = 0.85$, $T_{(a)LP1} = 0.15$ s, $K_{(p)LP1} = 0.85$ and $T_{(p)LP1} = 0.15$ s. Finally, the parameters $e_{(c)\text{est max}}$, $K_{(c)P}$, $K_{(c)I}$, $T_{(c)} > 0$, $\psi_{(c)} > 0$, $\delta_{(c)\text{MFSMCS}} > 0$, $e_{(a)\text{est max}}$, $K_{(a)P}$, $K_{(a)I}$, $T_{(a)} > 0$, $\psi_{(a)} > 0$, $\delta_{(a)\text{MFSMCS}} > 0$, $e_{(p)\text{est max}}$, $K_{(p)P}$, $K_{(p)I}$, $T_{(p)} > 0$, $\psi_{(p)} > 0$ and $\delta_{(p)\text{MFSMCS}} > 0$ are established according to steps 3–6 of the MIMO MFSMCS2 controller design and tuning presented in this section in the SISO case:

$$\chi = \begin{bmatrix} 3.5\ 0.001\ 0.005\ 115\ 0.001\ 0.001\ 3.5\ 0.0001\ 0.002\ 68\ 0.001\ 0.001 \\ 3.5\ 0.001\ 0.002\ 101\ 0.001\ 0.001 \end{bmatrix}^{T}. \quad (4.66)$$

The details related to the MIMO MFSMCS2 algorithm and the programs of steps 1–6 are implemented in MFSMCS2.m Matlab program for tower crane system control, which is included in the accompanying Matlab & Simulink programs. The simulations of the control system with the MIMO MFSMCS2 are conducted using the same Simulink diagram, i.e., the CS _ MFC _ SMC2 _ MIMO.mdl Simulink diagram for the tower crane system. The Simulink diagram makes use of the state-space model of the tower crane system viewed as a controlled process, which is described in Chapter 1 and is implemented in the process _ model.m S-function.

The simulation results obtained for control system with MIMO MFSMCS2 controller with the parameters in (4.66) and the user-chosen parameters $\alpha_{(c)}$, $\alpha_{(a)}$, $\alpha_{(p)}$, $K_{(c)LP1}$, $T_{(c)LP1}$, $K_{(a)LP1}$, $T_{(a)LP1}$, $K_{(p)LP1}$ and $T_{(p)LP1}$ are illustrated in Figure 4.21 for cart position control, Figure 4.22 for arm angular position control and Figure 4.23 for payload position control.

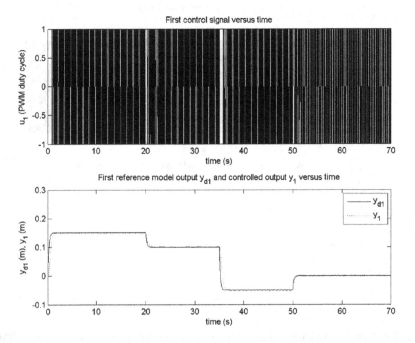

FIGURE 4.21 Simulation results of cart position control system with MIMO MFSMCS2 controller.

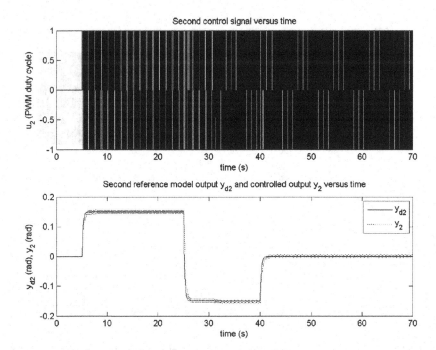

FIGURE 4.22 Simulation results of arm angular position control system with MIMO MFSMCS2 controller.

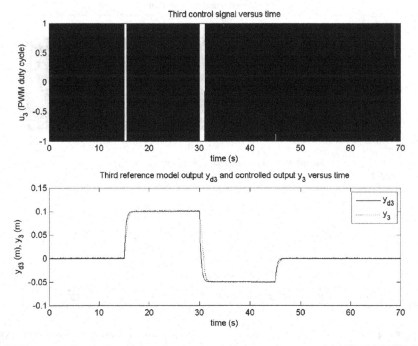

FIGURE 4.23 Simulation results of payload position control system with MIMO MFSMCS2 controller.

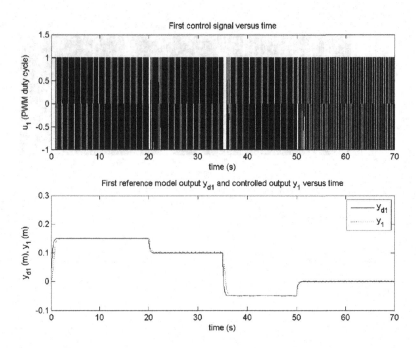

FIGURE 4.24 Experimental results of cart position control system with MIMO MFSMCS2 controller.

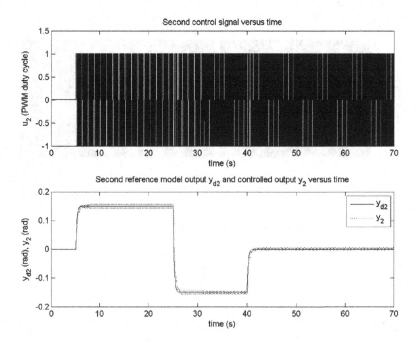

FIGURE 4.25 Experimental results of arm angular position control system with MIMO MFSMCS2 controller.

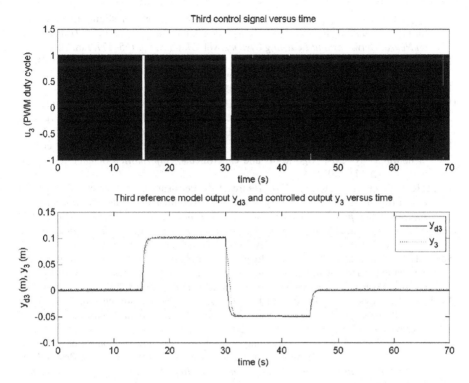

FIGURE 4.26 Experimental results of payload position control system with MIMO MFSMCS2 controller.

The experimental results obtained for the control system with MIMO MFSMCS2 controller with parameters in (4.66) and the user-chosen parameter $\alpha_{(c)}$, $\alpha_{(a)}$, $\alpha_{(p)}$, $K_{(c)LP1}$, $T_{(c)LP1}$, $K_{(a)LP1}$, $T_{(a)LP1}$, $K_{(p)LP1}$ and $T_{(p)LP1}$ are illustrated in Figure 4.24 for cart position control, Figure 4.25 for arm angular position control and Figure 4.26 for payload position control.

REFERENCES

[Cao20] L. Cao, S.-L. Gao and D.-Y. Zhao, "Data-driven model-free sliding mode learning control for a class of discrete-time nonlinear systems," *Transactions of the Institute of Measurement and Control*, DOI: 10.1177/0142331220921022, 2020.

[Che12] W. Chen, Y.-Q. Chen and C.-P. Yeh, "Robust iterative learning control via continuous sliding-mode technique with validation on an SRV02 rotary plant," *Mechatronics*, vol. 22, no. 5, pp. 588–593, Aug. 2012.

[Ebr18] N. Ebrahimi, S. Ozgoli and A. Ramezani, "Model-free sliding mode control, theory and application," *Proceedings of the Institution of Mechanical Engineers Part I Journal of Systems and Control Engineering*, vol. 232, no. 10, 095965181878059, Jun. 2018.

[Ebr20] N. Ebrahimi, S. Ozgoli and A. Ramezani, "Model free sliding mode controller for blood glucose control: Towards artificial pancreas without need to mathematical model of the system," *Computer Methods and Programs in Biomedicine*, vol. 195, p. 105663, Oct. 2020.

[Fan16] Q. Fan and G. Yang, "Adaptive actor-critic design-based integral sliding-mode control for partially unknown nonlinear systems with input disturbances," *IEEE Transactions on Neural Networks and Learning Systems*, vol. 27, no. 1, pp. 165–177, Jan. 2016.

[Li16] S.-Z. Li, H.-P. Wang, Y. Tian, A. Aitouch and J. Klein, "Direct power control of DFIG wind turbine systems based on an intelligent proportional-integral sliding mode control," *ISA Transactions*, vol. 64, pp. 431–439, Sep. 2016.

[Pre14] R.-E. Precup, M.-B. Radac, E.M. Petriu, C.-A. Dragos and S. Preitl, "Model-free tuning solution for sliding mode control of servo systems," in *Proceedings of 8th Annual IEEE International Systems Conference*, Ottawa, ON, Canada, 2014, pp. 30–35.

[Pre17] R.-E. Precup, M.-B. Radac and R.-C. Roman, "Model-free sliding mode control of nonlinear systems: Algorithms and experiments," *Information Sciences*, vol. 381, pp. 176–192, Mar. 2017.

[Pre21] R.-E. Precup, R.-C. Roman, E.-L. Hedrea, E.M. Petriu and C.-A. Dragos, "Data-driven model-free sliding mode and fuzzy control with experimental validation," *International Journal of Computers, Communications & Control*, vol. 16, no. 1, pp. 1–17, Feb. 2021.

[Rom18] R.-C. Roman, *Tehnici de tip model-free de acordare a parametrilor regulatoarelor automate* (in Romanian), PhD thesis, Editura Politehnica, Timisoara, 2018.

[Sch18] E. Schulken and A. Crassidis, "Model-free sliding mode control algorithms including application to a real-world quadrotor," in *Proceedings of 5th International Conference of Control, Dynamic Systems, and Robotics*, Niagara Falls, Canada, 2018, pp. 112-1–112-9.

[Vrk18] S. Vrkalovic, E.-C. Lunca and I.-D. Borlea, "Model-free sliding mode and fuzzy controllers for reverse osmosis desalination plants," *International Journal of Artificial Intelligence*, vol. 16, no. 2, pp. 208–222, Oct. 2018.

[Wan03] X.-S. Wang, S.-Y. Su and H. Hong, "Adaptive sliding inverse control of a class of nonlinear systems preceded by unknown non-symmetrical dead-zone," in *Proceedings of 2003 IEEE International Symposium on Intelligent Control*, Huston, TX, 2003, pp. 16–21.

[Wan15] H. Wang, X. Ye, Y. Tian and N. Christov, "Attitude control of a quadrotor using model free based sliding model controller," in *Proceedings of 20th International Conference on Control Systems and Science*, Bucharest, Romania, 2015, pp. 149–154.

[Wan16] X. Wang, X. Li, J. Wang, X. Fang and X. Zhu, "Data-driven model-free adaptive sliding mode control for the multi degree-of-freedom robotic exoskeleton", *Information Sciences*, vol. 327, pp. 246–257, Jan. 2016.

[Wan17] H.-P. Wang, Y. Tian, S.-S. Han and X.-K. Wang, "ZMP theory-based gait planning and model-free trajectory tracking control of lower limb carrying exoskeleton system," *Studies in Informatics and Control*, vol. 26, no. 2, pp. 161–170, Jun. 2017.

5 Model-Free Adaptive Controllers

5.1 INTRODUCTION

The model-free adaptive control (MFAC) technique and its associated algorithms implemented as controllers have been proposed by Zhong-Sheng Hou, who has published two recent synthesis papers on MFAC. In [Hou13a], a synthesis is performed on the transition from model-based control to data-driven control, and the need for data-driven control algorithms is justified. In [Hou17], a study is performed on model-based control and data-driven techniques, focusing on the improvements made to MFAC, which are also pointed out in [Rom18].

MFAC, as an important data-driven technique, uses the input/output data of the control process in the controller design. MFAC guarantees the stability of the control system structure and the convergence of the tracking error by fulfilling a set of reset conditions. In the design of the MFAC algorithms implemented as MFAC controllers, no additional external testing signals are needed, and no training stages as those specific to neural networks from nonlinear adaptive control system structures are needed, which leads to controllers characterized by low-cost design and implementation [Hou11a, Hou11b, Rom16, Rom18].

The MFAC-type control algorithms for discrete-time systems are classified in three versions: the Compact Form Dynamic Linearization (CFDL) with the algorithm known as MFAC-CFDL, the partial form dynamic linearization (PFDL) with the algorithm known as MFAC-PFDL, and the full form dynamic linearization (FFDL) with the algorithm known as MFAC-FFDL. According to [Hou11a, Hou11b], CFDL is the most used version of MFAC algorithms. As pointed out in [Rom18], nonparametric mathematical models of the processes can be used in the form of process responses to various input signals (control inputs) to tune the parameters using MFAC algorithms.

The three versions of the MFAC algorithms are based on linearized dynamic models of the process using a dynamic linearization technique plus an approximation of derivatives known as pseudo partial derivative (PPD).

A major challenge for designing and implementing the model-free controllers is the online method for estimating the unknown parameters of the system. This is named as *parameter estimator*. In Chapter 3, three different parameter estimators are proposed, where two of them defined in Sections 3.2.3 and 3.2.4 are model-based algorithms. These two solutions depend on regressor parameters, and hence, the persistent excitation (PE) requirement is a must. In addition, the parameter estimator suggested in Section 3.2.2 needs time integration over the previous control inputs in a fixed time window. Although that method can be considered as a model-free

DOI: 10.1201/9781003143444-5

parameter estimation, and the requirement of PE is revoked, it is not an adaptive algorithm, and it may be susceptible to divergence if there are changes in the system dynamics or external disturbances [Rom18]. In this regard, in Section 5.2.3 of the current chapter, a continuous-time adaptive model-free control algorithm with model-free online parameter estimators is given. Since the presented parameter estimators are model-free ones, there is no need for defining the regressor parameter, and thus, the PE requirement is revoked. This affects the convenience of the controller implementation. Moreover, since the parameter estimators are proposed in the format of robust adaptive laws, the whole system would be robustly adapted with the changes in the dynamic system as well as the external disturbances.

5.2 THEORY

5.2.1 The MFAC Algorithm for Discrete-Time Dynamic Systems

Definition 5.1

The MFAC algorithms for discrete-time dynamic systems are developed considering the nonlinear multiinput-multioutput (MIMO) discrete-time model of the process with the following expression [Hou11a; Hou11b]:

$$\mathbf{y}(k+1) = \mathbf{f}\big(\mathbf{y}(k),\ldots,\mathbf{y}(k-n_y),\mathbf{u}(k),\ldots,\mathbf{u}(k-n_u)\big), \qquad (5.1)$$

where $\mathbf{y}(k+1) = [y_1(k+1) \ldots y_n(k+1)]^T \in \mathfrak{R}^n$ is the output vector of the process, $\mathbf{u}(k) \in \mathfrak{R}^n$ is the input (control input or control signal) vector of the process, $n \in \mathbf{Z}, n \geq 0$ is the number of inputs of the process, equal to the number of the outputs of the process, n is known, $\mathbf{y}(k-n_y) \in \mathfrak{R}^n$ is the output vector at the time moment $k-n_y$, $\mathbf{u}(k-n_u) \in \mathfrak{R}^n$ is the control signal vector at the time moment $k-n_u$ and $\mathbf{f}: \mathfrak{R}^{n(n_u+n_y+2)} \to \mathfrak{R}^n$ is a nonlinear vectorial function of vectorial variable, which is unknown [Hou11a, Hou11b, Rom16, Rom18].

5.2.1.1 The MFAC-CFDL Algorithms
Remark 5.1

The function in (5.1) is Lipschitz generalized, namely, $\|\Delta\mathbf{y}(k+1)\| \leq b\|\Delta\mathbf{u}(k)\|$ for each k fixed and $\|\Delta\mathbf{u}(k)\| \neq 0$, where the incremented vectors of the outputs and inputs have the expressions $\Delta\mathbf{y}(k+1) = \mathbf{y}(k+1) - \mathbf{y}(k)$, $\Delta\mathbf{u}(k) = \mathbf{u}(k) - \mathbf{u}(k-1)$ and $b = \mathrm{const} > 0$.

Proposition 5.1

For the nonlinear system in (5.1) that fulfills the condition in Remark 5.1, a PPD matrix $\Phi(k) \in \mathfrak{R}^{n \times n}$ exists, such that the nonlinear MIMO model of the control process in discrete time in (5.1) can be transformed into the following CFDL model:

$$\Delta\mathbf{y}(k+1) = \Phi(k)\Delta\mathbf{u}(k), \qquad (5.2)$$

where $\Phi(k) = [\varphi_{ij}(k)]_{i,j=1...n}$, $\|\Phi(k)\| \le b$. It is considered that the PPD matrix $\Phi(k)$ is a diagonal dominant such that

$$|\varphi_{ij}(k)| \le b_1, \ b_2 \le |\varphi_{ii}(k)| \le ab_2, \ i = 1...n, \ j = 1...n, \ i \ne j, \ a \ge 1, \ b_2 > b_1(2a+1), \quad (5.3)$$

where $a = \text{const} > 0$, $b_1 = \text{const} > 0$, $b_2 = \text{const} > 0$, and the signs of all elements of the matrix $\Phi(k)$ are constant at all discrete-time moments.

Definition 5.2

The development of the MFAC algorithm aims to solve the following optimization problem [Hou11a, Rom18]:

$$\mathbf{u}^*(k) = \arg\min_{\mathbf{u}(k)} J(\mathbf{u}(k)),$$

$$J(\mathbf{u}(k)) = \|\mathbf{y}_d(k+1) - \mathbf{y}(k+1)\|^2 + \lambda \|\mathbf{u}(k) - \mathbf{u}(k-1)\|^2, \quad (5.4)$$

where $\mathbf{y}_d(k+1) = [y_{d1}(k+1) \ ... \ y_{dn}(k+1)]^T \in \Re^n$ is the reference input vector, and $\lambda = \text{const} > 0$ is a weighting parameter used in the design process.

Lemma 5.1

Referring to *Proposition 5.1*, the estimate of the PPD matrix $\Phi(k)$ is obtained from the recurrent equation

$$\hat{\Phi}(k) = \hat{\Phi}(k-1) + \frac{\eta[\Delta\mathbf{y}(k) - \hat{\Phi}(k-1)\Delta\mathbf{u}(k-1)]\Delta\mathbf{u}^T(k-1)}{\mu + \|\Delta\mathbf{u}(k-1)\|^2}, \quad (5.5)$$

where $\eta \in (0,1)$ is a constant parameter, and $\mu = \text{const} > 0$ is a parameter introduced to avoid zero denominator in (5.5). According to [Hou11a], the PPD matrix is updated at each time moment k using the input/output data measured during the real-time operation of the control system structure [Rom18]. The reset conditions of matrix $\hat{\Phi}(k)$ are

$$\hat{\varphi}_{ii}(k) = \hat{\varphi}_{ii}(1), \text{ if } \begin{cases} |\hat{\varphi}_{ii}(k)| < b_2 \text{ or} \\ |\hat{\varphi}_{ii}(k)| > ab_2 \text{ or} \\ \text{sgn}(\hat{\varphi}_{ii}(k)) \ne \text{sgn}(\hat{\varphi}_{ii}(1)) \end{cases},$$

$$\hat{\varphi}_{ij}(k) = \hat{\varphi}_{ij}(1), \text{ if } \begin{cases} |\hat{\varphi}_{ij}(k)| > b_1 \text{ or} \\ \text{sgn}(\hat{\varphi}_{ij}(k)) \ne \text{sgn}(\hat{\varphi}_{ij}(1)) \end{cases}, \ i \ne j, \quad (5.6)$$

where $\hat{\varphi}_{ij}(1)$ is the initial value of $\hat{\varphi}_{ij}(k)$ with $i = 1...n$ and $j = 1...n$.

Theorem 5.1

Recalling *Definition 5.2* and *Lemma 5.1*, the expression of the control related to the MFAC-CFDL algorithms is

$$\mathbf{u}(k) = \mathbf{u}(k-1) + \frac{\rho \hat{\boldsymbol{\Phi}}^T(k)[\mathbf{y}_d(k+1) - \mathbf{y}(k)]}{\lambda + \|\hat{\boldsymbol{\Phi}}(k)\|^2}, \tag{5.7}$$

where $\rho = \text{const} > 0$ is an additional parameter.

Remark 5.2

Theorem 5.1, together with the estimation mechanism in (5.5), the reset conditions specified in (5.6) and the control law in (5.7), guarantees the stability of the control system structure with MFAC-CFDL algorithm [Hou11a, Rom18].

Remark 5.3

In the MIMO case, the calculus of the norm from the previous equations is performed using the Euclidean norm (or the 2-norm) defined as follows:

$$\|\hat{\boldsymbol{\Phi}}\|_2 = \max_{\|\mathbf{x}\|_2 = 1} \|\hat{\boldsymbol{\Phi}}\mathbf{x}\|_2 = \sqrt{\gamma_{\max}}, \tag{5.8}$$

where $\gamma_{\max}$ is the maximum eigenvalue of matrix $\hat{\boldsymbol{\Phi}}^T\hat{\boldsymbol{\Phi}}$. The formula (5.8) is used in the initial implementation in [Hou11a] in the calculus of the 2-norm within the MFAC algorithm.

Proposition 5.2

The block diagram of the control system structure with MFAC-CFDL algorithm is presented in Figure 5.1, where the control error vector $\mathbf{e}(k) \in \mathfrak{R}^n$ is defined as follows:

$$\mathbf{e}(k) = \mathbf{y}_d(k) - \mathbf{y}(k) = [e_1(k) \ ... e_n(k)]^T,$$

$$e_i(k) = y_{di}(k) - y_i(k), \quad i = 1...n. \tag{5.9}$$

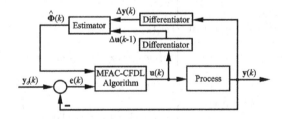

FIGURE 5.1   MIMO control system structure with MFAC-CFDL algorithm. (Adapted from Rom14, Rom15, Rom16, Rom18 and Rom19.)

Proposition 5.3

Based on the results presented in this section, the following **design and tuning steps for the control system structure with MFAC-CFDL algorithm** are carried out [Rom18]:

Step 1. Choosing the initial value of the estimate of the PPD matrix $\Phi(k)$ in order to fulfill the performance specifications of the control system structure and impose its desired behavior.

Step 2. Choosing the values of the lower limit b_2 and the upper limit ab_2 of the elements in the main diagonal and b_1 in the case of elements on the secondary diagonal in (5.3) such that $\hat{\Phi}(k)$ will vary during the experiments according to the constraint imposed in terms of the reset conditions of the matrix $\hat{\Phi}(k)$ in (5.6).

Step 3. Establishing the parameter values of $\eta \in (0,1)$, $\lambda > 0$, $\mu > 0$ and $\rho > 0$ using the experience of the control systems designer.

5.2.1.2 The MFAC-PFDL Algorithms
Remark 5.4

In the PFDL version of the MFAC algorithm, the function in (5.1) is also Lipschitz generalized, namely $\|\Delta y(k+1)\| \leq b\|\Delta U(k)\|$ for each k fixed and $\|\Delta U(k)\| \neq 0$, where $\Delta U(k) = [\Delta u(k) \ldots \Delta u(k-L+1)]^T$, $\Delta u(k-i) = u(k-i) - u(k-i-1)$, $i = 0 \ldots L-1$, $\Delta y(k+1) = y(k+1) - y(k)$, $u(k) = 0$ for $k \leq 0$ and $b = \text{const} > 0$, and $L = \text{const} > 0$ is a parameter that specifies the input time horizon for which the discrete-time nonlinear system is considered to be linearized [Hou11b, Rom18].

Definition 5.3

For the nonlinear system in (5.1) that fulfills the above requirements, where $\|\Delta U(k)\| \neq 0$ for each k, in the MFAC-CFDL algorithm, the following PFDL-type formulation similar to the CFDL model in (5.2) is used:

$$\Delta y(k+1) = \Phi^T(k)\Delta U(k), \tag{5.10}$$

where

$$\Phi(k) = \left[\Phi_1(k) \quad \Phi_2(k) \ldots \Phi_L(k) \right]^T, \| \Phi(k) \| \leq b,$$
$$\Phi_m(k) = [\varphi_{ij}(k)]_{i,j=1\ldots n}, \quad m = 1 \ldots L. \tag{5.11}$$

Remark 5.5

In the SISO case, the PPD matrix $\Phi(k)$ becomes a vector having the general shape of a column matrix $\Phi(k) = [\varphi_1(k) \ \varphi_2(k) \ \ldots \ \varphi_L(k)]^T \in \mathfrak{R}^{n \times 1}$. The design of the MFAC algorithm aims to solve the optimization problem defined in (5.3).

Lemma 5.2

The estimate $\hat{\Phi}(k)$ of the PPD matrix $\Phi(k)$ is obtained in terms of the recurrent equation

$$\hat{\Phi}(k) = \hat{\Phi}(k-1) + \frac{\eta \Delta U(k-1)[\Delta y(k) - \hat{\Phi}^T(k-1)\Delta U(k-1)]}{\mu + \|\Delta U(k-1)\|^2}, \qquad (5.12)$$

where $\eta \in (0,1)$ is a constant parameter, and $\mu = \text{const} > 0$ is a parameter introduced to avoid zero denominator in (5.12). Proceeding similarly to the MFAC-CFDL algorithm, the PPD matrix is updated at each time moment k using the input/output data measured online during the real-time operation of the control system structure [Rom18]. The reset conditions of matrix $\hat{\Phi}(k)$ are

$$\hat{\Phi}(k) = \hat{\Phi}(1), \text{ if } \begin{cases} \|\hat{\Phi}(k)\| \leq \beta \text{ or} \\ \text{sgn}(\hat{\varphi}_1(k)) \neq \text{sgn}(\hat{\varphi}_1(1)) \end{cases}, \qquad (5.13)$$

where $\hat{\varphi}_1(1)$ is the initial value of $\hat{\varphi}_1(k)$, and parameter $\beta = \text{const} > 0$ is chosen of low value.

Theorem 5.2

According to [Hou11a] and [Hou11b], the value of parameter L must be chosen between 1 and the approximate order of the unknown process. If $L = 1$, MFAC-PFDL becomes MFAC-CFDL [Rom18]. The expression of the control law specific to the MFAC-PFDL algorithm is presented as follows in the SISO formulation, which is relatively easily understandable to match the MIMO controller implementations in this paper as sets of separately designed and tuned SISO controllers:

$$u(k) = u(k-1) + \frac{\rho_1 \hat{\varphi}_1(k)[y_d(k+1) - y(k)]}{\lambda + |\hat{\varphi}_1(k)|^2} - \frac{\hat{\varphi}_1(k) \sum_{i=2}^{L} [\rho_i \hat{\varphi}_i(k) \Delta u(k-i+1)]}{\lambda + |\hat{\varphi}_1(k)|^2}, \qquad (5.14)$$

where $\rho = [\rho_1 \ \ldots \ \rho_L]^T$ is a vector of constant parameters, with $\rho_i > 0$, $i = 1 \ldots L$, and as in the MFAC-CFDL algorithm, $\lambda = \text{const} > 0$ is a weighting parameter in the objective function in (5.4).

Remark 5.6

Theorem 5.2, together with the estimation mechanism $\hat{\Phi}(k)$ given in equation (5.12), the reset conditions specified in (5.13) and with the control law in (5.14), guarantees the stability of the control system structure with MFAC-PFDL algorithm [Hou11b, Rom18].

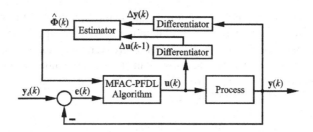

FIGURE 5.2   MIMO control system structure with MFAC-PFDL algorithm. (Adapted from Rom14, Rom15, Rom16, Rom18 and Rom19.)

Proposition 5.4

The block diagram of the control system structure with MFAC-PFDL algorithm is presented in Figure 5.2, where the control error vector $\mathbf{e}(k) \in \Re^n$ is defined in (5.9).

Proposition 5.5

Based on the results presented in this section, the following **design and tuning steps for the control system structure with MFAC-PFDL algorithm** are carried out [Rom18]:

Step 1. Establishing the value of the parameter $L > 0$.

Step 2. Choosing the initial value of $\hat{\mathbf{\Phi}}(1)$ of the estimate of the PPD matrix $\mathbf{\Phi}(k)$ in order to fulfill the performance specifications of the control system structure and impose its desired behavior.

Step 3. Choosing a small value for the lower limit $\beta = \text{const} > 0$ in (5.13) such that $\hat{\mathbf{\Phi}}(k)$ will vary during the experiments according to the constraint imposed in terms of the reset conditions of the matrix $\hat{\mathbf{\Phi}}(k)$ in (5.13).

Step 4. Establishing the parameter values of $\eta \in (0,1)$, $\lambda > 0$, $\mu > 0$ and $\rho = [\rho_1 \ \dots \ \rho_L]^T$, with $\rho_i > 0$, $i = 1 \dots L$, using the experience of the control system designer.

5.2.2   THE GENERIC STRUCTURE OF SINGLE-INTEGRATOR AND DOUBLE-INTEGRATOR COMPLETELY UNKNOWN NONLINEAR DYNAMIC SYSTEMS

Definition 5.4

Let us consider the dynamics of a single-integrator nonlinear system with the model

$$\dot{\mathbf{x}} = \mathbf{f}_0(\mathbf{x}, \mathbf{u}),$$
$$\mathbf{y} = \mathbf{x},$$

(5.15)

where $\mathbf{x} \in \Re^n$ is the system state vector, $\mathbf{u} \in \Re^n$ is the system input vector (or the control signal vector or the control input vector), $\mathbf{y} \in \Re^n$ is the system output vector,

$\mathbf{f}_0 : \mathfrak{R}^m \times \mathfrak{R}^n \to \mathfrak{R}^n$ is the vector function including the dynamics of the system (both linear and nonlinear terms), and the initial conditions are not specified. In our case, it is assumed that $\mathbf{f}_0$ is completely unknown. Moreover, since we are interested in a single-integrator system, the number of system states, the number of system outputs and the number of system inputs are equal to n. Then, adding and subtracting u from the right-hand side of (5.15) will lead to

$$\dot{\mathbf{x}} = \mathbf{f} + \mathbf{u}, \tag{5.16}$$

where $\mathbf{f} = \mathbf{f}_0 - \mathbf{u}$. Here, $\mathbf{f} \in \mathfrak{R}^n$ as a vector is considered as the new representation of the system dynamics. Furthermore, $\mathbf{f}$ can be represented by a linear term and a nonlinear term as follows [Saf18a, Saf18b]:

$$\mathbf{f} = \mathbf{A}\,\mathbf{x} + \mathbf{g}, \tag{5.17}$$

where $\mathbf{A} = \mathbf{A}^T \in \mathfrak{R}^{n \times n}$ is a diagonal system matrix and $\mathbf{g} \in \mathfrak{R}^n$ is a vector including all nonlinearities of the system. Here, $\mathbf{A}$ and $\mathbf{g}$ are completely unknown parameters and variables. It will be shown throughout this section that the representation of $\mathbf{f}$ as in (5.17) will enable to derive an online method for updating the main tuning gains of the model-free controller. Finally, the generic structure of a completely unknown single-integrator nonlinear dynamic system is defined as follows [Saf18a, Saf18b]:

$$\dot{\mathbf{x}} = \mathbf{A}\,\mathbf{x} + \mathbf{g} + \mathbf{u},$$
$$\mathbf{y} = \mathbf{x}. \tag{5.18}$$

Definition 5.5

In a double-integrator dynamic system, the number of system states is larger than the number of system inputs and outputs. In this regard and following *Definition 5.1*, the dynamics of a double-integrator nonlinear dynamic system is considered as

$$\dot{\mathbf{x}}_i = \mathbf{x}_j,$$
$$\dot{\mathbf{x}}_j = \mathbf{f}_j(\mathbf{x}, \mathbf{u}), \tag{5.19}$$

for $i = 1\ldots k$ and $j = k+1\ldots n$, with $k < n$, where $\mathbf{x} = \left[(\mathbf{x}_i)^T \quad (\mathbf{x}_j)^T \right]^T \in \mathfrak{R}^n$ includes all of the system states, $\mathbf{y} = \mathbf{x}_i \in \mathfrak{R}^k$ is the system output vector, and $\mathbf{u} \in \mathfrak{R}^m$ is the system input vector (or the control signal vector). Here, n is the number of system states, and m is the number of system inputs (for $m < n$), and k is the number of system outputs. In this regard, a virtual control variable $\mathbf{v} \in \mathfrak{R}^r$ for $r = n - k$ is inserted as follows:

$$\mathbf{v} = \mathbf{k}_p(\mathbf{y}_d - \mathbf{y}), \tag{5.20}$$

for $\mathbf{y}_d \in \mathfrak{R}^r$ as the reference input vector that includes the reference inputs (the setpoints or the desired outputs) for the actual outputs of the system, and $\mathbf{k}_p \in \mathfrak{R}^{r \times r}$ is a diagonal positive definite matrix accommodating all the controller gains for each system output variable. Furthermore, $\mathbf{v}$ is considered as the reference input vector, namely, its elements are the set-points for the remaining states of system that do not appear in the system output vector, i.e., $\mathbf{x}_j^d = \mathbf{v}$. Finally, the following generic model is considered for representing dynamics of the *inner loop* of the double-integrator system considered in (5.19):

$$\dot{\mathbf{x}}_j = \mathbf{A}\,\mathbf{x}_j + \mathbf{g} + \mathbf{u},$$
$$\mathbf{y}_j = \mathbf{x}_j. \tag{5.21}$$

The model in (5.21) is in the same format of that given in (5.18) for single-integrator dynamic systems. Consequently, this generic structure would be utilized throughout this section for proposing and implementing the MFAC algorithm on both single-integrator and double-integrator dynamic systems.

Remark 5.7

In (5.18), although the parameters and matrices $\mathbf{A}$ and $\mathbf{g}$ are unknown, it is assumed that they and their first time-derivatives are bounded. It means that the following inequalities hold for the elements of these matrices: $|\mathbf{A}| \leq \mathbf{U_A} \in \mathfrak{R}^{n \times n}, |\dot{\mathbf{A}}| \leq \mathbf{U_{\dot{A}}} \in \mathbf{R}^{n \times n}$, $|\mathbf{g}| \leq \mathbf{U_g} \in \mathfrak{R}^{n \times 1}$ and $|\dot{\mathbf{g}}| \leq \mathbf{U_{\dot{g}}} \in \mathfrak{R}^{n \times 1}$. The upper bounds $\mathbf{U_A}, U_{\dot{A}}, U_g$ and $U_{\dot{g}}$ are considered to be completely unknown.

5.2.3 MODEL-FREE ADAPTIVE CONTROLLER ALGORITHM FOR CONTINUOUS-TIME DYNAMIC SYSTEMS

Definition 5.6

For dynamic system defined in *Definitions 5.1* and *5.2*, the tracking error $\mathbf{e} \in \mathfrak{R}^n$ is defined as

$$\mathbf{e} = \mathbf{y}_d - \mathbf{y}, \tag{5.22}$$

where $\mathbf{y}_d \in \mathfrak{R}^n$ is the reference input vector whose elements are the reference inputs (or the set-points or the desired outputs). Moreover, the generalized tracking error $\sigma \in \mathfrak{R}^n$ is defined as follows:

$$\sigma = \mathbf{e} + \xi, \tag{5.23}$$

where $\xi = \int \mathbf{e}\,dt$. Here, the time-integral of the tracking error is added to σ in order to eliminate the error in the steady-state phase of the system operation. The objective of model-free controller is to control the system so that it converges σ to zero as time goes to infinity.

Definition 5.7

A vectorizing function $\upsilon(.)$ is to generate a vector $\mathbf{v_M} \in \mathfrak{R}^n$ using the diagonal elements of matrix $\mathbf{M} \in \mathfrak{R}^{n \times n}$, as follows:

$$\mathbf{v_M} = \upsilon(\mathbf{M}), \tag{5.24}$$

where $\mathbf{v_M}(i) = \mathbf{M}(i,i)$. In an opposite direction, the function $\mathbf{M}(.)$ is defined to produce a diagonal matrix $\mathbf{M_v} \in \mathfrak{R}^{n \times n}$ from all elements of the vector $\mathbf{v} \in \mathfrak{R}^n$, as follows:

$$\mathbf{M_v} = \mathbf{M}(\mathbf{v}), \tag{5.25}$$

where $\mathbf{M_v}(i,i) = \mathbf{v}(i)$.

Lemma 5.3

According to the *separation principle* [Ata99], by combining a stable observer into a stable controller, we can have a stable dynamic system that includes both the observer and the controller.

Theorem 5.3

In order to reach the tracking objective defined in *Definition 5.6* for the dynamic systems defined in *Definitions 5.4* and *5.5*, the model-free adaptive controller is proposed as follows [Saf18a, Saf18b]:

$$\mathbf{u} = \dot{\mathbf{y}}_d - \hat{\mathbf{A}}\,\mathbf{x} - \hat{\mathbf{g}} - \xi + \left(\mathbf{I}_n + 2\mathbf{P}^{-1}\mathbf{Q} + \hat{\mathbf{A}}\right)\sigma - \frac{1}{4}\mathbf{P}\,\sigma, \tag{5.26}$$

where $\mathbf{I}_n \in \mathfrak{R}^{n \times n}$ is an identity matrix, and $\mathbf{P} = \mathbf{P}^T \in \mathfrak{R}^{n \times n}$ is a positive definite matrix as the main controller gain defined by

$$\dot{\mathbf{P}} = 2\,\mathbf{P}\,\hat{\mathbf{A}} + 2\,\mathbf{Q} - \mathbf{P}\,\mathbf{P}, \tag{5.27}$$

where $\mathbf{Q} \in \mathfrak{R}^{n \times n}$ is a constant positive definite matrix and with incorporating the following two adaptive laws for linear and nonlinear terms of the dynamic system:

$$\dot{\hat{\mathbf{g}}} = -\Gamma_1\,\mathbf{P}\,\sigma - \rho_1\,\Gamma_1\,\hat{\mathbf{g}},$$
$$\dot{\mathbf{v}}_{\hat{\mathbf{A}}} = -\Gamma_2\,\mathbf{P}\,\mathbf{M}_\sigma(x - \sigma) - \rho_2\,\Gamma_2\,\mathbf{v}_{\hat{\mathbf{A}}}. \tag{5.28}$$

Here, $\Gamma_1 = \Gamma_1^T \in \mathfrak{R}^{n \times n}$ and $\Gamma_2 = \Gamma_2^T \in \mathfrak{R}^{n \times n}$ are two positive definite matrices acting as learning rates of the above adaptive laws; ρ_1 and ρ_2 are two positive leakage scalar gains providing the robustness to the adaptive laws in (5.28).

Proof (the proof presented here is following the proof presented in [Saf18b]). For the parameter estimation errors $\tilde{\mathbf{A}} = \mathbf{A} - \hat{\mathbf{A}}$ and $\tilde{\mathbf{g}} = \mathbf{g} - \hat{\mathbf{g}}$, and the generalized tracking error σ, the following Lyapunov function candidate is defined:

$$V = \frac{1}{2}\sigma^T \mathbf{P}\,\sigma + \frac{1}{2}\tilde{\mathbf{g}}\,\Gamma_1^{-1}\,\tilde{\mathbf{g}} + \frac{1}{2}\mathbf{v}_{\tilde{\mathbf{A}}}\,\Gamma_2^{-1}\,\mathbf{v}_{\tilde{\mathbf{A}}}. \tag{5.29}$$

Here, $\mathbf{v}_{\tilde{\mathbf{A}}} = \mathbf{v}(\tilde{\mathbf{A}})$ is considered according to *Definition 5.7.* Furthermore, the first time derivative of V is

$$\dot{V} = \sigma^T \mathbf{P}\,\dot{\sigma} + \frac{1}{2}\sigma^T \dot{\mathbf{P}}\,\sigma + \tilde{\mathbf{g}}\,\Gamma_1^{-1}\,\dot{\tilde{\mathbf{g}}} + \mathbf{v}_{\tilde{\mathbf{A}}}\,\Gamma_2^{-1}\,\dot{\mathbf{v}}_{\tilde{\mathbf{A}}}. \tag{5.30}$$

Similarly, here we have $\dot{\mathbf{v}}_{\tilde{\mathbf{A}}} = \mathbf{v}(\dot{\tilde{\mathbf{A}}})$. Moreover, replacing σ from (5.23), (5.22) and (5.18) into (5.30) leads to

$$\dot{V} = \sigma^T \mathbf{P}(\dot{\mathbf{y}}_d - \mathbf{A}\,\mathbf{x} - \mathbf{g} - \mathbf{u} + \mathbf{e}) + \frac{1}{2}\sigma^T \dot{\mathbf{P}}\,\sigma + \tilde{\mathbf{g}}^T\,\Gamma_1^{-1}\,\dot{\tilde{\mathbf{g}}} + \mathbf{v}_{\tilde{\mathbf{A}}}^T\,\Gamma_2^{-1}\,\dot{\mathbf{v}}_{\tilde{\mathbf{A}}}. \tag{5.31}$$

At this point, let us add and subtract $\sigma^T \mathbf{P}\mathbf{A}\,\sigma$ in the right-hand side of (5.31). This helps to arrange the term within the parenthesis of the first term in the right-hand side of (5.31), in a way that is appropriate for the rest of the proof. The result is

$$\dot{V} = \sigma^T \mathbf{P}(\dot{\mathbf{y}}_d - \mathbf{A}(\mathbf{x} - \sigma) - \mathbf{g} - \mathbf{u} + \mathbf{e} - \mathbf{A}\,\sigma) + \frac{1}{2}\sigma^T \dot{\mathbf{P}}\,\sigma + \tilde{\mathbf{g}}^T\,\Gamma_1^{-1}\,\dot{\tilde{\mathbf{g}}} + \mathbf{v}_{\tilde{\mathbf{A}}}^T\,\Gamma_2^{-1}\,\dot{\mathbf{v}}_{\tilde{\mathbf{A}}}. \tag{5.32}$$

Besides, we want to make a relation between the estimation errors $\tilde{\mathbf{A}}$ and $\tilde{\mathbf{g}}$ with the generalized tracking error σ. In this regard, we want to add and subtract the term $\sigma^T \mathbf{P}(\hat{\mathbf{A}}(\mathbf{x} - \sigma) + \hat{\mathbf{g}})$ in the right-hand side of (5.32). Thus, we have

$$\dot{V} = \sigma^T \mathbf{P}\left(\dot{\mathbf{y}}_d - \hat{\mathbf{A}}(\mathbf{x} - \sigma) - \hat{\mathbf{g}} - \mathbf{u} + \mathbf{e} - \mathbf{A}\,\sigma\right)$$
$$+ \frac{1}{2}\sigma^T \dot{\mathbf{P}}\,\sigma + \left(\tilde{\mathbf{g}}^T\,\Gamma_1^{-1}\,\dot{\tilde{\mathbf{g}}} - \sigma^T \mathbf{P}\mathbf{g} + \sigma^T \mathbf{P}\,\hat{\mathbf{g}}\right) \tag{5.33}$$
$$+ (\mathbf{v}_{\tilde{\mathbf{A}}}^T\,\Gamma_2^{-1}\,\dot{\mathbf{v}}_{\tilde{\mathbf{A}}} - \sigma^T \mathbf{P}\mathbf{A}(\mathbf{x} - \sigma) + \sigma^T \mathbf{P}\,\hat{\mathbf{A}}(\mathbf{x} - \sigma)).$$

Note that in the above method, the unknown parameters and vectors $\mathbf{A}$ and $\mathbf{g}$ are replaced with the estimated ones $\tilde{\mathbf{A}}$ and $\tilde{\mathbf{g}}$ within the first term. In addition, the variable σ is included within the third and fourth terms of the right-hand side of (5.33), in relation with $\hat{\mathbf{g}}$ and $\hat{\mathbf{A}}$, respectively. Furthermore, by recalling the definitions of $\tilde{\mathbf{A}}$ and $\tilde{\mathbf{g}}$, (5.33) will be simplified as follows:

$$\dot{V} = \sigma^T \mathbf{P}(\dot{\mathbf{y}}_d - \hat{\mathbf{A}}(\mathbf{x} - \sigma) - \hat{\mathbf{g}} - \mathbf{u} + \mathbf{e} - \mathbf{A}\,\sigma) + \frac{1}{2}\sigma^T \dot{\mathbf{P}}\,\sigma + (\tilde{\mathbf{g}}^T\,\Gamma_1^{-1}\,\dot{\tilde{\mathbf{g}}}$$
$$- \sigma^T \mathbf{P}\mathbf{g} + \sigma^T \mathbf{P}\,\hat{\mathbf{g}}) + (\mathbf{v}_{\tilde{\mathbf{A}}}^T\,\Gamma_2^{-1}\,\dot{\mathbf{v}}_{\tilde{\mathbf{A}}} - \sigma^T \mathbf{P}\,\tilde{\mathbf{A}}(\mathbf{x} - \sigma)). \tag{5.34}$$

Now, considering $\dot{\tilde{\mathbf{g}}} = \dot{\mathbf{g}} - \dot{\hat{\mathbf{g}}}$ and $\dot{\mathbf{v}}_{\tilde{\mathbf{A}}} = \dot{\mathbf{v}}_{\mathbf{A}} - \dot{\mathbf{v}}_{\hat{\mathbf{A}}}$ leads to

$$\dot{V} = \sigma^T \ \mathbf{P}(\dot{\mathbf{y}}_d - \hat{\mathbf{A}}(\mathbf{x} - \sigma) - \hat{\mathbf{g}} - \mathbf{u} + \mathbf{e} - \mathbf{A} \ \sigma) + \frac{1}{2}\sigma^T \ \dot{\mathbf{P}} \ \sigma + (\tilde{\mathbf{g}}^T \ \Gamma_1^{-1} \ \dot{\mathbf{g}}$$

$$- \tilde{\mathbf{g}}^T \ \Gamma_1^{-1}\dot{\hat{\mathbf{g}}} - \sigma^T \ \mathbf{P} \ \tilde{\mathbf{g}}) + (\mathbf{v}_{\tilde{\mathbf{A}}}^T \ \Gamma_2^{-1} \ \dot{\mathbf{v}}_{\mathbf{A}} - \mathbf{v}_{\tilde{\mathbf{A}}}^T\Gamma_2^{-1}\dot{\mathbf{v}}_{\hat{\mathbf{A}}} - \sigma^T \ \mathbf{P} \ \tilde{\mathbf{A}}(\mathbf{x} - \sigma)). \tag{5.35}$$

Moreover, a simple rearrangement leads to

$$\sigma^T \ \mathbf{P} \ \tilde{\mathbf{A}}(\mathbf{x} - \sigma) = \mathbf{v}_{\tilde{\mathbf{A}}}^T \ \mathbf{P} \ \mathbf{M}_\sigma(\mathbf{x} - \sigma), \tag{5.36}$$

and

$$\sigma^T \ \mathbf{P} \ \tilde{\mathbf{g}} = \tilde{\mathbf{g}}^T \ \mathbf{P} \ \sigma. \tag{5.37}$$

The first rearrangement helps us to have $\mathbf{v}_{\tilde{\mathbf{A}}}$ in the last term of the right-hand side of (5.35). This reaches at

$$\dot{V} = \sigma^T \ \mathbf{P}(\dot{\mathbf{y}}_d - \hat{\mathbf{A}}(\mathbf{x} - \sigma) - \hat{\mathbf{g}} - \mathbf{u} + \mathbf{e} - \mathbf{A} \ \sigma) + \frac{1}{2}\sigma^T \ \dot{\mathbf{P}} \ \sigma$$

$$+ \tilde{\mathbf{g}}^T(\Gamma_1^{-1} \ \dot{\mathbf{g}} - \Gamma_1^{-1}\dot{\hat{\mathbf{g}}} - \mathbf{P} \ \sigma) + \mathbf{v}_{\tilde{\mathbf{A}}}^T(\Gamma_2^{-1} \ \dot{\mathbf{v}}_{\mathbf{A}} - \Gamma_2^{-1}\dot{\mathbf{v}}_{\hat{\mathbf{A}}} - \mathbf{P} \ \mathbf{M}_\sigma(\mathbf{x} - \sigma)). \tag{5.38}$$

At this point, we want to construct a term including $\dot{\hat{\mathbf{g}}}$ in the third term at the right-hand side of (5.38), as well as a term including $\mathbf{v}_{\hat{\mathbf{A}}}$ into its fourth term. To achieve this, let us add and subtract the following terms s_1 and s_2 in (5.38):

$$s_1 = \frac{1}{4\rho_1}(\Gamma_1^{-1}\dot{\mathbf{g}} + \rho_1\mathbf{g})^T(\Gamma_1^{-1}\dot{\mathbf{g}} + \rho_1\mathbf{g}) + \rho_1\tilde{\mathbf{g}}^T\tilde{\mathbf{g}} + \rho_1\tilde{\mathbf{g}}^T\hat{\mathbf{g}},$$

$$s_2 = \frac{1}{4\rho_2}(\Gamma_2^{-1}\dot{\mathbf{v}}_{\mathbf{A}} + \rho_2\mathbf{v}_{\mathbf{A}})^T(\Gamma_2^{-1}\dot{\mathbf{v}}_{\mathbf{A}} + \rho_2\mathbf{v}_{\mathbf{A}}) + \rho_2\mathbf{v}_{\tilde{\mathbf{A}}}^T\mathbf{v}_{\tilde{\mathbf{A}}} + \rho_2\mathbf{v}_{\tilde{\mathbf{A}}}^T\mathbf{v}_{\hat{\mathbf{A}}}. \tag{5.39}$$

Thus, this leads to

$$\dot{V} = \sigma^T\mathbf{P}(\dot{\mathbf{y}}_d - \hat{\mathbf{A}}(\mathbf{x} - \sigma) - \hat{\mathbf{g}} - \mathbf{u} + \mathbf{e} - \mathbf{A} \ \sigma) + \frac{1}{2}\sigma^T \ \dot{\mathbf{P}} \ \sigma$$

$$+ \tilde{\mathbf{g}}^T(\rho_1\hat{\mathbf{g}} - \Gamma_1^{-1}\dot{\hat{\mathbf{g}}} - \mathbf{P} \ \sigma) + s_3 + \mathbf{v}_{\tilde{\mathbf{A}}}^T(\rho_2\mathbf{v}_{\hat{\mathbf{A}}} - \Gamma_2^{-1}\dot{\mathbf{v}}_{\hat{\mathbf{A}}} - \mathbf{P} \ \mathbf{M}_\sigma(\mathbf{x} - \sigma)) + s_4. \tag{5.40}$$

The following notations are employed in (5.40):

$$s_3 = \frac{1}{4\rho_1}(\Gamma_1^{-1}\dot{\mathbf{g}} + \rho_1\mathbf{g})^T(\Gamma_1^{-1}\dot{\mathbf{g}} + \rho_1\mathbf{g}) - \begin{pmatrix} \frac{1}{4\rho_1}(\Gamma_1^{-1}\dot{\mathbf{g}} + \rho_1\mathbf{g})^T(\Gamma_1^{-1}\dot{\mathbf{g}} + \rho_1\mathbf{g}) \\ \\ +\rho_1\tilde{\mathbf{g}}^T\tilde{\mathbf{g}} - 2\sqrt{\rho_1}\tilde{\mathbf{g}}^T\frac{1}{2\sqrt{\rho_1}}(\rho_1\tilde{\mathbf{g}} + \rho_1\hat{\mathbf{g}} + \Gamma_1^{-1}\dot{\mathbf{g}}) \end{pmatrix}, \tag{5.41}$$

and

$$s_4 = \frac{1}{4\rho_2}\left(\Gamma_2^{-1}\dot{\mathbf{v}}_A + \rho_2\mathbf{v}_A\right)^T\left(\Gamma_2^{-1}\dot{\mathbf{v}}_A + \rho_2\mathbf{v}_A\right)$$

$$- \left(\begin{array}{c} \dfrac{1}{4\rho_2}\left(\Gamma_2^{-1}\dot{\mathbf{v}}_A + \rho_2\mathbf{v}_A\right)^T\left(\Gamma_2^{-1}\dot{\mathbf{v}}_A + \rho_2\mathbf{v}_A\right) \\[2mm] + \rho_2\mathbf{v}_{\tilde{A}}{}^T\mathbf{v}_{\tilde{A}} - 2\sqrt{\rho_2}\mathbf{v}_{\tilde{A}}\dfrac{1}{2\sqrt{\rho_2}}\left(\rho_2\mathbf{v}_{\tilde{A}} + \rho_2\mathbf{v}_{\hat{A}} + \Gamma_2^{-1}\dot{\mathbf{v}}_A\right) \end{array}\right). \tag{5.42}$$

Note that the terms s_3 and s_4 are formulated in a way to produce squared terms. In this regard, these terms can be presented as follows:

$$s_3 = -\omega_1 + \delta_1,$$
$$s_4 = -\omega_2 + \delta_2, \tag{5.43}$$

where

$$\omega_1 = \left(\frac{1}{2\sqrt{\rho_1}}\left(\Gamma_1^{-1}\dot{\mathbf{g}} + \rho_1\mathbf{g}\right) - \sqrt{\rho_1}\tilde{\mathbf{g}}\right)^T\left(\frac{1}{2\sqrt{\rho_1}}\left(\Gamma_1^{-1}\dot{\mathbf{g}} + \rho_1\mathbf{g} - \sqrt{\rho_1}\tilde{\mathbf{g}}\right)\right),$$

$$\omega_2 = \left(\frac{1}{2\sqrt{\rho_2}}\left(\Gamma_2^{-1}\dot{\mathbf{v}}_A + \rho_2\mathbf{v}_A\right) - \sqrt{\rho_2}\mathbf{v}_{\tilde{A}}\right)^T\left(\frac{1}{2\sqrt{\rho_2}}\left(\Gamma_2^{-1}\dot{\mathbf{v}}_A + \rho_2\mathbf{v}_A\right) - \sqrt{\rho_2}\mathbf{v}_{\tilde{A}}\right). \tag{5.44}$$

and

$$\delta_1 = \frac{1}{4\rho_1}\left(\Gamma_1^{-1}\dot{\mathbf{g}} + \rho_1\mathbf{g}\right)^T\left(\Gamma_1^{-1}\dot{\mathbf{g}} + \rho_1\mathbf{g}\right),$$

$$\delta_2 = \frac{1}{4\rho_2}\left(\Gamma_2^{-1}\dot{\mathbf{v}}_A + \rho_2\mathbf{v}_A\right)^T\left(\Gamma_2^{-1}\dot{\mathbf{v}}_A + \rho_2\mathbf{v}_A\right). \tag{5.45}$$

On the other hand, utilizing the adaptive parameter estimators given in (5.28), the third and fifth terms of the right-hand side of (5.40) will be canceled. Hence, the result is

$$\dot{V} = \sigma^T\,\mathbf{P}(\dot{\mathbf{y}}_d - \hat{\mathbf{A}}(\mathbf{x} - \sigma) - \hat{\mathbf{g}} - \mathbf{u} + \mathbf{e} - \mathbf{A}\,\sigma) + \frac{1}{2}\sigma^T\,\dot{\mathbf{P}}\,\sigma$$
$$- \omega_1 + \delta_1 - \omega_2 + \delta_2. \tag{5.46}$$

Moreover, according to *Remark 5.7*, we have

$$\delta_1 \leq \frac{1}{4\rho_1}\left(\Gamma_1^{-1}\mathbf{U}_{\dot{\mathbf{g}}} + \rho_1\mathbf{U}_{\mathbf{g}}\right)^T\left(\Gamma_1^{-1}\mathbf{U}_{\dot{\mathbf{g}}} + \rho_1\mathbf{U}_{\mathbf{g}}\right) = \tau_1,$$

$$\delta_2 \leq \frac{1}{4\rho_2}\left(\Gamma_2^{-1}\mathbf{v}_{U_{\dot{A}}} + \rho_2\mathbf{v}_{U_A}\right)^T\left(\Gamma_2^{-1}\mathbf{v}_{U_{\dot{A}}} + \rho_2\mathbf{v}_{U_A}\right) = \tau_2. \tag{5.47}$$

Hence, incorporating the inequalities in (5.47) into (5.46) leads to

$$\dot{V} = \sigma^T \, \mathbf{P}(\dot{\mathbf{y}}_d - \hat{\mathbf{A}}(\mathbf{x} - \sigma) - \hat{\mathbf{g}} - \mathbf{u} + \mathbf{e} - \mathbf{A} \, \sigma) + \frac{1}{2}\sigma^T \, \dot{\mathbf{P}} \, \sigma - \omega_3 + \tau_3, \qquad (5.48)$$

where $\omega_3 = \omega_1 + \omega_2$ and $\tau_3 = \tau_1 + \tau_2$. Note that the variables τ_1, τ_2 and consequently τ_3 are constant positive, while ω_1, ω_2 and hence ω_3 are time-varying positive. Finally, utilizing the model-free adaptive controller proposed in (5.26) into (5.48) yields

$$\dot{V} \leq -\sigma^T \left(2 \, \mathbf{Q} + \mathbf{P} \, \mathbf{A} - \frac{1}{4}\mathbf{P} \, \mathbf{P} - \frac{1}{2}\dot{\mathbf{P}} \right)\sigma - \omega_3 + \tau_3. \qquad (5.49)$$

Replacing next $\dot{\mathbf{P}}$ from

$$\dot{\mathbf{P}} = 2 \, \mathbf{P} \, \mathbf{A} + 2 \, \mathbf{Q} - \mathbf{P} \, \mathbf{P} \qquad (5.50)$$

into (5.35) leads to

$$\dot{V} \leq -\sigma^T \left(\mathbf{Q} + \frac{1}{4}\mathbf{P} \, \mathbf{P} \right)\sigma - \omega_3 + \tau_3. \qquad (5.51)$$

Consequently, we have

$$\dot{V} \leq -\omega_4 + \tau_3, \qquad (5.52)$$

where

$$\omega_4 = \sigma^T \left(\mathbf{Q} + \frac{1}{4}\mathbf{P} \, \mathbf{P} \right)\sigma + \omega_3. \qquad (5.53)$$

Now, by recalling the *LaSalle-Yoshizawa* theorem formulated in *Lemma 3.3*, we conclude that V is UUB. Then, it is realized that σ, $\mathbf{v}_{\tilde{\mathbf{A}}}$ and $\tilde{\mathbf{g}}$ will converge to small bounded sets around the origin. In addition, according to *Lemma 5.3*, we can replace $\mathbf{A}$ in (5.50) with $\hat{\mathbf{A}}$ estimated in accordance with (5.28). So, the dynamic Riccati equation (DRE) in (5.27) is achieved. This completes the proof.

Remark 5.8

The values of $\dot{\mathbf{y}}_d$ in (5.26) are computed using the sliding mode differentiator specified in *Remark 3.3*.

Remark 5.9

Recalling the *proof* of *Theorem 5.3*, the values of the elements of Γ_1 and Γ_2 should be adjusted large enough, while the values of ρ_1 and ρ_2 are chosen small enough to provide stability and robustness of the model-free adaptive controller and the corresponding adaptive laws in *Theorem 5.3* [Saf18b].

Remark 5.10

Recalling (5.28), it is highlighted that no regressor parameters are utilized in the first terms on the right-hand sides of the adaptive laws. Only the modified tracking error is included. Compared to the mode-free control algorithms suggested in the context of reinforcement learning, where the construction of regressor parameters is a must for incorporating artificial neural networks, the implementation of the model-free adaptive controller in *Theorem 5.3* is more convenient. Moreover, since no regressor parameter is utilized, no condition for PE of the input signals is imposed on the model-free adaptive controller presented in this section.

Remark 5.11

The DRE presented in (5.27) is utilized for updating the main gains of the model-free controller, i.e., $\mathbf{P}$. The tuning parameters of this update law is $\mathbf{Q}$, which is set small enough to avoid divergence of the solution [Saf18b]. In addition, since having $\mathbf{P}$ to be a positive definite matrix is a requirement of the model-free controller in *Theorem 5.3*, it should be shown that for all values of $\hat{\mathbf{A}}$, this requirement is satisfied. The following lemma is discussed on this matter.

Lemma 5.4

In order to satisfy the positive definite requirement of $\mathbf{P}$ in the model-free adaptive controller defined in *Theorem 5.3*, the equation in (5.27) should have the following solution for each of the diagonal elements of $\mathbf{P}$ at each time step $t \in [0, \infty)$ of the controller operation:

$$\mathbf{P}(i,i) = P_0{}^i + \frac{1}{w^i}, \quad i = 1 \ldots n. \tag{5.54}$$

In this solution, the following notations are employed:

$$P_0{}^i = \hat{\mathbf{A}}(i,i) - \sqrt{\Delta}, \quad \Delta = (\hat{\mathbf{A}}(i,i))^2 + 2\,\mathbf{Q}(i,i), \tag{5.55}$$

and

$$w^i = \frac{1}{2\sqrt{\Delta}} + \left(w_0^i - \frac{1}{2\sqrt{\Delta}}\right)\exp\left(-\int_0^t 2\sqrt{\Delta}\,ds\right), \quad w_0^i = \frac{1}{1 - P_0{}^i}. \tag{5.56}$$

It can be shown that by choosing $\mathbf{Q}(i,i)$ large enough, the value of $\mathbf{P}(i,i)$ in (5.54) will always be positive.

Proof. For the proof of this lemma, refer to the proof presented for *Lemma 8.2*, by assuming $h = 1$.

5.3   EXAMPLE AND APPLICATION

The tower crane system described in Chapter 1 is considered as a controlled process to exemplify the MFAC algorithms in the CFDL and PFDL versions further used as MFAC-CFDL and MFAC-PFDL algorithms exemplified using SISO and MIMO control systems. Three separate SISO control systems for the three controlled outputs, namely cart position, arm angular position and payload position, will be first considered, and the MIMO control system to control all these three tower crane system outputs will be further presented.

5.3.1   SISO CONTROL SYSTEMS

For all SISO loops with MFAC algorithms presented in Section 5.2, the performance is assessed using the following objective function:

$$J_{(\psi)}(\chi_{(\psi)}) = \frac{1}{2} E \left\{ \sum_{k=1}^{N} \left[e_{(\psi)}(k, \chi_{(\psi)}) \right]^2 \right\}, \tag{5.57}$$

where the subscript $\psi \in \{c, a, p\}$ indicates the a specific controlled motion of the tower crane system, namely c – for the cart position (i.e., $y_{(c)} = y_1$), a – for the arm angular position (i.e., $y_{(a)} = y_2$) and p – for the payload position (i.e., $y_{(p)} = y_3$), the control error dynamics $\varepsilon_{(\psi)} = y^*_{(\psi)} - y_{(\psi)}$ and $\chi_{(\psi)}$ is the parameter vector of the controller. The mathematical expectation $E\{\Xi\}$ is taken with respect to the stochastic probability distribution of the disturbance inputs applied to the process and thus affects the control system behavior. The sampling period is $T_s = 0.01$ s, and the time horizon is 70 s, resulting in $N = 70 / 0.01 = 7,000$ s samples.

The MFAC algorithms in the CFDL and PFDL versions are applied, one for each controlled output of the process, by following the steps of the SISO MFAC algorithm in Section 5.2. The description of the steps is given in relation with the Matlab & Simulink programs and schemes.

The reference input vectors are obtained by first applying the following step signals as scalar reference inputs (or set-points), the same ones as in Chapters 3 and 4:

$$y^*_{(c)}(k) = 0.15 \text{ if } k \in [0, 2,000], 0.1 \text{ if } k \in (2,000, 3,500], -0.05 \text{ if } k \in (3,500, 5,000],$$

$$0 \text{ if } k \in (5,000, 7,000] \tag{5.58}$$

for cart position control,

$$y^*_{(a)}(k) = 0 \text{ if } k \in [0, 500], 0.15 \text{ if } k \in (500, 2,500], -0.15 \text{ if } k \in (2,500, 4,000],$$

$$0 \text{ if } k \in (4,000, 7,000] \tag{5.59}$$

for arm angular position control and

$$y_{(p)}^*(k) = 0 \text{ if } k \in [0,1,500], 0.1 \text{ if } k \in (1,500,3,000], -0.05 \text{ if } k \in (3,000,4,500],$$

$$0 \text{ if } k \in (4,500,7,000] \tag{5.60}$$

for payload position control. The above signals are next filtered to obtain the desired reference trajectory $y_{d(\psi)} = y_{(\psi)}^* H_{(\psi)}$ using the filters

$$H_{(c)}(s) = \frac{1}{1+0.2s}, \tag{5.61}$$

$$H_{(a)}(s) = \frac{1}{1+0.2s}, \tag{5.62}$$

$$H_{(p)}(s) = \frac{1}{1+0.2s}, \tag{5.63}$$

for cart position, angular position and payload position, with the corresponding discrete transfer functions as follows:

$$H_{(c)}(z^{-1}) = \frac{0.0488}{1-0.9512z^{-1}}, \tag{5.64}$$

$$H_{(a)}(z^{-1}) = \frac{0.0465}{1-0.9535z^{-1}}, \tag{5.65}$$

$$H_{(p)}(z^{-1}) = \frac{0.0328}{1-0.9672z^{-1}}. \tag{5.66}$$

5.3.1.1 The MFAC-CFDL Algorithms

Three MFAC-CFDL algorithms are separately used as controllers in the three control loops, where the parameter vector of the MFAC-CFDL algorithms is

$$\chi_{(\psi)} = [\eta \ \lambda \ \mu \ \rho \ \hat{\phi}(1)]^T, \tag{5.67}$$

with $\psi \in \{c, a, p\}$, i.e., one parameter for each controller.

The controllers are nonlinear since saturation-type nonlinearities are inserted on the three controller outputs to match the power electronics of the actuators, which operate on the basis of the pulse width modulation (PWM) principle. As shown in Chapter 1, the three control signals (or control inputs) are within −1 and 1, ensuring the necessary connection to the tower crane system subsystem, which models the controlled process in the control system structures developed in this chapter and the next chapters, also includes three zero-order hold (ZOH) blocks to enable digital control. The tower crane system subsystem belongs to the Process.mdl Simulink diagram, which is included in the accompanying Matlab & Simulink programs given in Chapter 1.

First the value of the initial value of the estimate of the PPD matrix $\hat{\varphi}(1)$ is chosen according to step 1:

$$\hat{\varphi}_{(c)}(1) = 3.5, \tag{5.68}$$

$$\hat{\varphi}_{(a)}(1) = 12, \tag{5.69}$$

$$\hat{\varphi}_{(p)}(1) = 0.89. \tag{5.70}$$

Next the parameters of lower and the upper limits will be considered as $1{,}000\,\hat{\varphi}_{(\psi)}(1)$ for the upper limit and $0.001\hat{\varphi}_{(\psi)}(1)$ for the lower limit and are imposed according to step 2 ensuring the conditions (5.3) and (5.6). After that, the rest of the parameters, i.e., $\eta \in (0,1)$, $\lambda > 0$, $\mu > 0$, and $\rho > 0$ are chosen as follows according to step 3 resulting the parameter vector of the MFAC-CFDL algorithm:

$$\chi_{(c)} = [0.6 \ 3.5 \ 28 \ 45 \ 3.5]^T, \tag{5.71}$$

$$\chi_{(a)} = [0.9 \ 0.01 \ 1 \ 990 \ 12]^T, \tag{5.72}$$

$$\chi_{(p)} = [0.81 \ 0.71 \ 14.9 \ 25.2 \ 0.89]^T. \tag{5.73}$$

The details of steps 1–3 are implemented in the MFAC _ CFDL _ c.m Matlab program for cart position control, the MFAC _ CFDL _ a.m Matlab program for arm angular position control and the MFAC _ CFDL _ p.m Matlab program for payload position control, which are included in the accompanying Matlab & Simulink programs. The digital simulations of the control systems with the three MFAC-CFDL algorithms are conducted using the CS _ MFAC _ CFDL _ SISO _ c.mdl Simulink diagram for cart position control, the CS _ MFAC _ CFDL _ SISO _ a.mdl Simulink diagram for arm angular position control and the CS _ MFAC _ CFDL _ SISO _ p.mdl Simulink diagram for payload position control. These three Simulink diagrams make use of the state-space model of the tower crane system viewed as a controlled process, which is described in Chapter 1 and is implemented in the process _ model.m S-function.

The simulation results obtained for the control systems with MFAC-CFDL with the parameters in (5.71–5.73) with $1{,}000\,\hat{\varphi}(1)_{(\psi)}$ as the upper limit and $0.001\hat{\varphi}_{(\psi)}(1)$ as the lower limit are illustrated in Figure 5.3 for cart position control, Figure 5.4 for arm angular position control and Figure 5.5 for payload position control.

The experimental results obtained for the control systems with MFAC-CFDL algorithms with the parameters in (5.71–5.73) with $1{,}000\,\hat{\varphi}(1)_{(\psi)}$ as the upper limit and $0.001\hat{\varphi}(1)_{(\psi)}$ as the lower limit are illustrated in Figure 5.6 for cart position control, Figure 5.7 for arm angular position control and Figure 5.8 for payload position control.

5.3.1.2 The MFAC-PFDL Algorithms

Three MFAC-PFDL algorithms are separately implemented as controllers in three SISO control system structures. According to step 1 of MFAC-PFDL algorithm, the

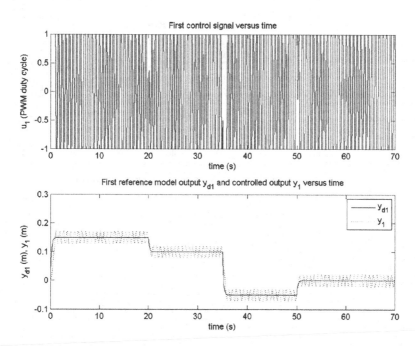

FIGURE 5.3 Simulation results of cart position control system with MFAC-CFDL algorithm.

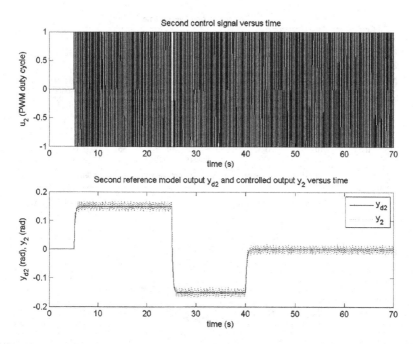

FIGURE 5.4 Simulation results of arm angular position control system with MFAC-CFDL algorithm.

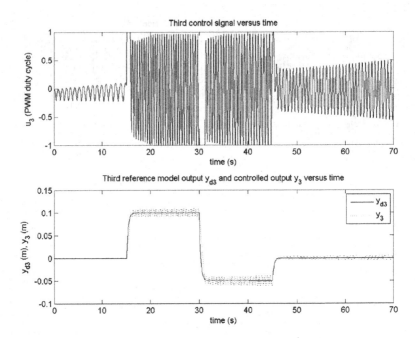

FIGURE 5.5 Simulation results of payload position control system with MFAC-CFDL algorithm.

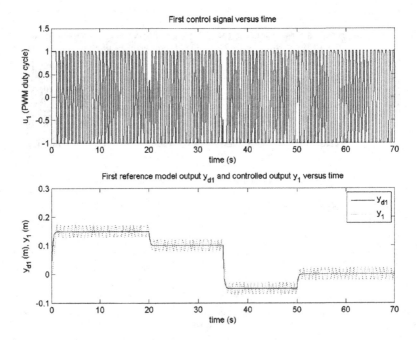

FIGURE 5.6 Experimental results of cart position control system with MFAC-CFDL algorithm.

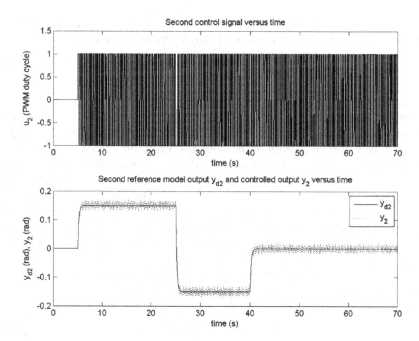

FIGURE 5.7 Experimental results of arm angular position control system with MFAC-CFDL algorithm.

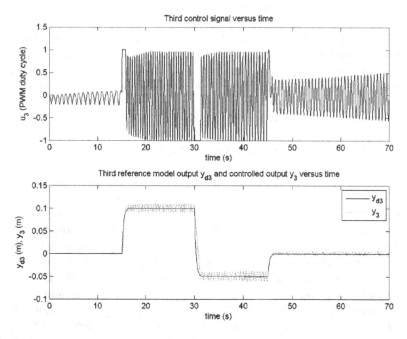

FIGURE 5.8 Experimental results of payload position control system with MFAC-CFDL algorithm.

first parameter that must be chosen in MFAC-PFDL algorithm is the value of $L = 3$, next the parameter vector of the MFAC-PFDL algorithm becomes

$$\chi_{(\psi)} = [\eta \ \lambda \ \mu \ \rho_1 \ \rho_2 \ \rho_3 \ \hat{\phi}_1(1) \ \hat{\phi}_2(1) \ \hat{\phi}_3(1)]^T, \tag{5.74}$$

where $\psi \in \{c,a,p\}$, i.e., one parameter for each controller.

Similar as in the previous version of MFAC algorithm, the controllers are nonlinear since saturation-type nonlinearities are inserted on the three controller outputs to match the power electronics of the actuators, which operate on the basis of the PWM principle. As shown in Chapter 1, the three control signals (or control inputs) are within −1 and 1, ensuring the necessary connection to the tower crane system subsystem, which models the controlled process in the control system structures developed in this chapter and the next chapters and also includes three ZOH blocks to enable digital control. The tower crane system subsystem belongs to the Process. mdl Simulink diagram, which is included in the accompanying Matlab & Simulink programs given in Chapter 1.

First, the value of the initial value of the estimate of the PPD matrix $\hat{\phi}_{(\psi)}(1) = [\hat{\phi}_1(1) \ \hat{\phi}_2(1) \ \hat{\phi}_3(1)]^T$ is chosen according to step 2 as follows:

$$\hat{\phi}_{(c)}(1) = [1.8 \ 0.91 \ 1.9]^T, \tag{5.75}$$

$$\hat{\phi}_{(a)}(1) = [3.2 \ 1 \ 0.6]^T, \tag{5.76}$$

$$\hat{\phi}_{(p)}(1) = [1.9 \ 0.91 \ 1.9]^T. \tag{5.77}$$

Next, the parameter of the lower limit $\beta = 0.001\hat{\phi}_{(\psi)}(1)$ is imposed according to step 3 ensuring the condition (5.13). After that, the rest of the parameters, namely $\eta \in (0,1)$, $\lambda > 0$, $\mu > 0$ and $\rho = [\rho_1 \ \rho_2 \ \rho_3]^T$, with $\rho_1 > 0$, $\rho_2 > 0$ and $\rho_3 > 0$, are chosen according to step 4 resulting in the parameter vector of the MFAC-PFDL algorithm

$$\chi_{(c)} = [0.5 \ 0.01 \ 0.05 \ 450 \ 0.3 \ 3.3 \ 1.8 \ 0.91 \ 1.9]^T, \tag{5.78}$$

$$\chi_{(a)} = [0.99 \ 0.02 \ 0.49 \ 150 \ 0.3 \ 10.4 \ 3.2 \ 1 \ 0.6]^T, \tag{5.79}$$

$$\chi_{(p)} = [0.5 \ 0.01 \ 0.05 \ 260 \ 0.3 \ 12.3 \ 1.9 \ 0.91 \ 1.9]^T. \tag{5.80}$$

The details of steps 1–3 are implemented in the MFAC _ PFDL _ c.m Matlab program for cart position control, the MFAC _ PFDL _ a.m Matlab program for arm angular position control and the MFAC _ PFDL _ p.m Matlab program for payload position control, which are included in the accompanying Matlab & Simulink programs. The simulations of the control systems with the three MFAC-PFDL algorithms are conducted using the CS _ MFAC _ PFDL _ SISO _ c.mdl Simulink diagram for cart position control, the CS _ MFAC _ PFDL _ SISO _ a.mdl Simulink diagram for arm angular position control and the CS _ MFAC _ PFDL _ SISO _ p. mdl Simulink diagram for payload position control. These three Simulink diagrams make use of the state-space model of the tower crane system viewed as a controlled process, which is described in Chapter 1 and is implemented in the process _ model.m S-function.

The simulation results obtained for the control systems with MFAC-PFDL algorithms with the parameters in (5.78–5.80) with $\beta = 0.001\hat{\phi}_{(\psi)}(1)$ as the lower limit are illustrated in Figure 5.9 for cart position control, Figure 5.10 for arm angular position control and Figure 5.11 for payload position control.

The experimental results obtained for the control systems with MFAC-PFDL algorithms with the parameters in (5.78–5.80) with $\beta = 0.001\hat{\phi}_{(\psi)}(1)$ as the lower limit are illustrated in Figure 5.12 for cart position control, Figure 5.13 for arm angular position control and Figure 5.14 for payload position control.

5.3.2 MIMO CONTROL SYSTEMS

Two types of MIMO control system structures are considered, namely one that consists of three SISO control loops that are running in parallel for MFAC-CFDL and MFAC-PFDL algorithms, and a MIMO control system structure based on a MFAC-CFDL controller considering the MFAC algorithms presented in Section 5.2. The performance is measured using the following objective function:

$$J(\chi) = \frac{1}{2}E\left\{\sum_{k=1}^{N}\left[\mathbf{e}^{T}(k,\chi)\mathbf{e}(k,\chi)\right]\right\}, \tag{5.81}$$

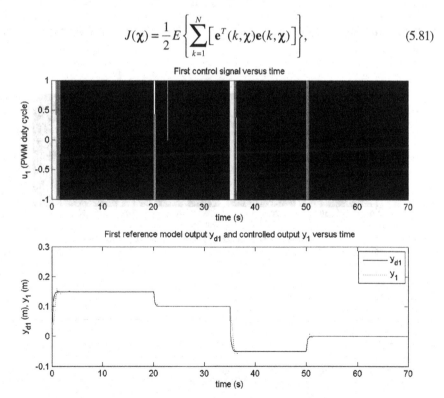

FIGURE 5.9 Simulation results of cart position control system with MFAC-PFDL algorithm.

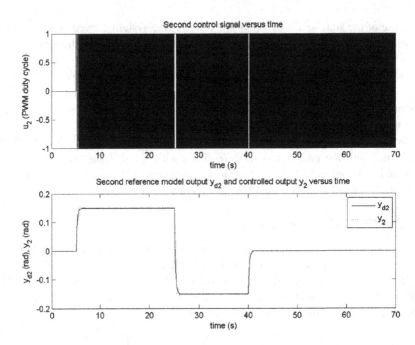

FIGURE 5.10   Simulation results of arm angular position control system with MFAC-PFDL algorithm.

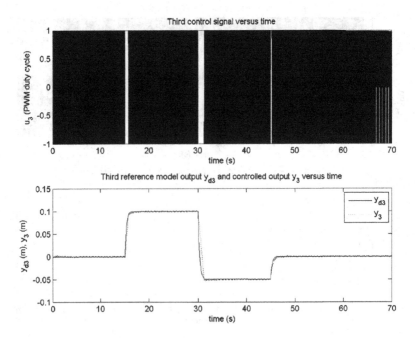

FIGURE 5.11   Simulation results of payload position control system with MFAC-PFDL algorithm.

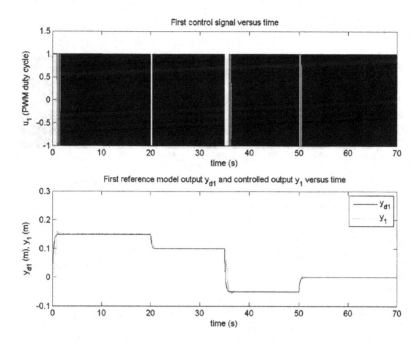

FIGURE 5.12 Experimental results of cart position control system with MFAC-PFDL algorithm.

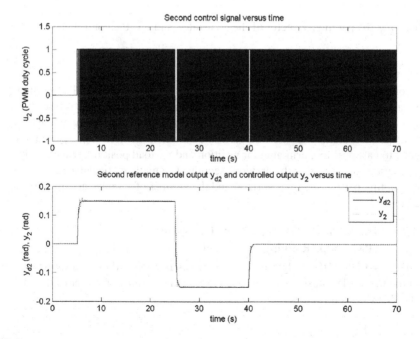

FIGURE 5.13 Experimental results of arm angular position control system with MFAC-PFDL algorithm.

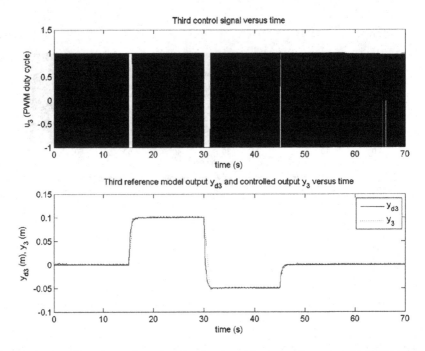

FIGURE 5.14 Experimental results of payload position control system with MFAC-PFDL algorithm.

where χ is the parameter vector of the controller, and $\mathbf{e}$ is the control error vector

$$\mathbf{e} = \mathbf{y}_d - \mathbf{y} = [e_1 \ e_2 \ e_3]^T = [y_{d1} - y_1 \ y_{d2} - y_2 \ y_{d3} - y_3]^T, \quad (5.82)$$

the control errors e_1, e_2 and e_3 are related to cart position, arm angular position and payload position, the controlled outputs y_1, y_2 and y_3 are the cart position, the arm angular position and the payload position, respectively, and the reference inputs (i.e., the set-points obtained as the output of the reference models) y_{d1}, y_{d2} and y_{d3} are related to cart position, arm angular position and payload position, respectively, and are generated according to the previous section. The mathematical expectation $E\{\Xi\}$, the sampling period, the time horizon and the number of samples are the same as in the SISO case.

5.3.2.1 The MFAC-CFDL Algorithms Using Three SISO Loops Running in Parallel

A MFAC-CFDL MIMO algorithm that consists of three SISO algorithms that are running in parallel is used as a controller, where the parameter vector of the MFAC-CFDL MIMO algorithm is

$$\chi = [\chi_{(c)} \ \chi_{(a)} \ \chi_{(p)}]^T$$

$$= [\eta_c \ \lambda_c \ \mu_c \ \rho_c \ \hat{\varphi}_c(1) \ \eta_a \ \lambda_a \ \mu_a \ \rho_a \ \hat{\varphi}_a(1) \ \eta_p \ \lambda_p \ \mu_p \ \rho_p \ \hat{\varphi}_p(1)]^T. \quad (5.83)$$

First, the value of the initial value of the estimates of the PPD element $\hat{\varphi}_{(\psi)}(1) = [\hat{\varphi}_c(1) \;\; \hat{\varphi}_a(1) \;\; \hat{\varphi}_p(1)]^T$ is chosen according to step 1

$$\hat{\varphi}_{(c)}(1) = 3.5, \tag{5.84}$$

$$\hat{\varphi}_{(a)}(1) = 12, \tag{5.85}$$

$$\hat{\varphi}_{(p)}(1) = 0.89. \tag{5.86}$$

Next, the parameters of lower and the upper limits will be considered as $1{,}000\,\hat{\varphi}_{(\psi)}(1)$ for the upper limit and $0.001\hat{\varphi}_{(\psi)}(1)$ for the lower limit and are imposed according to step 2 by ensuring the conditions (5.3) and (5.6). Finally, the remaining parameters, $\eta \in (0,1)$, $\lambda > 0$, $\mu > 0$ and $\rho > 0$, are chosen according to step 3 resulting in the parameter vector of the MFAC-CFDL algorithm leading to the parameter vector

$$\begin{aligned} \chi &= [\chi_{(c)} \;\; \chi_{(a)} \;\; \chi_{(p)}]^T \\ &= [0.6 \; 3.5 \; 28 \; 45 \; 3.5 \; 0.9 \; 0.01 \; 1 \; 990 \; 12 \; 0.81 \; 0.71 \; 14.9 \; 25.2 \; 0.89]^T. \end{aligned} \tag{5.87}$$

The details of steps 1–3 are implemented in the MFAC _ CFDL _ 3SISO.m Matlab program where the Matlab & Simulink programs are included. The experiment MFAC-CFDL algorithm is conducted by digital simulations using the CS _ MFAC _ CFDL _ 3SISO.mdl Simulink diagram. The Simulink diagram makes use of the state-space model of the tower crane system viewed as a controlled process, which is described in Chapter 1 and is implemented in the process _ model.m S-function.

The simulation results obtained for the MIMO control system MFAC-CFDL algorithm with the parameters in (5.87) with $1{,}000\,\hat{\varphi}(1)_{(\psi)}$ as the upper limit and $0.001\hat{\varphi}_{(\psi)}(1)$ as the lower limit are illustrated in Figure 5.15 for cart position control, Figure 5.16 for arm angular position control and Figure 5.17 for payload position control.

The experimental results obtained for the MIMO control system with MFAC-CFDL algorithm with the parameters in (5.87) with $1{,}000\hat{\varphi}(1)_{(\psi)}$ as the upper limit and $0.001\hat{\varphi}(1)_{(\psi)}$ as the lower limit are illustrated in Figure 5.18 for cart position control, Figure 5.19 for arm angular position control and Figure 5.20 for payload position control.

5.3.2.2 The MFAC-CFDL Algorithms Using a Single Loop

A MFAC-CFDL MIMO algorithm using a single MIMO MFAC controller is further used, where the parameter vector of this MFAC-CFDL MIMO algorithm is

$$\chi = \left[\eta \; \lambda \; \mu \; \rho \; \hat{\varphi}_{11}(1) \; \hat{\varphi}_{12}(1) \; \hat{\varphi}_{13}(1) \; \hat{\varphi}_{21}(1) \; \hat{\varphi}_{22}(1) \; \hat{\varphi}_{23}(1) \; \hat{\varphi}_{31}(1) \; \hat{\varphi}_{32}(1) \; \hat{\varphi}_{33}(1) \right]^T. \tag{5.88}$$

First, the value of the initial value of the estimates of the PPD matrix

$$\hat{\varphi}(1) = \begin{bmatrix} \hat{\varphi}_{11}(1) & \hat{\varphi}_{12}(1) & \hat{\varphi}_{13}(1) \\ \hat{\varphi}_{21}(1) & \hat{\varphi}_{22}(1) & \hat{\varphi}_{23}(1) \\ \hat{\varphi}_{31}(1) & \hat{\varphi}_{32}(1) & \hat{\varphi}_{33}(1) \end{bmatrix}$$ is chosen as follows according to step 1:

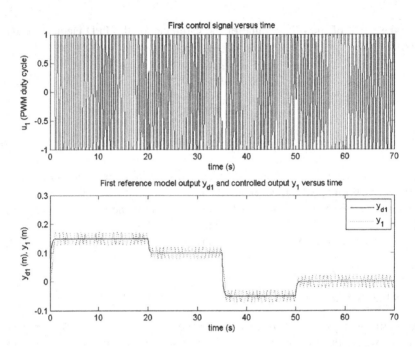

FIGURE 5.15 Simulation results of cart position control system with MFAC-CFDL algorithm with three SISO loops running in parallel.

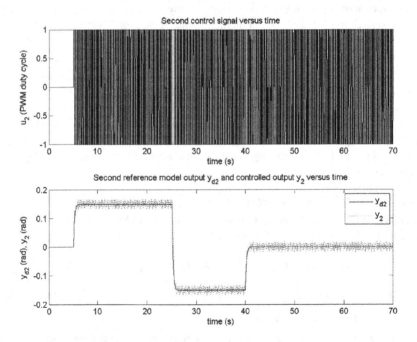

FIGURE 5.16 Simulation results of arm angular position control system with MFAC-CFDL algorithm with three SISO loops running in parallel.

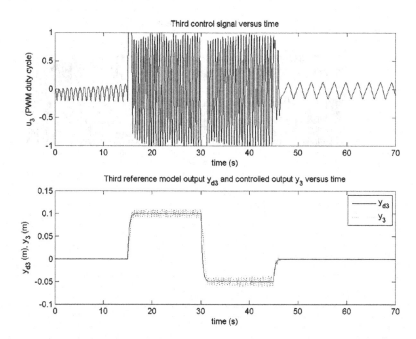

FIGURE 5.17 Simulation results of payload position control system with MFAC-CFDL algorithm with three SISO loops running in parallel.

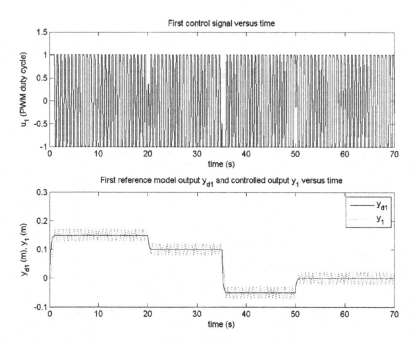

FIGURE 5.18 Experimental results of cart position control system with MFAC-CFDL algorithm with three SISO loops running in parallel.

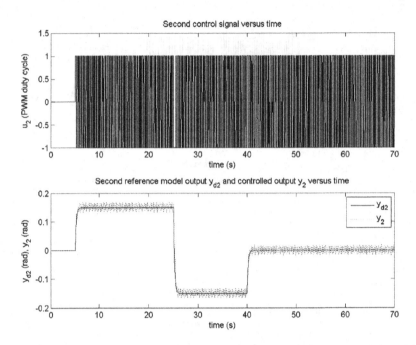

FIGURE 5.19 Experimental results of arm angular position control system with MFAC-CFDL algorithm with three SISO loops running in parallel.

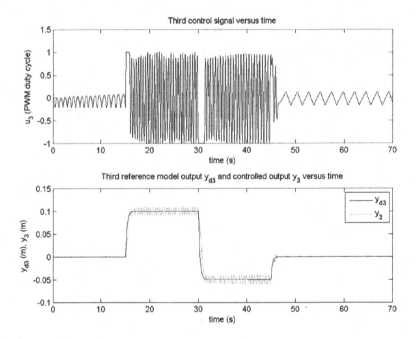

FIGURE 5.20 Experimental results of payload position control system with MFAC-CFDL algorithm with three SISO loops running in parallel.

$$\hat{\varphi}(1) = \begin{bmatrix} 3.5 & 0 & 0 \\ 0 & 12 & 0 \\ 0 & 0 & 0.89 \end{bmatrix}. \tag{5.89}$$

Next, the parameters will be considered as $1,000\,\hat{\varphi}(1)$ for the upper limit and $0.001\,\hat{\varphi}(1)$ for the lower limit and are imposed according to step 2 by ensuring the conditions (5.3) and (5.6). Finally, the remaining parameters, i.e., $\eta \in (0,1)$, $\lambda > 0$, $\mu > 0$ and $\rho > 0$, are chosen according to step 3 resulting in the parameter vector of the MFAC-CFDL algorithm

$$\chi = [0.6\ 0.11\ 0.19\ 951\ 3.5\ 0\ 0\ 0\ 12\ 0\ 0\ 0\ 0.89]^T. \tag{5.90}$$

The details of steps 1–3 are implemented in the MFAC _ CFDL.m Matlab program where the Matlab & Simulink programs are included. The experiment MFAC-CFDL algorithm is conducted by digital simulations using the CS _ MFAC _ CFDL _ MIMO.mdl Simulink diagram. The Simulink diagram makes use of the state-space model of the tower crane system viewed as a controlled process, which is described in Chapter 1 and is implemented in the process _ model.m S-function.

The simulation results obtained for the MIMO control system with MFAC-CFDL algorithm with the parameters in (5.90) with $1,000\,\hat{\varphi}(1)$ as the upper limit and $0.001\,\hat{\varphi}(1)$ as the lower limit are illustrated in Figure 5.21 for cart position control, Figure 5.22 for arm angular position control and Figure 5.23 for payload position control.

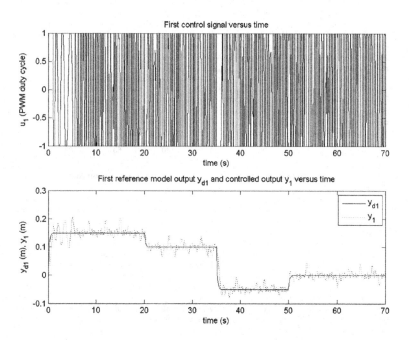

FIGURE 5.21 Simulation results of cart position control system with MFAC-CFDL algorithm using a single loop.

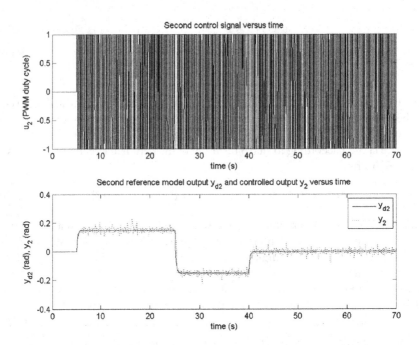

FIGURE 5.22   Simulation results of arm angular position control system with MFAC-CFDL algorithm using a single loop.

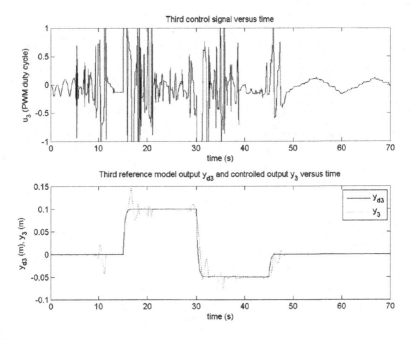

FIGURE 5.23   Simulation results of payload position control system with MFAC-CFDL algorithm using a single loop.

The experimental results obtained for the MIMO control system with MFAC-CFDL algorithm with the parameters in (5.90) with $1,000\,\hat{\varphi}(1)$ as the upper limit and $0.001\,\hat{\varphi}(1)$ as the lower limit are illustrated in Figure 5.24 for cart position control, Figure 5.25 for arm angular position control and Figure 5.26 for payload position control.

5.3.2.3 The MFAC-PFDL Algorithms Using Three SISO Loops Running in Parallel

Three MFAC-PFDL MIMO algorithms are implemented as three SISO controllers running in parallel. Step 1 of MFAC-PFDL algorithm mentions that the first parameter that must be chosen in MFAC-PFDL algorithm is the value of $L = 3$; after that, the parameter vector of the MFAC-PFDL algorithm is established

$$\chi = [\chi_c\ \chi_a\ \chi_p]^T$$

$$= \begin{bmatrix} \eta_c\ \lambda_c\ \mu_c\ \rho_{c1}\ \rho_{c2}\ \rho_{c3}\ \hat{\varphi}_{c1}(1)\ \hat{\varphi}_{c2}(1)\ \hat{\varphi}_{c3}1)\ \eta_a\ \lambda_a\ \mu_a \\ \rho_{a1}\ \rho_{a2}\ \rho_{a3}\ \hat{\varphi}_{a1}(1)\ \hat{\varphi}_{a2}(1)\ \hat{\varphi}_{a3}(1)\ \eta_p\ \lambda_p\ \mu_p\ \rho_{p1}\ \rho_{p2}\ \rho_{p3}\ \hat{\varphi}_{p1}(1)\ \hat{\varphi}_{p2}(1)\ \hat{\varphi}_{p3}(1) \end{bmatrix}^T.$$

(5.91)

The controller is nonlinear since saturation-type nonlinearities are inserted on the three controller outputs to match the power electronics of the actuators, which

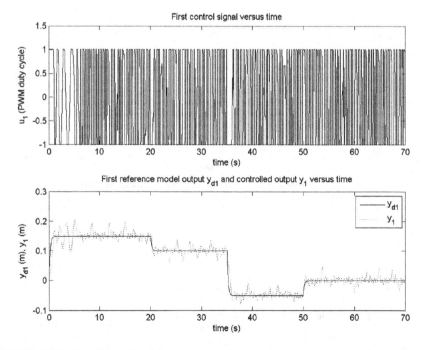

FIGURE 5.24 Experimental results of cart position control system with MFAC-CFDL algorithm using a single loop.

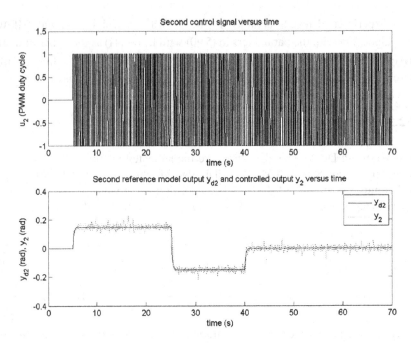

FIGURE 5.25 Experimental results of arm angular position control system with MFAC-CFDL algorithm using a single loop.

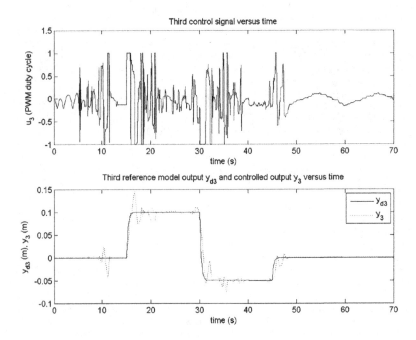

FIGURE 5.26 Experimental results of payload position control system with MFAC-CFDL algorithm using a single loop.

operate on the basis of the PWM principle. As shown in Chapter 1, the three control signals (or control inputs) are within −1 and 1, ensuring the necessary connection to the tower crane system subsystem, which models the controlled process in the control system structures developed in this chapter and the next chapters and also includes three ZOH blocks to enable digital control. The tower crane system subsystem belongs to the Process.mdl Simulink diagram, which is included in the accompanying Matlab & Simulink programs given in Chapter 1.

Next, the initial value of the estimate of the PPD matrix $\hat{\boldsymbol{\varphi}}_{(\psi)}(1) = \begin{bmatrix} \hat{\varphi}_{c1}(1) & \hat{\varphi}_{c2}(1) & \hat{\varphi}_{c3}(1) & \hat{\varphi}_{a1}(1) & \hat{\varphi}_{a2}(1) & \hat{\varphi}_{a3}(1) & \hat{\varphi}_{p1}(1) & \hat{\varphi}_{p2}(1) & \hat{\varphi}_{p3}(1) \end{bmatrix}^T$ is chosen according to step 2

$$\hat{\boldsymbol{\varphi}}_{(\psi)}(1) = [1.8 \ 0.91 \ 1.9 \ 3.2 \ 1 \ 0.6 \ 0.5 \ 1.8 \ 1]^T. \tag{5.92}$$

After that, the value of the parameter of the lower limit $\beta = 0.001\hat{\boldsymbol{\varphi}}_{(\psi)}(1)$ is imposed according to step 3 to ensure the condition (5.13). Next, the remaining parameters $\eta \in (0,1), \lambda > 0, \mu > 0$ and $\rho = [\rho_1 \ \rho_2 \ \rho_3]^T$, with $\rho_1 > 0, \rho_2 > 0$ and $\rho_3 > 0$, are chosen according to step 4 resulting in the parameter vector of the MFAC-PFDL algorithm

$$\boldsymbol{\chi} = [\chi_c \ \chi_a \ \chi_p]^T = [0.5 \ 0.01 \ 0.05 \ 450 \ 0.3 \ 3.3 \ 1.8 \ 0.91 \ 1.9 \ 0.99 \ 0.02 \ 0.49$$

$$150 \ 0.3 \ 10.4 \ 3.2 \ 1 \ 0.6 \ 0.1 \ 1.34 \ 8 \ 600 \ 0.25 \ 3.01 \ 0.5 \ 1.8 \ 1]^T. \tag{5.93}$$

The details of steps 1–3 are implemented in the MFAC _ PFDL _ 3SISO.m Matlab program where Matlab & Simulink programs are included. The simulation of the control system with the MFAC-PFDL algorithm is conducted using the CS _ MFAC _ PFDL _ 3SISO.mdl Simulink diagram for tower crane control.

The simulation results obtained for the MIMO control system with MFAC-PFDL algorithm with the parameters in (5.93) with $\beta = 0.001\hat{\boldsymbol{\varphi}}_{(\psi)}(1)$ as the lower limit are illustrated in Figure 5.27 for cart position control, Figure 5.28 for arm angular position control and Figure 5.29 for payload position control.

The experimental results obtained for the MIMO control system with MFAC-PFDL algorithm with the parameters in (5.93) with $\beta = 0.001\hat{\boldsymbol{\varphi}}_{(\psi)}(1)$ as the lower limit are illustrated in Figure 5.30 for cart position control, Figure 5.31 for arm angular position control and Figure 5.32 for payload position control.

5.3.2.4 Simulation and Experimental Results for the Continuous-Time MFAC Algorithm

This section gives the simulation and the experimental results concerning the MFAC algorithm given in Section 5.2.3 for continuous-time dynamic systems. The simulation and the experiments are conducted on the MIMO control system for tower crane position control, where three parallel control loops are implemented corresponding to the three dimensions of the process.

The tuning parameters of the MFAC controller for each dimension are set as $\mathbf{Q} = 1, \gamma_1 = 100, \gamma_2 = 1$ and $\rho_1 = \rho_2 = 1$. The plots for the control signals, tracking

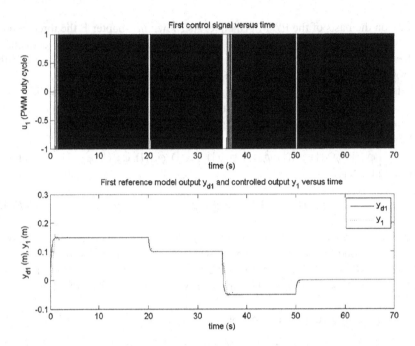

FIGURE 5.27 Simulation results of cart position control system with MFAC-PFDL algorithm with three SISO loops running in parallel.

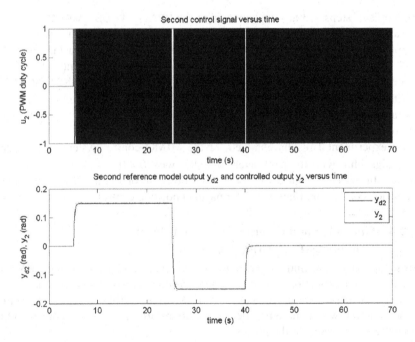

FIGURE 5.28 Simulation results of arm angular position control system with MFAC-PFDL algorithm with three SISO loops running in parallel.

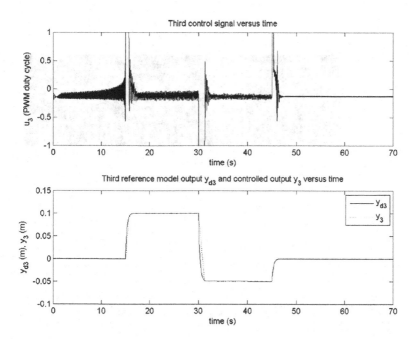

FIGURE 5.29 Simulation results of payload position control system with MFAC-PFDL algorithm with three SISO loops running in parallel.

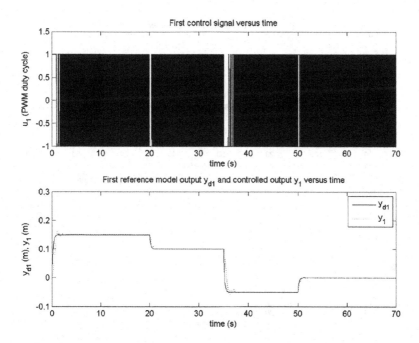

FIGURE 5.30 Experimental results of cart position control system with MFAC-PFDL algorithm with three SISO loops running in parallel.

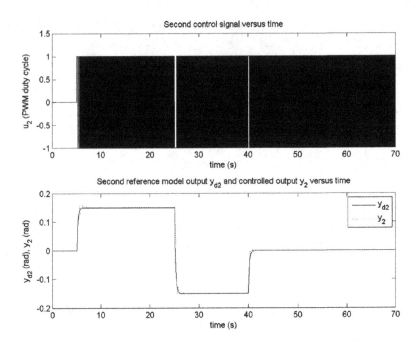

FIGURE 5.31 Experimental results of arm angular position control system with MFAC-PFDL algorithm with three SISO loops running in parallel.

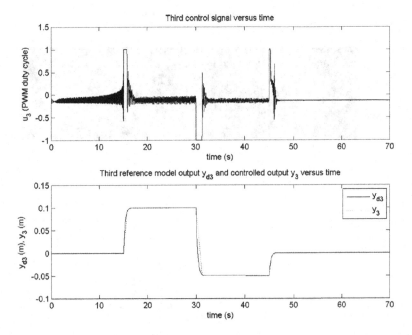

FIGURE 5.32 Experimental results of payload position control system with MFAC-PFDL algorithm with three SISO loops running in parallel.

performance and the online parameter estimations of the unknown linear and nonlinear terms as well as the updated values of the main controller gains (recalling the algorithm proposed in (5.26–5.28)) are illustrated in Figures 5.33–5.35. The corresponding value of the objective function (or the cost function) for this simulation is $J = 9.9365$.

The details of the controller design and tuning are given in the MFAC _ AMFC.m Matlab program. The simulation of the control system with the continuous-time MFAC algorithm is conducted using the CS _ MIMO _ AMFC _ R2007.mdl Simulink diagram.

The experimental results expressed as the control signals and the tracking performance (recalling the algorithm proposed in (5.26–5.28)) are illustrated in Figures 5.36–5.38. The corresponding value of the objective function (or the cost function) for this experiment is $J = 9.9426$.

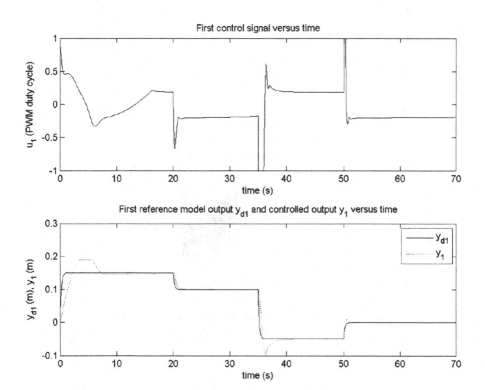

FIGURE 5.33 Simulation results: control signal and tracking performance for the first dimension of the process, using continuous MFAC.

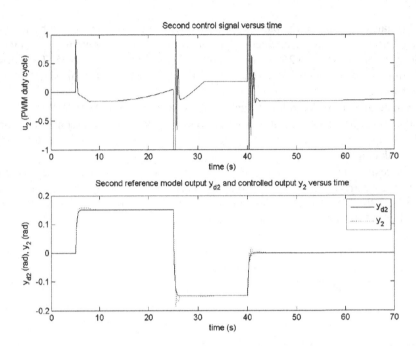

FIGURE 5.34 Simulation results: control signal and tracking performance for the second dimension of the process, using continuous MFAC.

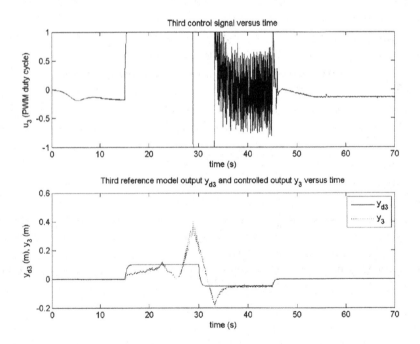

FIGURE 5.35 Simulation results: control signal and tracking performance for the third dimension of the process, using continuous MFAC.

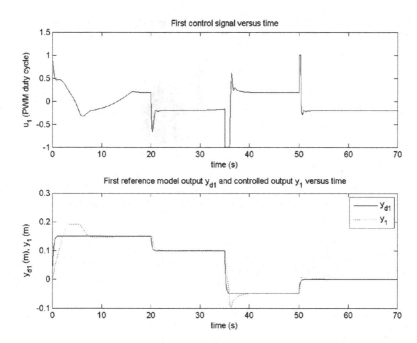

FIGURE 5.36 Experimental results: control signal and tracking performance for the first dimension of the process, using continuous MFAC.

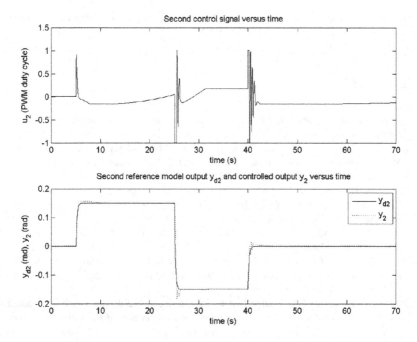

FIGURE 5.37 Experimental results: control signal and tracking performance for the second dimension of the process, using continuous MFAC.

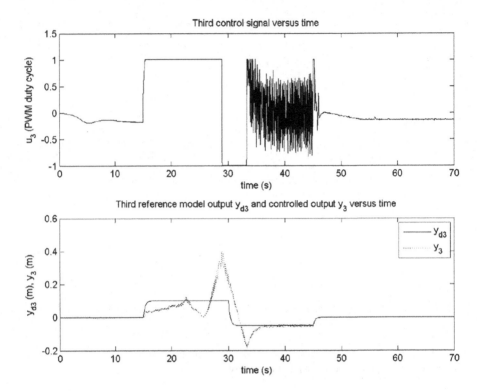

FIGURE 5.38 Experimental results: control signal and tracking performance for the third dimension of the process, using continuous MFAC.

REFERENCES

[Ata99] A. N. Atassi and H. Khalil, "A separation principle for the stabilization of a class of nonlinear systems," *IEEE Transactions on Automatic Control*, vol. 44, no. 9, pp. 1672–1687, Sep. 1999.

[Hou11a] Z.-S. Hou and S.-T. Jin, "Data-driven model-free adaptive control for a class of MIMO nonlinear discrete-time systems," *IEEE Transactions on Neural Networks*, vol. 22, no. 12, pp. 2173–2188, Dec. 2011.

[Hou11b] Z.-S. Hou and S.-T. Jin, "A novel data-driven control approach for a class of discrete-time nonlinear systems," *IEEE Transactions on Control Systems Technology*, vol. 19, no. 6, pp. 1549–1558, Nov. 2011.

[Hou13a] Z.-S. Hou and Z. Wang, "From model-based control to data-driven control: Survey, classification and perspective," *Information Sciences*, vol. 235, pp. 3–35, Jun. 2013.

[Hou17] Z.-S. Hou, R. Chi and H.-J. Gao, "An overview of dynamic linearization based data-driven control and applications," *IEEE Transactions on Industrial Electronics*, vol. 64, no. 5, pp. 4076–4090, May 2017.

[Rom14] R.-C. Roman, M.-B. Radac and R.-E. Precup, "Data-driven model-free adaptive control of twin rotor aerodynamic systems," in *Proceedings of IEEE 9th International Symposium on Applied Computational Intelligence and Informatics*, Timisoara, Romania, 2014, pp. 25–30.

[Rom15] R.-C. Roman, M.-B. Radac, R.-E. Precup and A.-I. Stinean, "Two data-driven control algorithms for a MIMO aerodynamic system with experimental validation," in *Proceedings of 19th International Conference on System Theory, Control and Computing Joint Conference*, Cheile Gradistei, Romania, 2015, pp. 736–741.

[Rom16] R.-C. Roman, M.-B. Radac and R.-E. Precup, "Multi-input-multi-output system experimental validation of model-free control and virtual reference feedback tuning techniques," *IET Control Theory & Applications*, vol. 10, no. 12, pp. 1395–1403, Aug. 2016.

[Rom18] R.-C. Roman, *Tehnici de tip model-free de acordare a parametrilor regulatoarelor automate* (in Romanian), PhD thesis, Editura Politehnica, Timisoara, 2018.

[Rom19] R.-C. Roman, R.-E. Precup, C.-A. Bojan-Dragos and A.-I. Szedlak-Stinean, "Combined model-free adaptive control with fuzzy component by virtual reference feedback tuning for tower crane systems," *Procedia Computer Science*, vol. 162, pp. 267–274, Oct. 2019.

[Saf18a] A. Safaei and M. N. Mahyuddin, "Adaptive model-free control based on an ultra-local model with model-free parameter estimations for a generic SISO system," *IEEE Access*, vol. 6, pp. 4266–4275, Jan. 2018.

[Saf18b] A. Safaei and M. N. Mahyuddin, "Optimal model-free control for a generic MIMO nonlinear system with application to autonomous mobile robots," *International Journal of Adaptive Control and Signal Processing*, vol. 32, no. 6, pp. 792–815, Jun. 2018.

6 Hybrid Model-Free and Model-Free Adaptive Virtual Reference Feedback Tuning Controllers

6.1 INTRODUCTION

A major shortcoming of the control algorithms developed based on the model-free control (MFC) techniques and model-free adaptive control (MFAC) techniques, regardless of the chosen variant (CFDL/PFDL/FFDL), is that they may have a lot of parameters to be tuned, and their tuning is difficult in the absence of dedicated steps for tuning the parameters. On the other hand, the major advantage of MFC algorithms is that the tuning of the parameters requires the real-time measured input/output data, a linear online approximation of the process and an estimator to update the linear approximation. Another advantage proven in [Rad14, Rom16a and Rom16b] is that the stability of the control system structure is guaranteed if certain inequality conditions are met [Rom18a]. In this regard, the two advantages of the MFAC algorithm are [Rom18a] as follows:

- the algorithm uses real-time measured input/output data, and
- it guarantees the stability of the control system structure according to [Hou11a] through the estimation mechanism leading to the PPD matrix, the reset conditions and the control law.

The virtual reference feedback tuning (VRFT) is a data-driven tuning technique, in which the input/output data related to the unknown open-loop process are measured, and using them, the controller parameters are calculated. The major disadvantage of VRFT is that it does not guarantee the stability of the control system structure [Rom18a].

Considering that MFC and VRFT and MFAC and VRFT have complementary characteristics, the two techniques were combined in [Rom16a, Rom17c and Rom18b] to benefit from the advantages of MFC and VRFT techniques, and the MFAC and VRFT techniques were combined in [Rom16c and Rom16d] to benefit from the advantages of both techniques. According to [Rom18a], the resulting new techniques are abbreviated as MFC-VRFT and MFAC-VRFT.

DOI: 10.1201/9781003143444-6

6.2 THEORY

6.2.1 THE VRFT TECHNIQUE

The formulation of VRFT for nonlinear systems is based on the output of the control system structure (closed-loop controlled process) and the output of a linear or nonlinear reference model. The VRFT technique for nonlinear systems uses a single open-loop experiment, in which a signal with a wide frequency range is applied as a stable nonlinear process input, and then the input/output signals are measured to be used in computing the controller parameters [Cam05, Cam06, Yan16, Rom16a, Rom16c, Rom16d, Rom18a].

Definition 6.1

For model reference tracking, the objective function used in VRFT for nonlinear systems is defined as follows [Yan16]:

$$J_{MR}(\chi) = \sum_{k=1}^{N} \| \mathbf{y}_\chi(k) - \mathbf{y}_d(k) \|^2, \tag{6.1}$$

with $\mathbf{y}_d(k) \in \mathfrak{R}^n$ – the reference model output vector, i.e., the desired output vector of the control system, and $\mathbf{y}_\chi(k) \in \mathfrak{R}^n$ – the (controlled) output vector of the nonlinear process, with the same number n of inputs and outputs, where $\mathbf{y}_\chi(k+1) \in \mathfrak{R}^n$ is obtained using the model

$$\mathbf{y}_\chi(k+1) = \mathbf{f}(\mathbf{y}_\chi(k),\ldots,\mathbf{y}_\chi(k - n_{yp}), \mathbf{u}_\chi(k),\ldots,\mathbf{u}_\chi(k - n_{up})), \tag{6.2}$$

which represents the local model, where $\mathbf{f} : \mathfrak{R}^{n(n_{yp}+n_{up}+2)} \to \mathfrak{R}^n$ is an unknown nonlinear vector function of vector variable, the control input (or the control signal) vector $\mathbf{u}_\chi(k) \in \mathfrak{R}^n$ is produced by the nonlinear controller, with fixed structure, called VRFT controller, $\chi \in \mathfrak{R}^{n_{\chi C}}$ is the parameter vector of the VRFT controller, $n_{\chi C}$ is the number of parameters of the VRFT controller, n_{yp} and n_{up} are the maximum known delay orders related to the controlled output and the control input, respectively, $\mathbf{C}_\chi : \mathfrak{R}^{n_{\chi C}+n(n_{uc}+n_{ec}+1)} \to \mathfrak{R}^n$ is a nonlinear vector function of a vector variable, and the control input vector is obtained from the model in terms of

$$\mathbf{u}_\chi(k) = \mathbf{C}_\chi(\chi, \mathbf{u}_\chi(k-1),\ldots,\mathbf{u}_\chi(k - n_{uc}), \mathbf{e}(k),\ldots,\mathbf{e}(k - n_{ec})), \tag{6.3}$$

where n_{uc} and n_{ec} are the maximum known delay orders related to the control input vector and the control error vector $\mathbf{e}(k) \in \mathfrak{R}^n$.

Definition 6.2

The following simplified notation is used or the VRFT nonlinear controller in the particular case $n_{uc} = 1$ and $n_{ec} = 0$, with $\mathbf{C}_\chi : \mathfrak{R}^{n_{\chi C}+2n} \to \mathfrak{R}^n$:

$$\mathbf{u}_\chi(k) = \mathbf{C}_\chi(\chi, \mathbf{u}_\chi(k-1), \mathbf{e}(k)). \tag{6.4}$$

The control error (or the tracking error) vector $\mathbf{e}(k) \in \mathfrak{R}^n$ has the expression

$$\mathbf{e}(k) = \mathbf{r}(k) - \mathbf{y}_\chi(k), \tag{6.5}$$

where $\mathbf{r}(k) \in \mathfrak{R}^n$ is the reference input vector applied to the control system structure. The reference model output vector $\mathbf{y}_d(k)$ is obtained as follows from the nonlinear reference model $\mathbf{m}_{RM}$ of order n_{ym} and n_{rm} selected by the control systems designer:

$$\mathbf{y}_d(k) = \mathbf{m}_{RM}(\mathbf{y}_d(k-1), \dots, \mathbf{y}_d(k - n_{ym}), \mathbf{r}(k-1), \dots, \mathbf{r}(k - n_{rm})). \tag{6.6}$$

According to (6.6), $\mathbf{m}_{RM} : \mathfrak{R}^{n(n_{ym} + n_{rm})} \to \mathfrak{R}^n$ is actually a nonlinear vector function of vector variable. Viewed as a square matrix $\mathbf{m} \in \mathfrak{R}^{n \times n}$, which is right-handed multiplied by $\mathbf{r}$ to give $\mathbf{y}_d$, i.e., $\mathbf{y}_d = \mathbf{m}\,\mathbf{r}$, the reference model (matrix) $\mathbf{m}$ should be nonsingular [Rom16a, Rom16c, Rom16d, Rom18a].

Remark 6.1

In the VRFT design, it is accepted that the input/output data pairs $\{\mathbf{u}_\chi(k), \mathbf{y}(k)\}$, with $k = 1 \dots N$, are available by measurement from the open-loop process. The virtual reference input vector $\bar{\mathbf{r}}(k)$ is calculated as

$$\bar{\mathbf{r}}(k) = \mathbf{m}_{RM}^{-1}(\mathbf{y}(k)), \tag{6.7}$$

so that the output of the reference model and the output of the control system have similar trajectories. The notation $\mathbf{m}_{RM}^{-1}(\mathbf{y}(k)) \in \mathfrak{R}^{n \times 1}$ is used in (6.7) to indicate that it represents a column matrix. In the context of considering the reference model characterize, as above, as the nonsingular square matrix $\mathbf{m} \in \mathfrak{R}^{n \times n}$ and $\mathbf{y}_d$ obtained as $\mathbf{y}_d = \mathbf{m}\,\mathbf{r}$, (6.7) is equivalent to $\bar{\mathbf{r}} = \mathbf{m}^{-1}\mathbf{y}$.

Definition 6.3

In the VRFT controller, the virtual control error (or the virtual tracking error) vector $\bar{\mathbf{e}}(k)$ has the expression

$$\bar{\mathbf{e}}(k) = \bar{\mathbf{r}}(k) - \mathbf{y}(k). \tag{6.8}$$

The VRFT controller works with the input vector $\bar{\mathbf{e}}(k)$ and elaborates the control input (or the control signal) vector $\mathbf{u}_\chi(k)$ with the objective of tracking $\mathbf{y}_d(k) \in \mathfrak{R}^n$. The parameters of this controller are calculated such that to ensure the minimization of the objective function [Yan16]

$$J_{\text{VRFT}}(\chi) = \frac{1}{N} \sum_{k=1}^{N} \left\| \mathbf{C}_\chi(\chi, \mathbf{u}_\chi(k-1), \bar{\mathbf{e}}(k)) - \mathbf{u}_\chi(k) \right\|^2. \tag{6.9}$$

Remark 6.2

In the case of single input-single output (SISO) VRFT, a time-varying filter is required to obtain approximately equal values of the objective functions $J_{MR}(\chi)$ in (6.1) and $J_{VRFT}(\chi)$ in (6.9). These objective functions are obtained by minimizing the sum of the squares of the errors at each sample. According to [Yan16], such a filter is not required in the multi input-multi output (MIMO) case. Values of the two objective functions can be obtained conveniently in the case of controllers with a large number of parameters such as, for example, artificial neural networks [Yan16, Esp11]. Details regarding the particularization of the results from this paragraph in the case of the nonlinear SISO VRFT technique are presented in [Rom16a, Rom16c and Rom16d].

In Sections 6.2.2 and 6.2.3, the MFC-VRFT and MFAC-VRFT hybrid techniques are developed using the VRFT technique in tuning the parameters of the MFC and MFAC algorithms. In Section 6.3, the validation of the VRFT technique is carried out on the tower crane system-type equipment with three degrees of freedom having three inputs and three outputs in the MIMO case, namely $n = 3$, and three processes controlled in the SISO case with $n = 1$.

6.2.2 THE FIRST-ORDER DISCRETE-TIME MODEL-FREE CONTROL-VRFT CONTROLLERS

This section presents the design and tuning of the first-order discrete-time MFC-VRFT controllers, namely it describes how the VRFT technique can be used in determining the parameters of the first-order discrete-time MFC algorithms with P/PI/PID components, i.e., the iP/iPI/iPID controllers.

Lemma 6.1

Starting with the control law of the first-order discrete-time MFC algorithm given in (3.69) for the first-order iP controller, in (3.76) for the first-order iPI controller and in (3.79) for the first-order iPID controller, replacing the control error vector $\mathbf{e}(k) \in \mathfrak{R}^n$ in (3.73) leads to the equivalent form [Rom16a, Rom17c, Rom18a]

$$\mathbf{u}(k) = \boldsymbol{\alpha}^{-1}(-\hat{\mathbf{F}}(k) + \mathbf{y}_d(k+1) - \mathbf{y}_d(k) - \mathbf{K}_1 \, (\mathbf{y}_d(k) - \mathbf{y}(k))) \qquad (6.10)$$

for the first-order discrete-time iP controller,

$$\mathbf{u}(k) = \boldsymbol{\alpha}^{-1}(-\hat{\mathbf{F}}(k) + \mathbf{y}_d(k+1) - \mathbf{y}_d(k) - \mathbf{K}_1 \, (\mathbf{y}_d(k) - \mathbf{y}(k)) - \mathbf{K}_2 \, (\mathbf{y}_d(k-1) - \mathbf{y}(k-1))),$$
$$(6.11)$$

for the first-order discrete-time iPI controller and

$$\mathbf{u}(k) = \boldsymbol{\alpha}^{-1}(-\hat{\mathbf{F}}(k) + \mathbf{y}_d(k+1) - \mathbf{y}_d(k) - \mathbf{K}_1 \, (\mathbf{y}_d(k) - \mathbf{y}(k))$$
$$- \mathbf{K}_2 \, (\mathbf{y}_d(k-1) - \mathbf{y}(k-1)) - \mathbf{K}_3 \, (\mathbf{y}_d(k-2) - \mathbf{y}(k-2))) \qquad (6.12)$$

for the first-order discrete-time iPID controller. The abbreviations and notations proposed in Chapter 3 are also employed in this chapter.

Lemma 6.2

Continuing *Lemma 6.1*, substituting the estimate $\hat{\mathbf{F}}(k) \in \mathfrak{R}^n$ of $\mathbf{F}(k)$ from (3.70) in (6.10–6.12), three new versions of the control law are obtained as follows:

$$\mathbf{u}(k) = \alpha^{-1}(\alpha\mathbf{u}(k-1) + \mathbf{y}_d(k+1) - \mathbf{y}_d(k) - \mathbf{y}(k) + \mathbf{y}(k-1) - \mathbf{K}_1 \, (\mathbf{y}_d(k) - \mathbf{y}(k)))) \quad (6.13)$$

for the first-order discrete-time iP controller,

$$\mathbf{u}(k) = \alpha^{-1}(\alpha\mathbf{u}(k-1) + \mathbf{y}_d(k+1) - \mathbf{y}_d(k) - \mathbf{y}(k) + \mathbf{y}(k-1)$$
$$- \mathbf{K}_1(\mathbf{y}_d(k) - \mathbf{y}(k)) - \mathbf{K}_2(\mathbf{y}_d(k-1) - \mathbf{y}(k-1))) \quad (6.14)$$

for the first-order discrete-time iPI controller and

$$\mathbf{u}(k) = \alpha^{-1}(\alpha\mathbf{u}(k-1) + \mathbf{y}_d(k+1) - \mathbf{y}_d(k) - \mathbf{y}(k) + \mathbf{y}(k-1) - \mathbf{K}_1(\mathbf{y}_d(k) - \mathbf{y}(k))$$
$$- \mathbf{K}_2(\mathbf{y}_d(k-1) - \mathbf{y}(k-1)) - \mathbf{K}_3(\mathbf{y}_d(k-2) - \mathbf{y}(k-2))) \quad (6.15)$$

for the first-order discrete-time iPID controller, where $\mathbf{y}(k) \in \mathfrak{R}^{n_y}$ is the controlled output (process output) vector in (3.72) for the first-order discrete-time iP controller, in (3.77) for the first-order discrete-time iPI controller and in (3.80) for the first-order discrete-time iPID controller, $\mathbf{y}_d(k) \in \mathfrak{R}^{n_y}$ is the reference input vector (i.e., the output vector of the reference model), α is a matrix, of constant value, chosen by the control system designer (in this case, it is considered to be fixed, but it can also be seen as a set of tuning parameters), and $\mathbf{K}_1$, $\mathbf{K}_2$ and $\mathbf{K}_3$ are square matrices.

Lemma 6.3

Substituting the control error vector from (3.73) in the expressions of the control law in (6.13) for the first-order discrete-time iP controller, in (6.14) for the first-order discrete-time iPI controller and in (6.15) for the first-order discrete-time iPID controller and adding and subtracting $\mathbf{y}(k)$ and $\mathbf{y}_d(k-1)$, the control laws in (6.13–6.15) become as follows [Rom16a, Rom17c, Rom18a]:

$$\mathbf{u}(k) = \mathbf{u}(k-1) + \alpha^{-1}((-\mathbf{I} - \mathbf{K}_1)\mathbf{e}(k) - \mathbf{e}(k-1) + \mathbf{y}_d(k-1) + \mathbf{y}_d(k+1) - 2\mathbf{y}(k)) \quad (6.16)$$

for the first-order discrete-time iP controller,

$$\mathbf{u}(k) = \mathbf{u}(k-1) + \alpha^{-1}((-\mathbf{I} - \mathbf{K}_1)\mathbf{e}(k) - (\mathbf{I} + \mathbf{K}_2)\mathbf{e}(k-1) + \mathbf{y}_d(k-1) + \mathbf{y}_d(k+1) - 2\mathbf{y}(k))$$
$$(6.17)$$

for the first-order discrete-time iPI controller and

$$
\begin{aligned}
\mathbf{u}(k) = \mathbf{u}(k-1) + \boldsymbol{\alpha}^{-1}((-\mathbf{I} - \mathbf{K}_1)\mathbf{e}(k) - (\mathbf{I} + \mathbf{K}_2)\mathbf{e}(k-1) \\
- \mathbf{K}_3\mathbf{e}(k-2) + \mathbf{y}_d(k-1) + \mathbf{y}_d(k+1) - 2\mathbf{y}(k))
\end{aligned}
\tag{6.18}
$$

for the first-order discrete-time iPID controller.

Theorem 6.1

Recalling Lemmas 6.1 to 6.3, three matrices $\mathbf{K}_1 = \mathrm{diag}(K_{11},\ldots,K_{1n}) \in \mathfrak{R}^{n \times n}$, $\mathbf{K}_2 = \mathrm{diag}(K_{21},\ldots,K_{2n}) \in \mathfrak{R}^{n \times n}$ and $\mathbf{K}_3 = \mathrm{diag}(K_{31},\ldots,K_{3n}) \in \mathfrak{R}^{n \times n}$ can be organized as the parameter vector χ, which is also a column matrix but with different dimensions

$$
\chi = \chi(1) = [K_{11} \ \ldots \ K_{1n}]^T \in \mathfrak{R}^n
\tag{6.19}
$$

for the first-order discrete-time iP controller,

$$
\chi = [\chi(1) \ \chi(2)]^T = [K_{11} \ \ldots \ K_{1n} \ K_{21} \ \ldots \ K_{2n}]^T \in \mathfrak{R}^{2n}
\tag{6.20}
$$

for the first-order discrete-time iPI controller and

$$
\chi = [\chi(1) \ \chi(2) \ \chi(3)]^T = [K_{11} \ \ldots \ K_{1n} \ K_{21} \ \ldots \ K_{2n} \ K_{31} \ \ldots \ K_{3n}]^T \in \mathfrak{R}^{3n}
\tag{6.21}
$$

for the first-order discrete-time iPID controller. The square matrix $\boldsymbol{\alpha} = \mathrm{diag}(\alpha_1,\ldots,\alpha_n) \in \mathfrak{R}^{n \times n}$ is also organized as the column matrix Λ_{MFC}

$$
\Lambda_{MFC} = [\alpha_1 \ \alpha_2 \ \ldots \ \alpha_n]^T \in \mathfrak{R}^{n \times 1}
\tag{6.22}
$$

and the notation $\boldsymbol{\theta}_1 \in \mathbf{R}^{4n}$

$$
\boldsymbol{\theta}_1 = [\boldsymbol{\theta}_1(1)^T \ \boldsymbol{\theta}_1(2)^T \ \boldsymbol{\theta}_1(3)^T \ \boldsymbol{\theta}_1(4)^T]^T = [\Lambda_{MFC}^T \ \mathbf{y}_d^T(k-1) \ \mathbf{y}_d^T(k+1) \ \mathbf{y}^T(k)]^T
\tag{6.23}
$$

will be used for the vector of untunable parameters. The control laws (6.16–6.18) are next expressed in terms of the following recurrent forms:

$$
\mathbf{u}(k) = \mathbf{g}(\mathbf{e}(k), \mathbf{e}(k-1), \mathbf{u}(k-1), \chi, \boldsymbol{\theta}_1)
\tag{6.24}
$$

for the first-order discrete-time iP controller considering χ in (6.19),

$$
\mathbf{u}(k) = \mathbf{g}(\mathbf{e}(k), \mathbf{e}(k-1), \mathbf{u}(k-1), \chi, \boldsymbol{\theta}_1)
\tag{6.25}
$$

for the first-order discrete-time iPI controller considering χ in (6.20) and

$$
\mathbf{u}(k) = \mathbf{g}(\mathbf{e}(k), \mathbf{e}(k-1), \mathbf{e}(k-2), \mathbf{u}(k-1), \chi, \boldsymbol{\theta}_1)
\tag{6.26}
$$

for the first-order discrete-time iPID controller considering χ in (6.21), and $\mathbf{g}$ is a nonlinear vector function of vector variable with the expression

$$\mathbf{g} : \mathfrak{R}^{8n} \to \mathfrak{R}^n, \ \mathbf{g}(\mathbf{e}(k), \mathbf{e}(k-1), \mathbf{u}(k-1), \chi, \theta_1)$$
$$= \mathbf{u}(k-1) + \alpha^{-1}[(-\mathbf{I} - \mathbf{K}_1)\mathbf{e}(k) - \mathbf{e}(k-1) + \theta_1(2) + \theta_1(3) - 2\theta_1(4)] \tag{6.27}$$

for the first-order discrete-time iP controller,

$$\mathbf{g} : \mathfrak{R}^{9n} \to \mathfrak{R}^n, \ \mathbf{g}(\mathbf{e}(k), \mathbf{e}(k-1), \mathbf{u}(k-1), \chi, \theta_1)$$
$$= \mathbf{u}(k-1) + \alpha^{-1}[(-\mathbf{I} - \mathbf{K}_1)\mathbf{e}(k) - (\mathbf{I} + \mathbf{K}_2)\mathbf{e}(k-1) + \theta_1(2) + \theta_1(3) - 2\theta_1(4)] \tag{6.28}$$

for the first-order discrete-time iPI controller and

$$\mathbf{g} : \mathfrak{R}^{10n} \to \mathfrak{R}^n, \ \mathbf{g}(\mathbf{e}(k), \mathbf{e}(k-1), \mathbf{e}(k-2), \mathbf{u}(k-1), \chi, \theta_1)$$
$$= \mathbf{u}(k-1) + \alpha^{-1}[(-\mathbf{I} - \mathbf{K}_1)\mathbf{e}(k) - (\mathbf{I} + \mathbf{K}_2)\mathbf{e}(k-1) \tag{6.29}$$
$$- \mathbf{K}_3\mathbf{e}(k-2) + \theta_1(2) + \theta_1(3) - 2\theta_1(4)]$$

for the first-order discrete-time iPID controller.

Remark 6.3

According to [Rom16a] and [Rom17c], in the design and tuning of the first-order discrete-time MFC-VRFT controller, it is considered that the MFC-specific control error (tracking error) vector $\mathbf{e}(k)$ is equivalent to the VRFT-specific virtual control error vector $\bar{\mathbf{e}}(k)$ in the nonlinear mathematical model (6.3) of the VRFT controller, where $\theta = [\chi^T \ \theta_1^T]^T$ is the extended vector of the parameters of the MFC-VRFT controller ($\theta \in \mathfrak{R}^{5n}$ for the first-order discrete-time iP controller, $\theta \in \mathfrak{R}^{6n}$ for the first-order discrete-time iPI controller and $\theta \in \mathfrak{R}^{7n}$ for the first-order discrete-time iPID controller), and the virtual reference input vector $\bar{\mathbf{r}}(k)$ within the VRFT controller is considered equivalent to the reference input vector $\mathbf{y}_d(k)$ of the MFC algorithm. Therefore, the parameters of the MFC controller within the control system structure will be tuned by VRFT based on the following relationships between the parameters and variables in the case of the VRFT controller presented in this section and the general one of the VRFT controller in Section 6.2.1:

$$n_{\chi C} \text{ in VRFT} = n \text{ in MFC-VRFT},$$

$$\chi \text{ in VRFT} = \chi \text{ in MFC-VRFT},$$

$$\bar{\mathbf{e}}(k) \text{ in VRFT} = \mathbf{e}(k) \text{ in MFC-VRFT}, \tag{6.30}$$

$$\bar{\mathbf{r}}(k) \text{ in VRFT} = \mathbf{y}_d(k) \text{ in MFC-VRFT}.$$

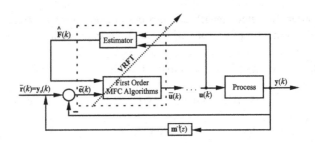

FIGURE 6.1 MIMO control system structure with discrete-time first-order MFC-VRFT controller. (Adapted from Rom16a, Rom17c and Rom18a.)

Proposition 6.1

The control system structure with MFC-VRFT controller is presented in Figure 6.1. In other words, the structure contains an MFC controller whose parameters are tuned by VRFT. In Figure 6.1, $\hat{\mathbf{F}}(k)$ indicates the vector $\mathbf{F}(k)$ of estimates specific to the MFC algorithm defined in (3.70), and $\mathbf{m}^{-1}(z)$ is the inverse of the discrete transfer function matrix of the reference model specific to VRFT.

Proposition 6.2

Based on the results presented in this section, the **following design and tuning steps** for the control system structure with discrete-time first-order MFC-VRFT controller are carried out [Rom18a]:

Step 1. Choosing the value of the design parameter $\alpha = \text{diag}(\alpha_1,...,\alpha_n) \in \mathfrak{R}^{n \times n}$ such that $\Delta \mathbf{y}(k+1) = \mathbf{y}(k+1) - \mathbf{y}(k)$ and $\alpha \mathbf{u}(k)$ will have the same order of magnitude.

Step 2. Choosing the initial signals $\mathbf{u}(k) \in \mathfrak{R}^n$ with a wide range of frequency specific to VRFT that should be applied to the open-loop controlled process to collect the input/output data pairs $\{\mathbf{u}(k), \mathbf{y}(k)\}$.

Step 3. Establishing the reference model matrix $\mathbf{m} \in \mathfrak{R}^{n \times n}$ that leads to the virtual reference input vector $\bar{\mathbf{r}}(k) \in \mathfrak{R}^n$ according to (6.7) so that the output of the control system performance specifications is achieved in terms of ensuring that the output of the reference model and the output of the control system (i.e., the closed-loop controlled process) have similar trajectories.

Step 4. Obtaining the parameter vector χ considering the stability conditions in (3.74) for the first-order discrete-time iP controller, (3.78) for the first-order discrete-time iPI controller and (3.81) for the first-order discrete-time iPID controller in terms of solving the optimization problem in (6.9).

6.2.3 THE SECOND-ORDER DISCRETE-TIME MODEL-FREE CONTROL-VRFT CONTROLLERS

This section gives details on the design and tuning of the second-order discrete-time MFC-VRFT controllers, namely how the VRFT technique can be used in determining the parameters of the second-order discrete-time MFC algorithms with P/PI/PID component, i.e., the iP/iPI/iPID controllers.

Lemma 6.4

Starting with the control law of the second-order discrete-time MFC algorithm given in (3.83) for the second-order iP controller, in (3.87) for the second-order iPI controller and in (3.90) for the second-order iPID controller, replacing $e(k) \in \Re^n$ in (3.73) yields the equivalent forms of the control law [Rom16a, Rom17c, Rom18a]

$$\mathbf{u}(k) = \alpha^{-1}(-\hat{\mathbf{F}}(k) + \mathbf{y}_d(k+1) - 2\mathbf{y}_d(k) + \mathbf{y}_d(k-1) - \mathbf{K}_1 (\mathbf{y}_d(k) - \mathbf{y}(k))) \quad (6.31)$$

for the second-order discrete-time iP controller,

$$\mathbf{u}(k) = \alpha^{-1}(-\hat{\mathbf{F}}(k) + \mathbf{y}_d(k+1) - 2\mathbf{y}_d(k) + \mathbf{y}_d(k-1)$$
$$- \mathbf{K}_1(\mathbf{y}_d(k) - \mathbf{y}(k)) - \mathbf{K}_2(\mathbf{y}_d(k-1) - \mathbf{y}(k-1))) \quad (6.32)$$

for the second-order discrete-time iPI controller, and

$$\mathbf{u}(k) = \alpha^{-1}(-\hat{\mathbf{F}}(k) + \mathbf{y}_d(k+1) - 2\mathbf{y}_d(k) + \mathbf{y}_d(k-1) - \mathbf{K}_1(\mathbf{y}_d(k) - \mathbf{y}(k))$$
$$- \mathbf{K}_2(\mathbf{y}_d(k-1) - \mathbf{y}(k-1)) - \mathbf{K}_3(\mathbf{y}_d(k-2) - \mathbf{y}(k-2))) \quad (6.33)$$

for the second-order discrete-time iPID controller.

Lemma 6.5

Substituting the estimate $\hat{\mathbf{F}}(k) \in \Re^n$ of $\mathbf{F}(k)$ specified in (3.84) in (6.31–6.33), the following versions of the control laws are obtained:

$$\mathbf{u}(k) = \alpha^{-1}(\alpha \, \mathbf{u}(k-1) + \mathbf{y}_d(k-1) - 2\mathbf{y}_d(k) + \mathbf{y}_d(k+1) - \mathbf{y}(k-2)$$
$$+ 2\mathbf{y}(k-1) - \mathbf{y}(k) - \mathbf{K}_1(\mathbf{y}_d(k) - \mathbf{y}(k))) \quad (6.34)$$

for the second-order discrete-time iP controller,

$$\mathbf{u}(k) = \alpha^{-1}(\alpha \, \mathbf{u}(k-1) + \mathbf{y}_d(k-1) - 2\mathbf{y}_d(k) + \mathbf{y}_d(k+1) - \mathbf{y}(k-2)$$
$$+ 2\mathbf{y}(k-1) - \mathbf{y}(k) - \mathbf{K}_1(\mathbf{y}_d(k) - \mathbf{y}(k)) - \mathbf{K}_2(\mathbf{y}_d(k-1) - \mathbf{y}(k-1))) \quad (6.35)$$

for the second-order discrete-time iPI controller and

$$\mathbf{u}(k) = \alpha^{-1}(\alpha \, \mathbf{u}(k-1) + \mathbf{y}_d(k-1) - 2\mathbf{y}_d(k) + \mathbf{y}_d(k+1) - \mathbf{y}(k-2) + 2\mathbf{y}(k-1) - \mathbf{y}(k)$$
$$- \mathbf{K}_1(\mathbf{y}_d(k) - \mathbf{y}(k)) - \mathbf{K}_2(\mathbf{y}_d(k-1) - \mathbf{y}(k-1)) - \mathbf{K}_3(\mathbf{y}_d(k-2) - \mathbf{y}(k-2)))$$
$$(6.36)$$

for the second-order discrete-time iPID controller, where $\mathbf{y}(k) \in \mathfrak{R}^n$ is the controlled output (the process output) vector that appears in (3.85) for the second-order discrete-time iP controller, in (3.88) for the second-order discrete-time iPI controller and in (3.91) for the second-order discrete-time iPID controller, and the rest of the vectors of matrices are described in the previous section.

Lemma 6.6

Substituting $\mathbf{e}(k)$ given in (3.73) in the expressions of the control laws in (6.34) for the second-order discrete-time iP controller, in (6.35) for the second-order discrete-time iPI controller and in (6.36) for the second-order discrete-time iPID controller and adding and subtracting $\mathbf{y}(k-1)$ and $\mathbf{y}(k)$, the control laws in (6.34), (6.35) and (6.36) become [Rom16a, Rom17c, Rom18a]

$$\mathbf{u}(k) = \mathbf{u}(k-1) + \alpha^{-1}((-\mathbf{I} - \mathbf{K}_1)\mathbf{e}(k) + \mathbf{e}(k-1) - \mathbf{y}_d(k)$$
$$+ \mathbf{y}_d(k+1) - \mathbf{y}(k-2) + 3\mathbf{y}(k-1) - 2\mathbf{y}(k)) \tag{6.37}$$

for the second-order discrete-time iP controller,

$$\mathbf{u}(k) = \mathbf{u}(k-1) + \alpha^{-1}((-\mathbf{I} - \mathbf{K}_1)\mathbf{e}(k) + (\mathbf{I} + \mathbf{K}_2)\mathbf{e}(k-1) - \mathbf{y}_d(k)$$
$$+ \mathbf{y}_d(k+1) - \mathbf{y}(k-2) + 3\mathbf{y}(k-1) - 2\mathbf{y}(k)) \tag{6.38}$$

for the second-order discrete-time iPI controller and

$$\mathbf{u}(k) = \mathbf{u}(k-1) + \alpha^{-1}((-\mathbf{I} - \mathbf{K}_1)\mathbf{e}(k) + (\mathbf{I} + \mathbf{K}_2)\mathbf{e}(k-1) - \mathbf{K}_3\mathbf{e}(k-2)$$
$$- \mathbf{y}_d(k) + \mathbf{y}_d(k+1) - \mathbf{y}(k-2) + 3\mathbf{y}(k-1) - 2\mathbf{y}(k)), \tag{6.39}$$

for the second-order discrete-time iPID controller.

Theorem 6.2

Based on *Lemmas 6.4–6.6*, three matrices $\mathbf{K}_1 = \text{diag}(K_{11},\dots,K_{1n}) \in \mathbf{R}^{n \times n}$, $\mathbf{K}_2 = \text{diag}(K_{21},\dots,K_{2n}) \in \mathbf{R}^{n \times n}$ and $\mathbf{K}_3 = \text{diag}(K_{31},\dots,K_{3n}) \in \mathbf{R}^{n \times n}$ are organized as the column matrix and also parameter vector:

$$\chi = [K_{11} \ \dots \ K_{1n}]^T \in \mathfrak{R}^n \tag{6.40}$$

for the second-order discrete-time iP controller,

$$\chi = [K_{11} \ \dots \ K_{1n} \ K_{21} \ \dots \ K_{2n}]^T \in \mathfrak{R}^{2n} \tag{6.41}$$

for the second-order discrete-time iPI controller and

$$\chi = [K_{11} \dots K_{1n} \ K_{21} \dots K_{2n} \ K_{31},\dots,K_{3n}]^T \in \mathfrak{R}^{3n} \tag{6.42}$$

for the second-order discrete-time iPID controller. The quadratic matrix $\alpha = \text{diag}$ $(\alpha_1,...,\alpha_n) \in \mathfrak{R}^{n \times n}$ is also organized as the column matrix Λ_{MFC}

$$\Lambda_{\text{MFC}} = [\alpha_1 \; \alpha_2 \; ... \; \alpha_n]^T \in \mathfrak{R}^n, \tag{6.43}$$

and the notation $\boldsymbol{\theta}_1 \in \mathfrak{R}^{6n}$

$$\boldsymbol{\theta}_1 = [\boldsymbol{\theta}_1(1)^T \; \boldsymbol{\theta}_1(2)^T \; \boldsymbol{\theta}_1(3)^T \; \boldsymbol{\theta}_1(4)^T \; \boldsymbol{\theta}_1(5)^T \; \boldsymbol{\theta}_1(6)^T]^T$$

$$= [\Lambda_{MFC}^T \; \mathbf{y}_d^T(k) \; \mathbf{y}_d^T(k+1) \; \mathbf{y}^T(k-2) \; \mathbf{y}^T(k-1) \; \mathbf{y}^T(k)]^T \tag{6.44}$$

will be used for the untunable parameters. The control laws in (6.37–6.39) are next transformed into the recurrent form

$$\mathbf{u}(k) = \mathbf{g}(\mathbf{e}(k), \mathbf{e}(k-1), \mathbf{u}(k-1), \chi, \boldsymbol{\theta}_1) \tag{6.45}$$

for the second-order discrete-time iP controller considering χ in (6.40),

$$\mathbf{u}(k) = \mathbf{g}(\mathbf{e}(k), \mathbf{e}(k-1), \mathbf{u}(k-1), \chi, \boldsymbol{\theta}_1) \tag{6.46}$$

for the second-order discrete-time iPI controller considering χ in (6.41) and

$$\mathbf{u}(k) = \mathbf{g}(\mathbf{e}(k), \mathbf{e}(k-1), \mathbf{e}(k-2), \mathbf{u}(k-1), \chi, \boldsymbol{\theta}_1) \tag{6.47}$$

for the second-order discrete-time iPID controller considering χ in (6.42), where $\mathbf{g}$ is a nonlinear vector function of vector variable with the expression

$$\mathbf{g} : \mathfrak{R}^{10n} \to \mathfrak{R}^n, \; \mathbf{g}(\mathbf{e}(k), \mathbf{e}(k-1), \mathbf{u}(k-1), \chi, \boldsymbol{\theta}_1)$$

$$= \mathbf{u}(k-1) + \alpha^{-1}[(-\mathbf{I} - \mathbf{K}_1)\mathbf{e}(k) + \mathbf{e}(k-1) - \boldsymbol{\theta}_1(2) + \boldsymbol{\theta}_1(3) - \boldsymbol{\theta}_1(4) + 3\boldsymbol{\theta}_1(5) - 2\boldsymbol{\theta}_1(6)] \tag{6.48}$$

for the second-order discrete-time iP controller,

$$\mathbf{g} : \mathfrak{R}^{11n} \to \mathfrak{R}^n, \; \mathbf{g}(\mathbf{e}(k), \mathbf{e}(k-1), \mathbf{u}(k-1), \chi, \boldsymbol{\theta}_1)$$

$$= \mathbf{u}(k-1) + \alpha^{-1}[(-\mathbf{I} - \mathbf{K}_1)\mathbf{e}(k) + (\mathbf{I} + \mathbf{K}_2)\mathbf{e}(k-1) \tag{6.49}$$

$$- \boldsymbol{\theta}_1(2) + \boldsymbol{\theta}_1(3) - \boldsymbol{\theta}_1(4) + 3\boldsymbol{\theta}_1(5) - 2\boldsymbol{\theta}_1(6)]$$

for the second-order discrete-time iPI controller and

$$\mathbf{g} : \mathfrak{R}^{12n} \to \mathfrak{R}^n, \; \mathbf{g}(\mathbf{e}(k), \mathbf{e}(k-1), \mathbf{u}(k-1), \chi, \boldsymbol{\theta}_1)$$

$$= \mathbf{u}(k-1) + \alpha^{-1}[(-\mathbf{I} - \mathbf{K}_1)\mathbf{e}(k) + (\mathbf{I} + \mathbf{K}_2)\mathbf{e}(k-1) - \mathbf{K}_3\mathbf{e}(k-2) \tag{6.50}$$

$$- \boldsymbol{\theta}_1(2) + \boldsymbol{\theta}_1(3) - \boldsymbol{\theta}_1(4) + 3\boldsymbol{\theta}_1(5) - 2\boldsymbol{\theta}_1(6)]$$

for the second-order discrete-time iPID controller.

Remark 6.4

According to [Rom16a] and [Rom17c], in the design and tuning of the second-order discrete-time MFC-VRFT controller, it is considered that the vector of MFC-specific control errors (tracking errors) $e(k)$ is equivalent to the VRFT-specific virtual error vector $\bar{e}(k)$, from the nonlinear mathematical model (6.3) of the VRFT controller, where $\theta = [\chi^T \; \theta_1^T]^T$ is the extended parameter vector of the MFC-VRFT controller ($\theta \in \mathfrak{R}^{7n}$ for the second-order discrete-time iP controller, $\theta \in \mathfrak{R}^{8n}$ for the second-order discrete-time iPI controller and $\theta \in \mathfrak{R}^{9n}$ for the second-order discrete-time iPID controller), and the virtual reference input vector $\bar{r}(k)$ within the VRFT controller is considered equivalent to the vector reference input vector $y_d(k)$ within the MFC controller. Therefore, the parameters of the MFC controller will be tuned by VRFT based on the following equivalences between the parameters and variables in the case of the VRFT controller presented in this section and the general one of the VRFT controller given in Section 6.2.1:

$$n_{\chi C} \text{ in VRFT} = n \text{ in MFC-VRFT},$$

$$\chi \text{ in VRFT} = \chi \text{ in MFC-VRFT},$$

$$\bar{e}(k) \text{ in VRFT} = e(k) \text{ in MFC-VRFT}, \qquad (6.51)$$

$$\bar{r}(k) \text{ in VRFT} = y_d(k) \text{ in MFC-VRFT}.$$

Proposition 6.3

In Figure 6.2, the control system structure with MFC-VRFT controller is presented. In other words, the structure contains an MFC controller whose parameters are tuned by VRFT. In Figure 6.2, $\hat{F}(k)$ is the vector of $F(k)$ estimate specific to the MFC algorithm defined in equation (3.84), and $m^{-1}(z)$ is the inverse of the discrete transfer function matrix of the reference model specific to VRFT.

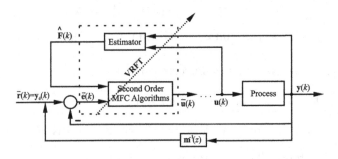

FIGURE 6.2 MIMO control system structure with discrete-time second-order MFC-VRFT controller. (Adapted from Rom16a, Rom17c and Rom18a.)

Proposition 6.4

Based on the results presented in this section, the **following design and tuning steps** for the control system structure with discrete-time second-order MFC-VRFT controller are carried out [Rom18a]:

Step 1. Choosing the value of the design parameter $\alpha = \mathrm{diag}(\alpha_1,...,\alpha_n) \in \mathfrak{R}^{n \times n}$ such that $\mathbf{y}(k) - 2\mathbf{y}(k-1) - \mathbf{y}(k-2)$ and $\alpha\,\mathbf{u}(k)$ will have the same order of magnitude.

Step 2. Choosing the initial signals $\mathbf{u}(k) \in \mathfrak{R}^n$ with a wide range of frequency specific to VRFT that should be applied to the open-loop controlled process to collect the input/output data pairs $\{\mathbf{u}(k), \mathbf{y}(k)\}$.

Step 3. Establishing the reference model matrix $\mathbf{m} \in \mathfrak{R}^{n \times n}$ that leads to the virtual reference input vector $\bar{\mathbf{r}}(k) \in \mathfrak{R}^n$ according to (6.7) so that the output of the control system performance specifications is achieved in terms of ensuring that the output of the reference model and the output of the control system (i.e., the closed-loop controlled process) have similar trajectories.

Step 4. Obtaining the parameter vector χ considering the stability conditions in (3.86) for the second-order discrete-time iP controller, (3.89) for the second-order discrete-time iPI controller and (3.92) for the second-order discrete-time iPID controller in terms of solving the optimization problem in (6.9).

6.2.4 THE MFAC-VRFT ALGORITHMS

This section presents the design and tuning of the MFAC-VRFT controllers, namely how the VRFT technique presented in Section 6.2.1 can be used to determine the values of the parameters of the MFAC controller in the CFDL version.

Remark 6.5

Recalling Chapter 5, the relations (5.5) and (5.7) specific to the MFAC algorithm in the CFDL version (i.e., the MFAC-CFDL algorithm) are rewritten as

$$\mathbf{u}(k) = \mathbf{u}(k-1) + \frac{\rho \hat{\Phi}^T(k)[\mathbf{y}_d(k+1) - \mathbf{y}(k)]}{\lambda + \|\hat{\Phi}(k)\|^2},$$

$$\hat{\Phi}(k) = \hat{\Phi}(k-1) + \frac{\eta[\Delta\mathbf{y}(k) - \hat{\Phi}(k-1)(\mathbf{u}(k-1) - \mathbf{u}(k-2))](\mathbf{u}^T(k-1) - \mathbf{u}^T(k-2))}{\mu + \|\mathbf{u}(k-1) - \mathbf{u}(k-2)\|^2},$$

$$(6.52)$$

where the first expression is the control law specific to the MFAC-CFDL algorithm given in (5.7), $\hat{\Phi}(k) \in \mathfrak{R}^{n \times n}$ is the estimate of the PPD matrix defined in (5.5), $\mathbf{y}_d(k) \in \mathfrak{R}^n$ is the reference input vector (the set-point vector as the output of the reference model), $\mathbf{y}(k) \in \mathfrak{R}^n$ is the controlled output (the controlled process output) vector, $\rho = \mathrm{const} > 0$ is an additional parameter used in the control law, $\lambda = \mathrm{const} > 0$

is a weighting parameter involved in the design of the MFAC controller, $\eta \in (0,1)$ is a constant parameter, and $\mu = \text{const} > 0$ is a parameter specific to the MFAC controller introduced to avoid the zero denominator in the second law in (6.52).

Definition 6.4

The PPD matrix $\hat{\Phi}(k) = [\hat{\varphi}_{ij}(k)]_{i,j=1...n}$ is organized as a column vector $\hat{\Lambda}(k)$

$$\hat{\Lambda}(k) = [\hat{\varphi}_{11}(k)\ \hat{\varphi}_{21}(k)...\hat{\varphi}_{n1}(k)\ \hat{\varphi}_{12}(k)\ \hat{\varphi}_{22}(k)...\hat{\varphi}_{n2}(k)\ \hat{\varphi}_{1n}(k)\ \hat{\varphi}_{2n}(k)...\hat{\varphi}_{nn}(k)]^T \in \Re^{n^2},$$
(6.53)

which will be used as a state vector within the mathematical state model of the MFAC algorithm in the CFDL version, and the next vector χ of parameters of the MFAC algorithm is defined with the expression

$$\chi = [\chi(1)\ \chi(2)\ \chi(3)\ \chi(4)]^T = [\rho\ \eta\ \lambda\ \mu]^T \in \Re^4,$$
(6.54)

used in the state-space mathematical model of the MFAC algorithm in the CFDL version.

Lemma 6.7

The control law and the estimate of the PPD matrix in (6.52) are reformulated as follows such that to allow expressing the state equations related to the MFAC algorithm in the CFDL version [Rom16c, Rom16d, Rom18a]:

$$\mathbf{u}(k) = \mathbf{g}(\hat{\Lambda}(k), \mathbf{u}(k-1), \mathbf{y}_d(k+1), \mathbf{y}(k), \chi),$$
$$\hat{\Lambda}(k) = \mathbf{h}(\hat{\Lambda}(k-1), \mathbf{u}(k-1), \mathbf{u}(k-2), \mathbf{y}(k), \mathbf{y}(k-1), \chi),$$
(6.55)

where the initial nonzero conditions related to (6.55) are $\hat{\Lambda}(1)$ and $\mathbf{u}(1)$, $\mathbf{g} : \Re^{n(n+3)+4} \to \Re^n$ and $\mathbf{h} : \Re^{n(n+4)+4} \to \Re^{n^2}$ are two nonlinear vector functions of vector variable with the expressions

$$\mathbf{g}(\hat{\Lambda}(k), \mathbf{u}(k-1), \mathbf{y}_d(k+1), \mathbf{y}(k), \chi) = \mathbf{u}(k-1) + \frac{\chi(1)\hat{\Lambda}^T(k)[\mathbf{y}_d(k+1) - \mathbf{y}(k)]}{\chi(3) + \|\hat{\Lambda}(k)\|^2},$$

$$\mathbf{h}(\hat{\Lambda}(k-1), \mathbf{u}(k-1), \mathbf{u}(k-2), \mathbf{y}(k), \mathbf{y}(k-1), \chi) = \hat{\Lambda}(k-1)$$

$$+ \frac{\chi(2)[\mathbf{y}(k) - \mathbf{y}(k-1) - \hat{\Lambda}(k-1)(\mathbf{u}(k-1) - \mathbf{u}(k-2))](\mathbf{u}^T(k-1) - \mathbf{u}^T(k-2))}{\chi(4) + \|\mathbf{u}(k-1) - \mathbf{u}(k-2)\|^2}.$$
(6.56)

Introducing an additional state vector $\mathbf{z}(k)$

$$\mathbf{z}(k) = \mathbf{u}(k-1), \tag{6.57}$$

defining the input vector $\mathbf{U}(k)$

$$\mathbf{U}(k) = [\mathbf{y}_d(k+1)^T \quad \mathbf{y}(k)^T \quad \mathbf{y}(k-1)^T]^T \in \mathfrak{R}^{3n} \tag{6.58}$$

and the state vector $\boldsymbol{\theta}(k)$

$$\boldsymbol{\theta}(k) = [\mathbf{u}(k)^T \quad \mathbf{z}(k)^T \quad \hat{\Lambda}(k)^T]^T \in \mathfrak{R}^{n(n+2)}, \tag{6.59}$$

the state equations are obtained from the state-space model related to the MFAC algorithm in the CFDL version:

$$\boldsymbol{\theta}(k) = \Gamma(\boldsymbol{\theta}(k-1), \mathbf{U}(k), \boldsymbol{\chi}), \tag{6.60}$$

where the initial condition related to (6.60) is $\boldsymbol{\theta}(1) = \boldsymbol{\theta}_1 \neq \mathbf{0}$, and $\Gamma : \mathfrak{R}^{n(n+5)+4} \to \mathfrak{R}^{n(n+2)}$ is a nonlinear function of vector variable with the expression

$$\Gamma(\boldsymbol{\theta}(k-1), \mathbf{U}(k), \boldsymbol{\chi}) = \begin{bmatrix} \mathbf{g}(\hat{\Lambda}(k-1), \mathbf{u}(k-1), \mathbf{z}(k-1), \mathbf{y}(k), \mathbf{y}(k-1), \mathbf{y}_d(k+1), \boldsymbol{\chi}) \\ \mathbf{u}(k-1) \\ \mathbf{h}(\hat{\Lambda}(k-1), \mathbf{z}(k), \mathbf{z}(k-1), \mathbf{y}(k), \mathbf{y}(k-1), \boldsymbol{\chi}) \end{bmatrix},$$
$$\tag{6.61}$$

where the parameter vector $\boldsymbol{\chi}$ is considered in [Rom16c, Rom16d, Rom18a] as an additional input vector that plays a disturbance role.

Theorem 6.3

Continuing *Lemma 6.7*, also using the additional notation in (6.57) and replacing the PPD matrix $\hat{\Phi}(k)$ considered as the output expressed through the second state equation in (6.55) in the first state equation in (6.55) and using the state vector in (6.53), the following expression of the state equations related to the MFAC MIMO algorithm is obtained:

$$\mathbf{u}(k) = \mathbf{g}(\hat{\Lambda}(k-1), \mathbf{u}(k-1), \mathbf{z}(k-1), \mathbf{y}(k), \mathbf{y}(k-1), \mathbf{y}_d(k+1), \boldsymbol{\chi}),$$

$$\mathbf{z}(k) = \mathbf{u}(k-1), \tag{6.62}$$

$$\hat{\Lambda}(k) = \mathbf{h}(\hat{\Lambda}(k-1), \mathbf{z}(k), \mathbf{z}(k-1), \mathbf{y}(k), \mathbf{y}(k-1), \boldsymbol{\chi}).$$

Starting with the initial conditions $\hat{\Lambda}(1), \mathbf{u}(1), \mathbf{z}(1) = \mathbf{u}(0)$ associated with the state equations in (6.62), these equations are expressed at the discrete-time moments 2, 3,..., k and the recurrent equation related to the control law results as follows:

$$\hat{\Lambda}(2) = \mathbf{h}(\hat{\Lambda}(1), \mathbf{u}(1), \mathbf{u}(0), \mathbf{y}(2), \mathbf{y}(1), \mathbf{\theta}),$$

$$\mathbf{u}(2) = \mathbf{g}(\hat{\Lambda}(1), \mathbf{u}(1), \mathbf{u}(0), \mathbf{y}(2), \mathbf{y}(1), \mathbf{y}^*(3), \mathbf{\theta}),$$

$$\mathbf{u}(3) = \mathbf{g}(\hat{\Lambda}(2), \mathbf{u}(2), \mathbf{u}(1), \mathbf{y}(3), \mathbf{y}(2), \mathbf{y}_d(4), \mathbf{\chi}) = \mathbf{g}(\mathbf{h}(\hat{\Lambda}(1), \mathbf{u}(1), \mathbf{u}(0), \mathbf{y}(2),$$

$$\mathbf{y}(1), \mathbf{\chi}), \mathbf{g}(\hat{\Lambda}(1), \mathbf{u}(1), \mathbf{u}(0), \mathbf{y}(2), \mathbf{y}(1), \mathbf{y}_d(3), \mathbf{\chi}), \mathbf{u}(1), \mathbf{y}(3), \mathbf{y}(2), \mathbf{y}_d(4), \mathbf{\chi})$$

$$= \mathbf{g}(\hat{\Lambda}(1), \mathbf{u}(1), \mathbf{u}(0), \mathbf{y}(3), \mathbf{y}(2), \mathbf{y}(1), \mathbf{y}_d(3), \mathbf{y}_d(4), \mathbf{\chi}), \qquad (6.63)$$

...

$$\mathbf{u}(k) = \mathbf{g}(\hat{\Lambda}(1), \mathbf{u}(1), \mathbf{u}(0), \mathbf{y}(k), \mathbf{y}(k-1), \ldots, \mathbf{y}(2), \mathbf{y}(1), \mathbf{y}_d(k+1), \mathbf{y}_d(k), \ldots,$$

$$\mathbf{y}_d(4), \mathbf{y}_d(3), \mathbf{\chi}) = \mathbf{g}(\hat{\Lambda}(1), \mathbf{u}(1), \mathbf{u}(0), \mathbf{y}_d(k+1) - \mathbf{y}(k), \mathbf{y}_d(k) - \mathbf{y}(k-1), \ldots,$$

$$\mathbf{y}_d(3) - \mathbf{y}(2), \mathbf{y}(1), \mathbf{\chi}).$$

Inserting the notation $\mathbf{e}(k) = \mathbf{y}_d(k+1) - \mathbf{y}(k) \in \mathfrak{R}^n$ and the extended parameter vector of the MFAC-VRFT controller $\mathbf{\chi}_{ext} = [\mathbf{\chi}^T \ \hat{\Lambda}(1)^T]^T \in \mathfrak{R}^{4+n^2}$, the control law at time k in (6.63) is expressed as follows in the nonlinear recurrent form:

$$\mathbf{u}_{\chi_{ext}}(k) = C_{\chi_{ext}}(\mathbf{\chi}_{ext}, \mathbf{u}(k-1), \ldots, \mathbf{u}(k-n_{uc}), \overline{\mathbf{e}}(k), \ldots, \overline{\mathbf{e}}(k-n_{ec})), \qquad (6.64)$$

where $C_{\chi_{ext}} : \mathfrak{R}^{4+n(n+n_{uc}+n_{ec}+2)} \to \mathfrak{R}^n$ is a nonlinear vector function of vector variable, $\mathbf{u}_{\chi_{ext}}(k) \in \mathfrak{R}^n$ is the control input vector elaborated by the nonlinear controller equivalent to the one described in (6.54) related to VRFT, $\overline{\mathbf{e}}(k)$ is the virtual control (or tracking error) vector, and uc and ec are the maximum known delay orders related to the control input and the control error, respectively.

Remark 6.6

If the vector of the virtual reference input vector $\overline{\mathbf{r}}(k) \in \mathfrak{R}^n$ specific to VRFT is considered equivalent to the reference vector at the one-step-ahead moment $\mathbf{y}_d(k+1) \in \mathfrak{R}^n$ specific to MFAC, i.e., $\overline{\mathbf{r}}(k) = \mathbf{y}_d(k+1) \in \mathfrak{R}^n$, then the parameter tuning of the MFAC algorithm within the control system structure is performed by VRFT based on the following equivalences between the parameters and variables specific to VRFT presented in this section and the general VRFT controller in Section 6.2.1:

$$n_{\chi C} \text{ in VRFT} = 4 \text{ in MFAC-VRFT},$$

$$\chi \text{ in VRFT} = \chi_{ext} \text{ in MFAC-VRFT},$$

$$\overline{e}(k) \text{ in VRFT} = e(k) \text{ in MFAC-VRFT}, \tag{6.65}$$

$$\overline{r}(k) \text{ in VRFT} = y_d(k+1) \text{ in MFAC-VRFT}.$$

Proposition 6.5

Figure 6.3 defines the control system structure with MFAC-VRFT controller [Rom16c, Rom16d, Rom18a], where $\mathbf{m}^{-1}(z)$ is the inverse of the discrete transfer function matrix of the reference model specific to VRFT.

Remark 6.7

According to the results presented in Section 6.2.1, in order to achieve the reference model tracking based on VRFT in the general case of nonlinear systems, the objective function in (6.1) [Yan16] is transformed into the following objective function in terms of choosing a reference model $\mathbf{m} = \mathbf{I}_n$:

$$J_{MR}(\chi) = \sum_{k=1}^{N} \left\| \mathbf{y}_\chi(k) - \mathbf{y}_d(k) \right\|^2. \tag{6.66}$$

The objective function J_{MR} in (6.66) represents the sum of N steps for $k = 1 \dots N$ of the objective function J_{MFAC} specific to MFAC in (5.3) for $\lambda = 0$. However, in practice, no causal controller can be achieved with $\mathbf{m} = \mathbf{I}_n$. Therefore, the choice of $\mathbf{m} \neq \mathbf{I}_n$ in VRFT is equivalent to $\lambda \neq 0$ in MFAC. The parameter λ is important because it influences the behavior of the control system structure with MFAC algorithms in the sense that a high value of λ will lead to the reduction of the overshoot. From the point

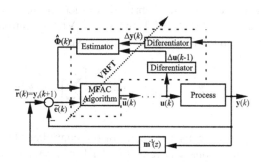

FIGURE 6.3 MIMO control system structure with MFAC-VRFT controller. (Adapted from Rom16c, Rom16d and Rom18a.)

of view of VRFT, this means choosing an appropriate reference model **m** through which the performance of the control system structure is imposed. Therefore, the calculus approach of the parameters $\hat{\Lambda}(1)$ and $\theta = [\rho \ \eta \ \lambda \ \mu]^T$ using the MFAC-VRFT controller is transposed into a simplified version using the reference model **m**. The above objective function is obtained by minimizing the sum of the errors' squares for each sample [Rom16c, Rom16d, Rom18a].

Proposition 6.6

Based on the results presented in this section, the following **design and tuning steps for the control system structure with MFAC-VRFT controller** are carried out [Rom18a]:

Step 1. Choosing the initial signals $\mathbf{u}(k) \in \mathfrak{R}^n$ with a wide range of frequency specific to VRFT that should be applied to the open-loop controlled process to collect the input/output data pairs $\{\mathbf{u}(k), \mathbf{y}(k)\}$.

Step 2. Establishing the reference model matrix $\mathbf{m} \in \mathfrak{R}^{n \times n}$ that leads to the virtual reference input vector $\bar{\mathbf{r}}(k) \in \mathfrak{R}^n$ according to (6.7) so that the output of the control system performance specifications is achieved in terms of ensuring that the output of the reference model and the output of the control system (i.e., the closed-loop controlled process) have similar trajectories.

Step 3. Obtaining the value $\chi_{ext} = [\chi^T \ \hat{\Lambda}(1)^T]^T \in \mathfrak{R}^{n^2+4}$, where $\hat{\Lambda}(1) \in \mathfrak{R}^{n^2}$ is defined in (6.53) and $\chi = [\rho \ \eta \ \lambda \ \mu]^T \in \mathbf{R}^4$ in terms of solving the optimization problem in (6.9) such that $\Phi(k) \in \mathbf{R}^{n \times n}$ will vary during the experiments according to the reset conditions in the context of (5.6).

6.3 EXAMPLE AND APPLICATION

6.3.1 SISO CONTROL SYSTEMS

The objective function $J_{(\psi)}(\chi_{(\psi)})$ used to measure the performance of each SISO control system structure with the hybrid first-order discrete-time MFC-VRFT, second-order discrete-time MFC-VRFT and MFAC-VRFT controllers presented in Section 6.2 is included in the following optimization problem:

$$\chi_{(\psi)}^* = \arg \min_{\chi_{(\psi)}} J(\chi_{(\psi)}), \ J_{(\psi)}(\chi_{(\psi)}) = \frac{1}{2} E \left\{ \sum_{k=1}^N \left[e_{(\psi)}(k, \chi_{(\psi)}) \right]^2 \right\}, \quad (6.67)$$

where $\chi_{(\psi)}^*$ is the optimal vector parameter of the controller obtained in terms of solving the optimization problem specific to VRFT, $e_{(\psi)} = y_{d(\psi)} - y_{(\psi)}$ is the control error, and the subscript $\psi \in \{c, a, p\}$ indicates both the controlled output and its corresponding controller, namely c – the cart position (i.e., $y_{(c)} = y_1$), a – the arm angular position (i.e., $y_{(a)} = y_2$) and p – the payload position (i.e., $y_{(p)} = y_3$). The mathematical expectation $E\{\Xi\}$ is taken with respect to the stochastic probability distribution of the disturbance inputs applied to the process and thus affects the control system behavior. The sampling period is $T_s = 0.01$ s, and the time horizon is 70 s, leading to a number of $N = 70 / 0.01 = 7,000$ s samples.

The hybrid first-order discrete-time MFC-VRFT, second-order discrete-time MFC-VRFT and MFAC-VRFT controllers are designed and tuned, one for each controlled output of the process, respecting the steps given in Section 6.2. The application of the steps is given as follows in relation with the Matlab & Simulink programs and schemes.

The reference input vector is obtained by first applying the three step signals:

$$y_c^*(k) = 0.15 \text{ if } k \in [0,2,000], 0.1 \text{ if } k \in (2,000,3,500], -0.05 \text{ if } k \in (3,500,5,000],$$

$$0 \text{ if } k \in (5,000,7,000]$$

(6.68)

for cart position control,

$$y_a^*(k) = 0 \text{ if } k \in [0,500], 0.15 \text{ if } k \in (500,2,500], -0.15 \text{ if } k \in (2,500,4,000],$$

$$0 \text{ if } k \in (4,000,7,000]$$

(6.69)

for arm angular position control and

$$y_p^*(k) = 0 \text{ if } k \in [0,1,500], 0.1 \text{ if } k \in (1,500,3,000], -0.05 \text{ if } k \in (3,000,4,500],$$

$$0 \text{ if } k \in (4,500,7,000],$$

(6.70)

for cart position control. The above step signals are filtered to obtain the desired reference trajectory $y_{d(\psi)} = y_{(\psi)}^* H_{(\psi)}$ that represents the reference input vector using the filters

$$H_{(c)}(s) = \frac{1}{1+0.2s}$$

(6.71)

for cart position control,

$$H_{(a)}(s) = \frac{1}{1+0.2s}$$

(6.72)

for arm angular position control and

$$H_{(p)}(s) = \frac{1}{1+0.2s}$$

(6.73)

for payload position control, with the corresponding discrete transfer functions:

$$H_{(c)}(z^{-1}) = \frac{0.0488}{1-0.9512z^{-1}},$$

(6.74)

$$H_{(a)}(z^{-1}) = \frac{0.0465}{1 - 0.9535z^{-1}}, \tag{6.75}$$

$$H_{(p)}(z^{-1}) = \frac{0.0328}{1 - 0.9672z^{-1}}. \tag{6.76}$$

For the sake of reducing the big volume of this chapter and enabling the relatively easy understanding of the application of the controllers, only the results of hybrid first-order discrete-time MFC-VRFT with P component, second-order discrete-time MFC-VRFT with P component and MFAC-VRFT in the CFDL version are presented as follows.

6.3.1.1 The Discrete-Time First-Order MFC-VRFT Controller with P Component

Three first-order discrete-time MFC-VRFT controllers with P component are used separately in three SISO control system structures, where the parameter vectors of the first-order discrete-time iP controllers are

$$\chi_{(\psi)} = K_1, \tag{6.77}$$

with $\psi \in \{c, a, p\}$, i.e., one parameter for each controller.

The controllers are nonlinear since saturation-type nonlinearities are inserted on the three controller outputs in order to match the power electronics of the actuators, which operate on the basis of the pulse width modulation (PWM) principle. As shown in Chapter 1, the three control signals (or control inputs) are within -1 and 1, ensuring the necessary connection to the tower crane system subsystem, which models the controlled process in the control system structures developed in this chapter and the next chapters and also includes three Zero-Order Hold (ZOH) blocks to enable digital control. The tower crane system subsystem belongs to the Process.mdl Simulink diagram, which is included in the accompanying Matlab & Simulink programs given in Chapter 1.

The parameter $\alpha = 0.0015$ is chosen according to step 1, then the parameters of the MFC-VRFT controller with P component are calculated using only the input/output data measured from the controlled process obtained by going through step 2 and solving the optimization problem in (6.9) using genetic algorithms implemented in terms of the Matlab predefined function. The discrete-time reference model is specified as transfer functions in (6.74) for cart position, (6.75) for arm angular position and (6.76) for payload according to step 3. According to step 4, to establish the parameter K_1 in (6.77), to ensure the condition (3.74), solving the optimization problem in (6.9) leads to

$$\chi_{(c)} = -0.05243, \tag{6.78}$$

$$\chi_{(a)} = -0.0915, \tag{6.79}$$

$$\chi_{(p)} = -0.045. \tag{6.80}$$

The details related to VRFT and the programs of steps 1–4 are implemented in script_initialization_MFC_VRFT_c.m (with its Simulink diagram in OL _ SISO _ PRBS _ c.mdl), initial _ data _ after _ PRBS _ MFC _ VRFT _ c.m, optimization _ MFC _ VRFT _ c.m, and script _ function _ MFC _ VRFT _ c.m (with its Simulink diagram in OL _ SISO _ MFC _ VRFT _ c.mdl), Matlab programs for cart position control, in script _ initialization_MFC _ VRFT _ a.m (with its Simulink diagram in OL _ SISO _ PRBS _ a.mdl), initial _ data _ after _ PRBS _ MFC _ VRFT _ a.m, optimization _ MFC _ VRFT _ a.m, and script _ function _ MFC _ VRFT _ a.m (with its Simulink diagram in OL _ SISO _ MFC _ VRFT _ a.mdl), Matlab programs for arm angular position control and in script _ initialization _ MFC _ VRFT _ p.m (with its Simulink diagram in OL _ SISO _ PRBS _ p.mdl), initial _ data _ after _ PRBS _ MFC _ VRFT _ p.m, optimization _ MFC _ VRFT _ p.m, and script _ function _ MFC _ VRFT _ p.m (with its Simulink diagram in OL _ SISO _ MFC _ VRFT _ p.mdl), Matlab programs for payload position control. The details of testing the first-order discrete-time MFC-VRFT controller with P component in closed loop are implemented in the MFC _ VRFT _ iPID _ 1st _ order _ c.m Matlab program for cart position control, the MFC _ VRFT _ iPID _ 1st _ order _ a.m Matlab program for arm angular position control and in the MFC _ VRFT _ iPID _ 1st _ order _ p.m Matlab program for payload position control. The digital simulations of the three control systems with first-order discrete-time iP MFC controllers with the parameters tuned via VRFT are conducted using the CS _ iP _ VRFT _ SISO _ c.mdl Simulink diagram for cart position control, the CS _ iP _ VRFT _ SISO _ a.mdl Simulink diagram for arm angular position control and the CS _ iP _ VRFT _ SISO _ p.mdl Simulink diagram for payload position control. These three Simulink diagrams make use of the state-space model of the tower crane system viewed as a controlled process, which is described in Chapter 1 and is implemented in the process _ model.m S-function.

The simulation results obtained for the control systems with first-order discrete-time iP controllers as MFC-VRFT controllers with the parameters in (6.78–6.80) and the user-chosen parameter α are illustrated in Figure 6.4 for cart position control, Figure 6.5 for arm angular position control and Figure 6.6 for payload position control.

The experimental results obtained for the control systems with first-order discrete-time iP controllers as MFC-VRFT controllers with the parameters in (6.78–6.80) and the user-chosen parameter α are illustrated in Figure 6.7 for cart position control, Figure 6.8 for arm angular position control and Figure 6.9 for payload position control.

6.3.1.2 The Discrete-Time Second-Order MFC-VRFT Controller with P Component

Three second-order discrete-time MFC-VRFT controllers with P component are designed and tuned separately in three SISO control system structures, where the parameter vectors of the second-order discrete-time iP controllers are

$$\chi_{(\psi)} = K_1, \tag{6.81}$$

with $\psi \in \{c, a, p\}$, i.e. one parameter for each controller.

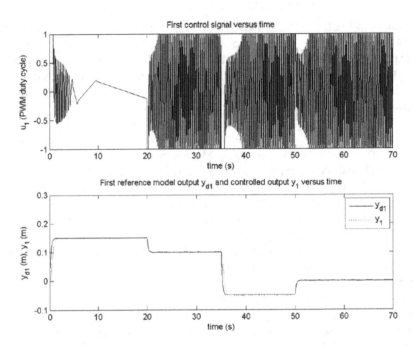

FIGURE 6.4 Simulation results of cart position control system with first-order discrete-time iP controller as MFC-VRFT controller.

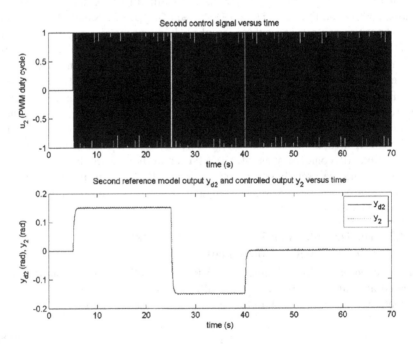

FIGURE 6.5 Simulation results of arm angular position control system with first-order discrete-time iP controller as MFC-VRFT controller.

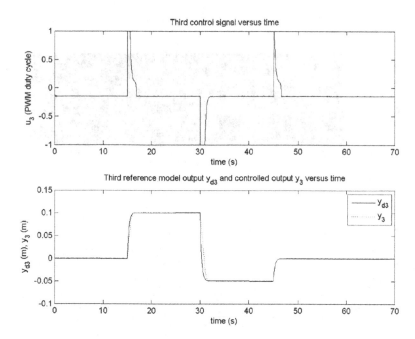

FIGURE 6.6 Simulation results of payload position control system with first-order discrete-time iP controller as MFC-VRFT controller.

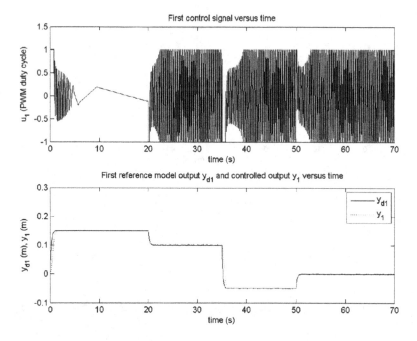

FIGURE 6.7 Experimental results of cart position control system with first-order discrete-time iP controller as MFC-VRFT controller.

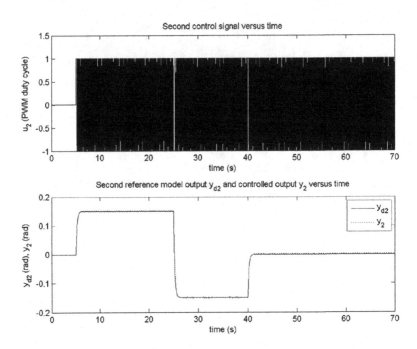

FIGURE 6.8 Experimental results of arm angular position control system with first-order discrete-time iP controller as MFC-VRFT controller.

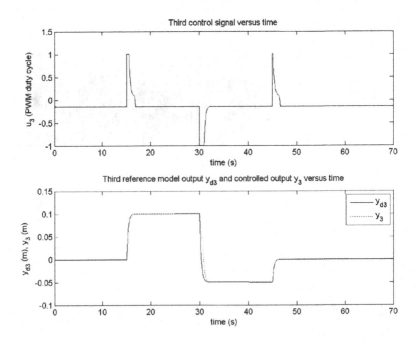

FIGURE 6.9 Experimental results of payload position control system with first-order discrete-time iP controller as MFC-VRFT controller.

The parameter $\alpha = 0.0015$ is chosen according to step 1, then the parameters of the MFC-VRFT controller with P component were calculated using only the input/output data measured from the controlled process obtained by going through step 2 and solving the optimization problem in (6.9) using genetic algorithms implemented in terms of the Matlab predefined function. According to step 4, to establish the parameter K_1 in (6.81), to ensure the condition (3.86), solving the optimization problem in (6.9) results in

$$\chi_{(c)} = -0.7698, \tag{6.82}$$

$$\chi_{(a)} = -0.914, \tag{6.83}$$

$$\chi_{(p)} = -0.98. \tag{6.84}$$

The details related to VRFT and the programs of steps 1–4 are implemented in script _ initialization _ MFC _ 2nd _ order _ VRFT _ c.m (with its Simulink diagram in OL _ SISO _ PRBS _ c.mdl), initial _ data _ after _ PRBS _ MFC _ 2nd _ order _ VRFT _ c.m, optimization _ MFC _ 2nd _ order _ VRFT _ c.m, and script _ function _ MFC _ 2nd _ order _ VRFT _ c.m (with its Simulink diagram in OL _ SISO _ MFC _ 2nd _ order _ VRFT _ c.mdl), Matlab programs for cart position control, in script _ initialization _ MFC _ 2nd _ order _ VRFT _ a.m (with its Simulink diagram in OL _ SISO _ PRBS _ a.mdl), initial _ data _ after _ PRBS _ MFC _ 2nd _ order _ VRFT _ a.m, optimization _ MFC _ 2nd _ order _ VRFT _ a.m, and script _ function _ MFC _ 2nd _ order _ VRFT _ a.m (with its Simulink diagram in OL _ SISO _ MFC _ 2nd _ order _ VRFT _ a.mdl), Matlab programs for arm angular position control and in script _ initialization _ MFC _ 2nd _ order _ VRFT _ p.m (with its Simulink diagram in OL _ SISO _ PRBS _ p.mdl), initial _ data _ after _ PRBS _ MFC _ 2nd _ order _ VRFT _ p.m, optimization _ MFC _ 2nd _ order _ VRFT _ p.m, and script _ function _ MFC _ 2nd _ order _ VRFT _ p.m (with its Simulink diagram in OL _ SISO _ MFC _ 2nd _ order _ VRFT _ p.mdl), Matlab programs for payload position control. The details of testing the second-order discrete-time MFC-VRFT controller with P component in closed loop are implemented in the MFC _ VRFT _ iPID _ 2nd _ order _ c.m Matlab program for cart position control, the MFC _ VRFT _ iPID _ 2nd _ order _ a.m Matlab program for arm angular position control and in the MFC _ VRFT _ iPID _ 2nd _ order _ p.m Matlab program for payload position control. The digital simulations of the control systems with the three second-order discrete-time iP MFC algorithms with the parameters tuned via VRFT are conducted using the CS _ iP _ 2nd _ order _ VRFT _ SISO _ c.mdl Simulink diagram for cart position control, the CS _ iP _ 2nd _ order _ VRFT _ SISO _ a.mdl Simulink diagram for arm angular position control and the CS _ iP _ 2nd _ order _ VRFT _ SISO _ p.mdl Simulink diagram for payload position control. These three Simulink diagrams make use of the state-space model of the tower crane system viewed as a controlled process, which is described in Chapter 1 and is implemented in the `process _ model.m` S-function.

The simulation results obtained for the control systems with second-order discrete-time iP controllers as MFC-VRFT controllers with the parameters in (6.82–6.84) and the user-chosen parameter α are illustrated in Figure 6.10 for cart position control, Figure 6.11 for arm angular position control and Figure 6.12 for payload position control.

The experimental results obtained for the control systems with second-order discrete-time iP controllers as MFC-VRFT controllers with the parameters in (6.82–6.84) and the user-chosen parameter α are illustrated in Figure 6.13 for cart position control, Figure 6.14 for arm angular position control and Figure 6.15 for payload position control.

6.3.1.3 The MFAC-VRFT Algorithms in Compact-Form Dynamic Linearization Version

Three MFAC-VRFT controllers in the CFDL version are separately used as controllers in the three SISO control system structures, where the parameter vectors of the MFAC-VRFT controllers are

$$\chi_{\text{ext}\,(\psi)} = [\eta \ \lambda \ \mu \ \rho \ \hat{\varphi}(1)]^T, \tag{6.85}$$

with $\psi \in \{c, a, p\}$, i.e. one parameter for each controller.

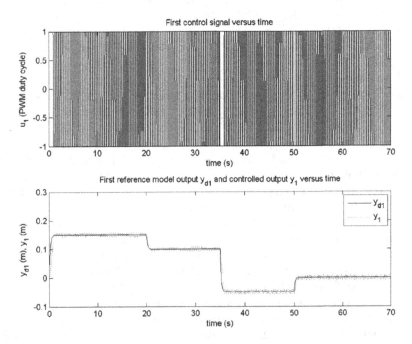

FIGURE 6.10 Simulation results of cart position control system with second-order discrete-time iP controller as MFC-VRFT controller.

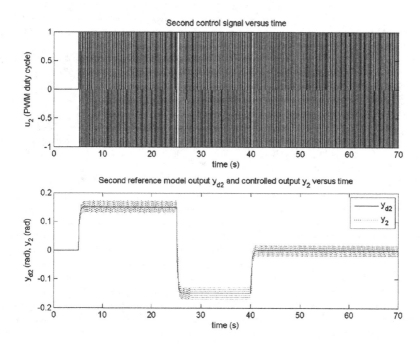

FIGURE 6.11 Simulation results of arm angular position control system with second-order discrete-time iP controller as MFC-VRFT controller.

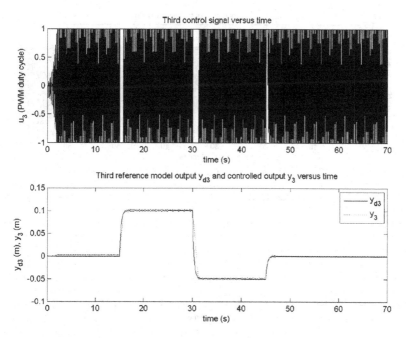

FIGURE 6.12 Simulation results of payload position control system with second-order discrete-time iP controller as MFC-VRFT controller.

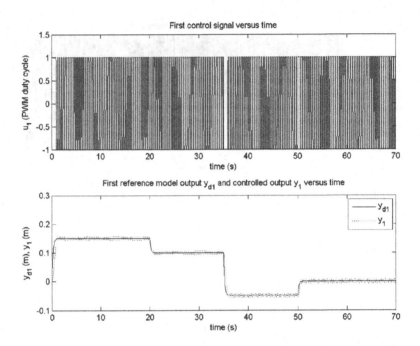

FIGURE 6.13 Experimental results of cart position control system with second-order discrete-time iP controller as MFC-VRFT controller.

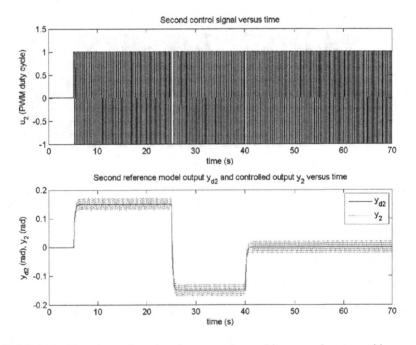

FIGURE 6.14 Experimental results of arm angular position control system with second-order discrete-time iP controller as MFC-VRFT controller.

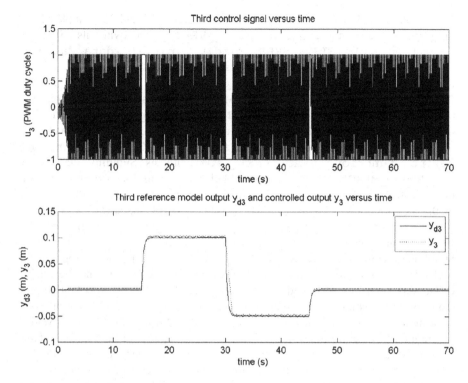

FIGURE 6.15 Experimental results of payload position control system with second-order discrete-time iP controller as MFC-VRFT controller.

The parameters of the MFAC-VRFT controllers in the CFDL version are calculated using only the input/output data measured from the controlled process obtained according to step 1, next the control systems designer must establish the discrete-time reference model **m** specified as transfer functions in (6.74) for cart position, (6.75) for arm angular position and (6.76) for payload that leads to the virtual reference input vector $\bar{r}(k)$ according to (6.7) in step 2. Finally the parameters $\chi_{\text{ext}(\psi)}$ of the MFAC-VRFT controllers in the CFDL version are obtained after solving the optimization problem in (6.9) such that $\hat{\Phi}(k)$ will vary during the experiments according to the reset constraints imposed in (5.6) according to step 3, where the parameters of lower and the upper limits will be considered as $1,000\,\hat{\Phi}(1)$ for the upper limit and $0.001\,\hat{\Phi}(1)$ for the lower limit and are imposed by ensuring conditions (5.4) and (5.6) and the parameters $\eta \in (0,1)$, $\lambda > 0$, $\mu > 0$, and $\rho > 0$ are also chosen fulfilling the MFAC-CFDL algorithm constraints. These steps lead to

$$\chi_{\text{ext}(c)} = [0.5771 \ \ 3.2334 \ \ 28.9407 \ \ 60.6864 \ \ 5.6469]^T, \tag{6.86}$$

$$\chi_{\text{ext}(a)} = [0.4442 \ \ 0.00145 \ \ 0.00887 \ \ 850 \ \ 18.8211]^T, \tag{6.87}$$

$$\chi_{\text{ext}(p)} = [0.6921 \ \ 2.8527 \ \ 4.1251 \ \ 34.3581 \ \ 7.1244]^T. \tag{6.88}$$

The details related to VRFT and the programs of steps 1, 2 and 3 are implemented in script_initialization_MFAC_VRFT_c.m (with its Simulink diagram in OL_SISO_PRBS_c.mdl), initial_data_after_PRBS_MFAC_VRFT_c.m, optimization_MFAC_VRFT_c.m, and script_function_MFAC_VRFT_c.m (with its Simulink diagram in OL_SISO_MFAC_VRFT_c.mdl), Matlab programs for cart position control, in script_initialization_MFAC_VRFT_a.m (with its Simulink diagram in OL_SISO_PRBS_a.mdl), initial_data_after_PRBS_MFAC_VRFT_a.m, optimization_MFAC_VRFT_a.m, and script_function_MFAC_VRFT_a.m (with its Simulink diagram in OL_SISO_MFAC_VRFT_a.mdl), Matlab programs for arm angular position control and in script_initialization_MFAC_VRFT_p.m (with its Simulink diagram in OL_SISO_PRBS_p.mdl), initial_data_after_PRBS_MFAC_VRFT_p.m, optimization_MFAC_VRFT_p.m, and script_function_MFAC_VRFT_p.m (with its Simulink diagram in OL_SISO_MFAC_VRFT_p.mdl), Matlab programs for payload position control. The details of the testing the MFAC-VRFT controllers in the CFDL version are implemented in the MFAC_CFDL_VRFT_c.m Matlab program for cart position control, the MFAC_CFDL_VRFT_a.m Matlab program for arm angular position control and in the MFAC_CFDL_VRFT_p.m Matlab program for payload position control. The digital simulations of the control systems with the three MFAC-VRFT controllers are carried out using the CS_MFAC_CFDL_VRFT_SISO_c.mdl Simulink diagram for cart position control, the CS_MFAC_CFDL_VRFT_SISO_a.mdl Simulink diagram for arm angular position control and the CS_MFAC_CFDL_VRFT_SISO_p.mdl Simulink diagram for payload position control. These three Simulink diagrams make use of the state-space model of the tower crane system viewed as a controlled process, which is described in Chapter 1 and is implemented in the process_model.m S-function.

The simulation results obtained for the control systems with MFAC-VRFT controllers in the CFDL version with the parameters in (6.86–6.88) with $1,000\hat{\varphi}(1)_{(\psi)}$ as the upper limit and $0.001\hat{\varphi}_{(\psi)}(1)$ as the lower limit are illustrated in Figure 6.16 for cart position control, Figure 6.17 for arm angular position control and Figure 6.18 for payload position control.

The experimental obtained for the control systems with MFAC-CFDL controllers with the parameters in (6.86–6.88) with $1,000\hat{\varphi}(1)_{(\psi)}$ as the upper limit and $0.001\hat{\varphi}(1)_{(\psi)}$ as the lower limit are illustrated in Figure 6.19 for cart position control, Figure 6.20 for arm angular position control and Figure 6.21 for payload position control.

6.3.2   MIMO CONTROL SYSTEMS

In this section, the validation of the hybrid first-order discrete-time MFC-VRFT, second-order discrete-time MFC-VRFT and MFAC-VRFT controllers is carried out on the tower crane system equipment with three inputs and three outputs, in this case for $n = 3$. The performance is assessed using the following optimization problem:

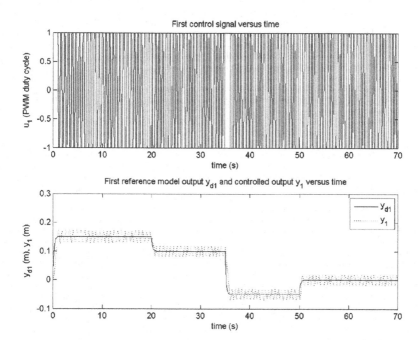

FIGURE 6.16 Simulation results of cart position control system with the MFAC-VRFT controller in the CFDL version.

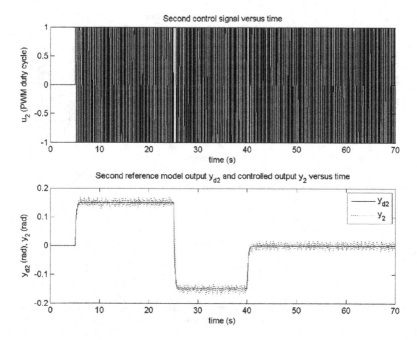

FIGURE 6.17 Simulation results of arm angular position control system with the MFAC-VRFT controller in the CFDL version.

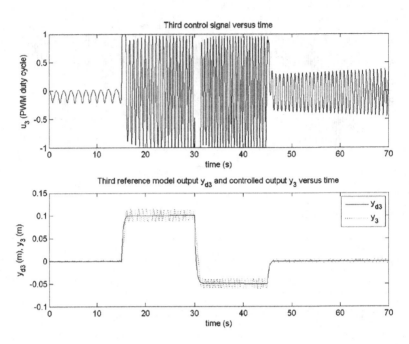

FIGURE 6.18 Simulation results of payload position control system with the MFAC-VRFT controller in the CFDL version.

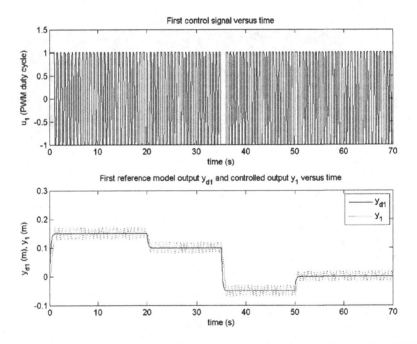

FIGURE 6.19 Experimental results of cart position control system with the MFAC-VRFT controller in the CFDL version.

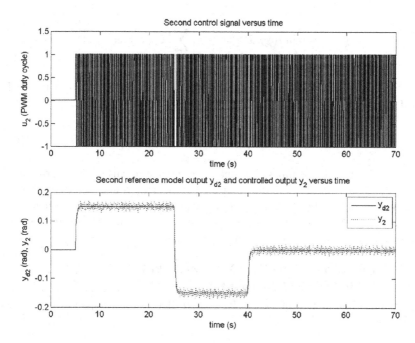

FIGURE 6.20 Experimental results of arm angular position control system with the MFAC-VRFT controllers in the CFDL version.

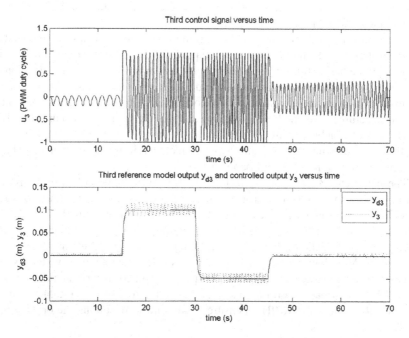

FIGURE 6.21 Experimental results of payload position control system with the MFAC-VRFT controllers in the CFDL version.

$$\chi^* = \arg \min_{\chi} J(\chi), \ J(\chi) = \frac{1}{2} E \left\{ \sum_{k=1}^{N} \left[\mathbf{e}^T(k,\chi)\mathbf{e}(k,\chi) \right] \right\}, \tag{6.89}$$

where χ^* is the optimal vector parameter of the controller obtained using the VRFT-based solving of the optimization problem, $\mathbf{e}$ is the control error vector

$$\mathbf{e} = \mathbf{y}_d - \mathbf{y} = [e_1 \ e_2 \ e_3]^T = [y_{d1} - y_1 \ y_{d2} - y_2 \ y_{d3} - y_3]^T, \tag{6.90}$$

and the control error dynamics e_1, e_2 and e_3 are related to cart position, arm angular position and payload position, the controlled outputs y_1, y_2 and y_3 are the cart position, the arm angular position and the payload position, respectively. The mathematical expectation $E\{\Xi\}$, the sampling period, the time horizon, the number of samples and the reference input vector are same as in the SISO case.

6.3.2.1 The Discrete-Time First-Order MFC-VRFT Controller with P Component

The discrete-time first-order MFC-VRFT controller with P component uses parameter vector

$$\chi = [\chi_{(c)1} \ \chi_{(a)1} \ \chi_{(p)1}]^T = [K_{11} \ K_{12} \ K_{13}]^T, \tag{6.91}$$

where $\mathbf{K}_1 = \text{diag}(K_{11}, K_{12}, K_{13})$, $\psi \in \{c, a, p\}$, i.e. three parameters for the MIMO controller, one parameter for each motion.

The controller is nonlinear since saturation-type nonlinearities are inserted on the controller outputs in order to match the power electronics of the actuators, which operate on the basis of the PWM principle. As shown in Chapter 1, the three control signals (or control inputs) are within -1 and 1, ensuring the necessary connection to the tower crane system subsystem, which models the controlled process in the control system structures developed in this chapter and the next chapters, also includes three ZOH blocks to enable digital control. The tower crane system subsystem belongs to the Process.mdl Simulink diagram, which is included in the accompanying Matlab & Simulink programs given in Chapter 1.

The parameter $\Lambda_{\text{MFC}} = [\alpha_1 \ \alpha_2 \ \alpha_3]^T = [0.0015 \ 0.0015 \ 0.0015]^T$ in (6.22) is chosen according to step 1, and then, the parameters of the discrete-time first-order MFC-VRFT controller with P component are calculated using only the input/output data measured from the controlled process obtained by going through step 2 and solving the optimization problem in (6.9) using genetic algorithms implemented in terms of the Matlab predefined function. The discrete-time reference model is specified as transfer functions in (6.74) for cart position, (6.75) for arm angular position and (6.76) for payload according to step 3. According to step 4, to establish parameters K_{11}, K_{12} and K_{13} in (6.91), to ensure condition (3.74), solving the optimization problem in (6.9) leads to

$$\chi = [\chi_{(c)1} \ \chi_{(a)1} \ \chi_{(p)1}]^T = [K_{11} \ K_{12} \ K_{13}]^T = [-0.7884 \ -0.1280 \ -0.0101]^T. \tag{6.92}$$

The details related to VRFT and the programs of steps 1–4 are implemented in script _ initialization _ MFC _ VRFT.m (with its Simulink diagram in OL _ MIMO _ PRBS.mdl), initial _ data _ after _ MFC _ VRFT _ PRBS.m, optimization _ MFC _ VRFT.m and script _ function _ MFC _ VRFT.m (with its Simulink diagram in OL _ MIMO _ MFC _ VRFT.mdl) Matlab programs. The details of testing the first-order discrete-time MFC-VRFT controller with P component in closed loop are implemented in the MFC _ VRFT _ iPID _ 1st _ order.m Matlab program for MIMO control. The simulations of three control systems with first-order discrete-time iP MFC controllers with the parameters tuned via VRFT are conducted using the CS _ iP _ VRFT _ MIMO. mdl Simulink diagram for MIMO control. This Simulink diagram makes use of the state-space model of the tower crane system viewed as a controlled process, which is described in Chapter 1 and is implemented in the process _ model.m S-function.

The simulation results obtained for the control system with discrete-time first-order MFC-VRFT MIMO controller with P component with the parameters in (6.92) and the user-chosen parameter α are illustrated in Figure 6.22 for cart position control, Figure 6.23 for arm angular position control and Figure 6.24 for payload position control.

The experimental results obtained for the discrete-time first-order MFC-VRFT MIMO controller with P component with the parameters in (6.92) and the user-chosen parameter α are illustrated in Figure 6.25 for cart position control, Figure 6.26 for arm angular position control and Figure 6.27 for payload position control.

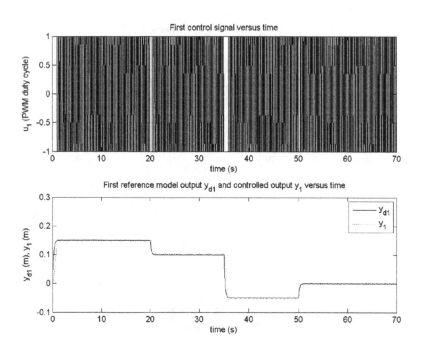

FIGURE 6.22 Simulation results of cart position control system with first-order discrete-time iP controller as MFC-VRFT MIMO controller.

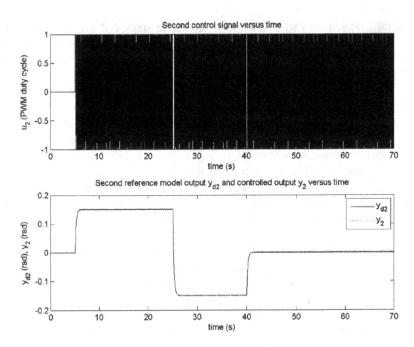

FIGURE 6.23 Simulation results of arm angular position control system with first-order discrete-time iP controller as MFC-VRFT MIMO controller.

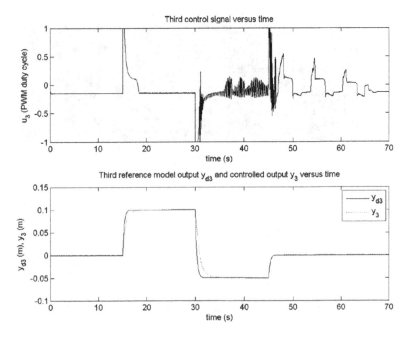

FIGURE 6.24 Simulation results of payload position control system with first-order discrete-time iP controller as MFC-VRFT MIMO controller.

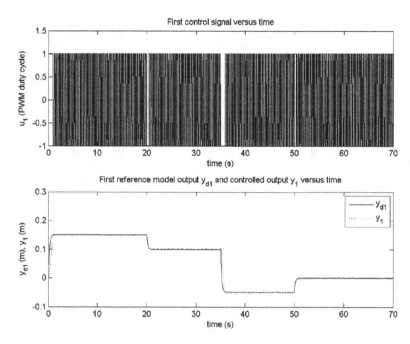

FIGURE 6.25 Experimental results of cart position control system with first-order discrete-time iP controller as MFC-VRFT MIMO controller.

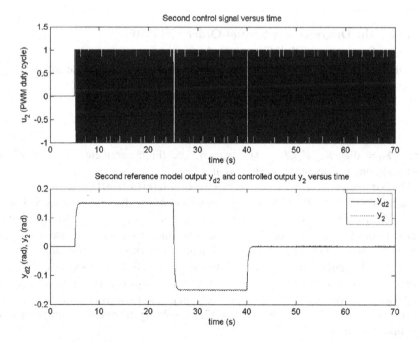

FIGURE 6.26 Experimental results of arm angular position control system with first-order discrete-time iP controller as MFC-VRFT MIMO controller.

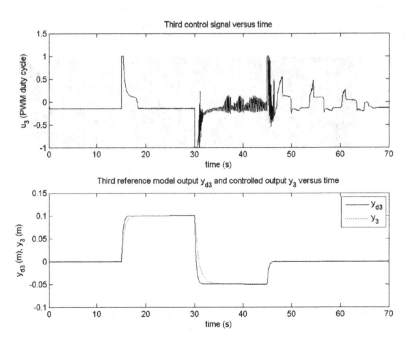

FIGURE 6.27 Experimental results of payload position control system with first-order discrete-time iP controller as MFC-VRFT MIMO controller.

6.3.2.2 The Discrete-Time Second-Order MFC-VRFT Controller with P Component

The discrete-time second-order MFC-VRFT controller with P component makes use of the following parameter vector in its MIMO implementation:

$$\chi = [\chi_{(c)1} \ \chi_{(a)1} \ \chi_{(p)1}]^T = [K_{11} \ K_{12} \ K_{13}]^T, \tag{6.93}$$

where $\mathbf{K}_1 = \mathrm{diag}(K_{11}, K_{12}, K_{13})$, $\psi \in \{c, a, p\}$, i.e., three parameters for the MIMO controller, one parameter for each motion.

The parameter $\Lambda_{\mathrm{MFC}} = [\alpha_1 \ \alpha_2 \ \alpha_3]^T = [0.0015 \ 0.0015 \ 0.0015]^T$ in (6.43) is chosen according to step 1, and then, the parameters of the discrete-time second-order MFC-VRFT controller with P component are calculated using only the input/output data measured from the controlled process obtained by going through step 2 and solving the optimization problem in (6.9) genetic algorithms implemented in terms of the Matlab predefined function. The discrete-time reference model is specified as transfer functions in (6.74) for cart position, (6.75) for arm angular position and (6.76) for payload according to step 3. According to step 4, to establish parameters K_{11}, K_{12} and K_{13} in (6.91) and to ensure condition (3.83), solving the optimization problem in equation (6.9) results in

$$\chi = [\chi_{(c)1} \ \chi_{(a)1} \ \chi_{(p)1}]^T = [K_{11} \ K_{12} \ K_{13}]^T = [-0.75 \ -0.89 \ -0.85]^T. \tag{6.94}$$

The details related to VRFT and the programs of steps 1–4 are implemented in script _ initialization _ MFC _ 2nd _ order _ VRFT.m (with its Simulink diagram in OL _ MIMO _ PRBS.mdl), initial _ data _ after _ PRBS _ MFC _ 2nd _ order _ VRFT.m, optimization _ MFC _ 2nd _ order _ VRFT.m and script _ function _ MFC _ 2nd _ order _ VRFT.m (with its Simulink diagram in OL _ MIMO _ MFC _ 2nd _ order _ VRFT.mdl) Matlab programs. The details of testing the second-order discrete-time MFC-VRFT controller with P component in the closed-loop control system are implemented in the MFC _ VRFT _ iPID _ 2nd _ order.m Matlab program for MIMO control. The simulations of three control systems with second-order discrete-time iP MFC controllers with the parameters tuned via VRFT are conducted using the CS _ iP _ 2nd _ order _ VRFT _ MIMO. mdl Simulink diagram for MIMO control. This Simulink diagram makes use of the state-space model of the tower crane system viewed as a controlled process, which is described in Chapter 1 and is implemented in the process _ model.m S-function.

The simulation results obtained for the control system with discrete-time second-order MFC-VRFT MIMO controller with P component with the parameters in (6.94) and the user-chosen parameter α are illustrated in Figure 6.28 for cart position control, Figure 6.29 for arm angular position control and Figure 6.30 for payload position control.

The experimental results obtained for the control system with discrete-time second-order MFC-VRFT MIMO controller with P component with the parameters in (6.94) and the user-chosen parameter α are illustrated in Figure 6.31 for cart position control, Figure 6.32 for arm angular position control and Figure 6.33 for payload position control.

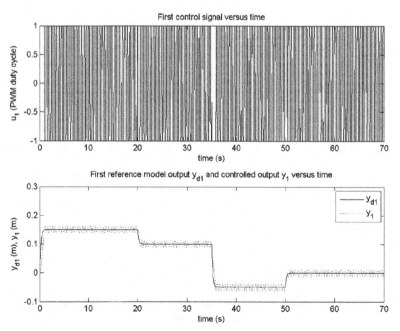

FIGURE 6.28 Simulation results of cart position control system with second-order discrete-time iP controller as MFC-VRFT MIMO controller.

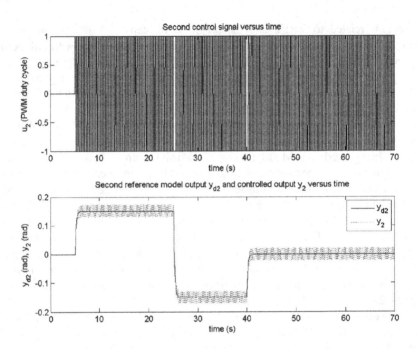

FIGURE 6.29 Simulation results of arm angular position control system with second-order discrete-time iP controller as MFC-VRFT MIMO controller.

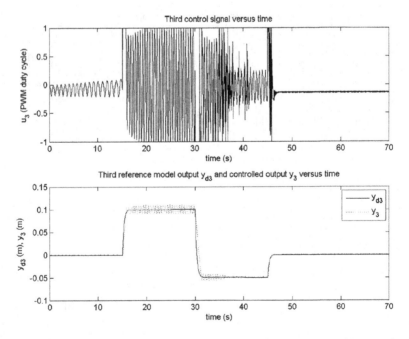

FIGURE 6.30 Simulation results of payload position control system with second-order discrete-time iP controller as MFC-VRFT MIMO controller.

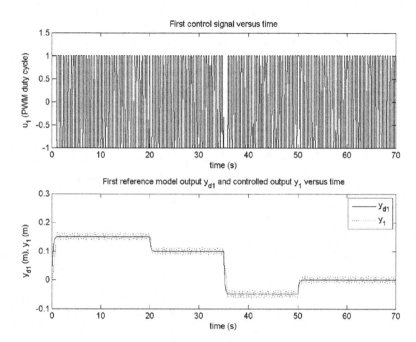

FIGURE 6.31 Experimental results of cart position control system with second-order discrete-time iP controller as MFC-VRFT MIMO controller.

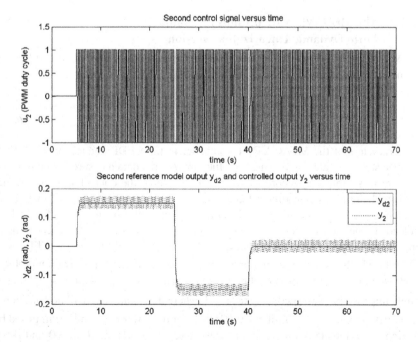

FIGURE 6.32 Experimental results of arm angular position control system with second-order discrete-time iP controller as MFC-VRFT MIMO controller.

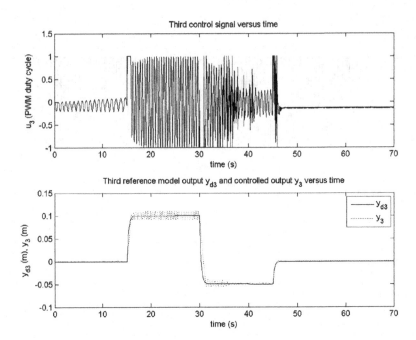

FIGURE 6.33 Experimental results of payload position control system with second-order discrete-time iP controller as MFC-VRFT MIMO controller.

6.3.2.3 The MFAC-VRFT Algorithms in Compact-Form Dynamic Linearization Version

The MFAC-VRFT controller in the CFDL version operates with the following parameter vector in its MIMO implementation:

$$\chi_{ext} = [\eta \ \lambda \ \mu \ \rho \ \hat{\phi}_{11}(1) \ \hat{\phi}_{12}(1) \ \hat{\phi}_{13}(1) \ \hat{\phi}_{21}(1) \ \hat{\phi}_{22}(1) \ \hat{\phi}_{23}(1) \ \hat{\phi}_{31}(1) \ \hat{\phi}_{32}(1) \ \hat{\phi}_{33}(1)]^T.$$
(6.95)

The parameters of the MFAC-VRFT controller in the CFDL version of the MIMO control system are calculated using only the input/output data measured from the controlled process obtained according to step 1, and next, the control system designer must establish the discrete-time reference model **m** specified as transfer functions in equation (6.74) for cart position, (6.75) for arm angular position and (6.76) for payload that leads to the virtual reference vector $\bar{r}(k)$ according to equation (6.7) according to step 2. Finally, the parameters χ_{ext} of the MFAC-VRFT controller in the CFDL version of the MIMO control system are obtained after solving the optimization problem in (6.9) such that $\hat{\Phi}(k)$ will vary during the experiments according to the reset constraints imposed by (5.6) according to step 3, where the parameters will be considered as $1,000\,\hat{\Phi}(1)$ for the upper limit and $0.001\,\hat{\Phi}(1)$ for the lower limit and are imposed by ensuring conditions (5.4) and (5.6), and parameters $\eta \in (0,1)$, $\lambda > 0$, $\mu > 0$ and $\rho > 0$ are also chosen fulfilling the MFAC-CFDL algorithm constraints, resulting in

$$\chi = [0.9999 \ 0.0883 \ 0.3190 \ 890.25 \ 2.4930 \ 0 \ 0 \ 0 \ 0.75 \ 0 \ 0 \ 0 \ 2.4435]^T. \quad (6.96)$$

The details related to VRFT and the programs of steps 1–3 are implemented in script _ initialization _ MFAC _ VRFT.m (with its Simulink diagram in OL _ MIMO _ PRBS.mdl), initial _ data _ after _ PRBS _ MFAC _ VRFT.m, optimization _ MFAC _ VRFT.m and script _ function _ MFAC _ VRFT.m (with its Simulink diagram in OL _ MIMO _ MFAC _ VRFT.mdl) Matlab programs. The details of testing the MFAC-VRFT controller in the CFDL version of the MIMO control system are implemented in the MFAC _ CFDL _ VRFT.m Matlab program for MIMO control. The simulations of the MFAC-VRFT controllers in the CFDL version of the MIMO control system with the parameters tuned via VRFT are conducted using the CS _ MFAC _ CFDL _ VRFT _ MIMO. mdl Simulink diagram for MIMO control. The Simulink diagram makes use of the state-space model of the tower crane system viewed as a controlled process, which is described in Chapter 1 and is implemented in the process _ model.m S-function.

The simulation results obtained for the control system with MFAC-VRFT controller in the CFDL version of the MIMO control system with the parameters in (6.96) with $1,000\,\hat{\Phi}(1)$ as the upper limit and $0.001\,\hat{\Phi}(1)$ as the lower limit are illustrated in Figure 6.34 for cart position control, Figure 6.35 for arm angular position control and Figure 6.36 for payload position control.

The experimental results obtained for the control system with MFAC-VRFT controller in the CFDL version of the MIMO control system with parameters in (6.96) with $1,000\,\hat{\Phi}(1)$ as the upper limit and $0.001\,\hat{\Phi}(1)$ as the lower limit are illustrated in Figure 6.37 for cart position control, Figure 6.38 for arm angular position control and Figure 6.39 for payload position control.

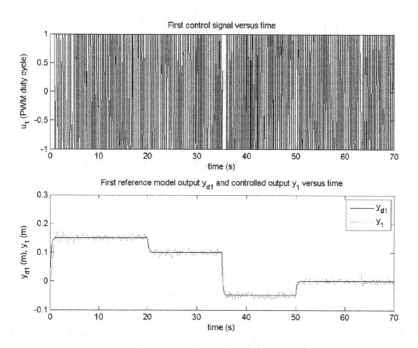

FIGURE 6.34 Simulation results of cart position control system with the MFAC-VRFT algorithm in the CFDL version as MFC-VRFT MIMO controller.

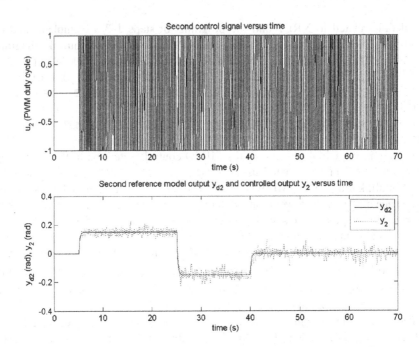

FIGURE 6.35 Simulation results of arm angular position control system with the MFAC-VRFT algorithm in the CFDL version as MFC-VRFT MIMO controller.

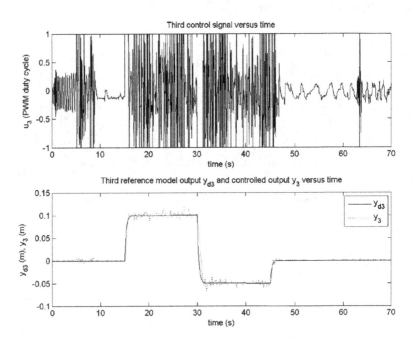

FIGURE 6.36 Simulation results of payload position control system with the MFAC-VRFT algorithm in the CFDL version as MFC-VRFT MIMO controller.

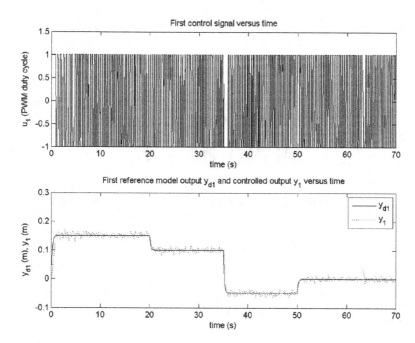

FIGURE 6.37 Experimental results of cart position control system with the MFAC-VRFT algorithm in the CFDL version as MFC-VRFT MIMO controller.

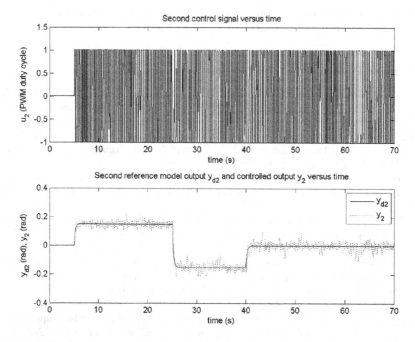

FIGURE 6.38 Experimental results of arm angular position control system with the MFAC-VRFT algorithm in the CFDL version as MFC-VRFT MIMO controller.

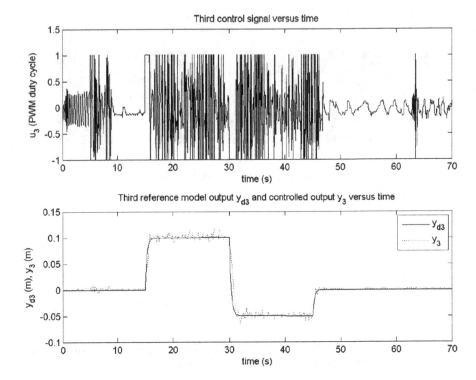

FIGURE 6.39 Experimental results of payload position control system with the MFAC-VRFT algorithm in the CFDL version as MFC-VRFT MIMO controller.

REFERENCES

[Cam05] M. C. Campi and S. M. Savaresi, "Virtual reference feedback tuning for non-linear systems," in *Proceedings of 44th IEEE Conference on Decision and Control and European Control Conference*, Seville, Spain, 2005, pp. 6608–6613.

[Cam06] M. C. Campi and S. M. Savaresi, "Direct nonlinear control design: the Virtual Reference Feedback Tuning (VRFT) approach," *IEEE Transactions on Automatic Control*, vol. 51, no. 1, pp. 14–27, Jan. 2006.

[Esp11] A. Esparza, A. Sala and P. Albertos, "Neural networks in virtual reference tuning," *Engineering Applications of Artificial Intelligence*, vol. 24, no. 6, pp. 983–995, Sep. 2011.

[Hou11a] Z. S. Hou and S. Jin, "Data-driven model-free adaptive control for a class of MIMO nonlinear discrete-time systems," *IEEE Transactions on Neural Networks*, vol. 22, no. 12, pp. 2173–2188, Dec. 2011.

[Rad14] M.-B. Radac, R.-C. Roman, R.-E. Precup and E. M. Petriu, "Data-driven model-free control of twin rotor aerodynamic systems: algorithms and experiments," in *Proceedings of 2014 IEEE International Symposium on Intelligent Control*, Antibes, France, 2014, pp. 1889–1894.

[Rom16a] R.-C. Roman, M.-B. Radac and R.-E. Precup, "Mixed MFC-VRFT approach for a multivariable aerodynamic system position control," in *Proceedings of 2016 IEEE International Conference on Systems, Man, and Cybernetics*, Budapest, Hungary, 2016, pp. 2615–2620.

[Rom16b] R.-C. Roman, M.-B. Radac and R.-E. Precup, "Multi-input-multi-output system experimental validation of model-free control and virtual reference feedback tuning techniques," *IET Control Theory & Applications*, vol. 10, no. 12, pp. 1395–1403, Aug. 2016.

[Rom16c] R.-C. Roman, M.-B. Radac, R.-E. Precup, and E. M. Petriu, "Data-driven model-free adaptive control tuned by virtual reference feedback tuning," *Acta Polytechnica Hungarica*, vol. 13, no. 1, pp. 83–96, Feb. 2016.

[Rom16d] R.-C. Roman, M.-B. Radac, R.-E. Precup and E. M. Petriu, "Virtual reference feedback tuning of MIMO data-driven model-free adaptive control algorithms," in *Proceedings of 7th Advanced Doctoral Conference on Computing, Electrical and Industrial Systems*, Caparica (Lisbon), Portugal, 2016, pp. 253–260.

[Rom17c] R.-C. Roman, M.-B. Radac, R.-E. Precup and E. M. Petriu, "Virtual reference feedback tuning of model-free control algorithms for servo systems," *Machines*, vol. 5, no. 4, pp. 1–15, Oct. 2017.

[Rom18a] R.-C. Roman, *Tehnici de tip model-free de acordare a parametrilor regulatoarelor automate* (in Romanian), PhD thesis, Editura Politehnica, Timisoara, 2018.

[Rom18b] R.-C. Roman, R.-E. Precup and R.-C. David, "Second order intelligent proportional-integral fuzzy control of twin rotor aerodynamic systems," *Procedia Computer Science*, vol. 139, pp. 372–380, Oct. 2018.

[Yan16] P. Yan, D. Liu, D. Wang and H. Ma, "Data-driven controller design for general MIMO nonlinear systems via virtual reference feedback tuning and neural networks," *Neurocomputing*, vol. 171, pp. 815–825, Jan. 2016.

7 Hybrid Model-Free and Model-Free Adaptive Fuzzy Controllers

7.1 A SHORT OVERVIEW OF FUZZY LOGIC AND CONTROL

As shown in [Pre99a], the "classical" engineering approaches to characterize real-world problems are essentially qualitative and quantitative ones, based on more or less accurate mathematical modeling. In such approaches, expressions such as "medium temperature," "large humidity," "small pressure" and "very large speed," related to the variables specific to the behavior of a **controlled process**, are subjected to relatively difficult quantitative interpretations. This happens because the "classical" automation handles variables and information processed with well-specified numerical values. In this regard, the controller design and tuning and next its implementation in the control equipment require an as accurate as possible quantitative modeling of the controlled process. Advanced control strategies (e.g., adaptive, predictive and variable structure ones) require even the permanent reassessment of the models and the values of the parameters that characterize these (parametric) models.

L. A. Zadeh set the basics of fuzzy set theory [Zad65], which initially seemed to be just mathematical entertainment. The boom in the 70s in computer science opened the first prospects for practical applications of the meanwhile built theory in the field of automatic control, and these first applications belong to **E. H. Mamdani** and his coauthors [Mam74, Mam75]. The reference application of fuzzy control involves some "special" controllers based on fuzzy set theory, referred to as **fuzzy controllers**, in cement kiln control [Hol82]. In the 80s, in Japan, USA and later Europe, the so-called **fuzzy boom** took place in the field of fuzzy control applications involving several domains ranging from the electrical household industry up to control of vehicles, transportation systems and robots. This is caused partly by the spectacular development of electronics technologies and computer systems that enabled the manufacturing of circuits with very high speed of information processing, dedicated (by construction and usage) to a certain purpose including fuzzy information processing, and also the development of computer-aided design programs, which allowed the control system designer to efficiently use a large amount of information concerning the controlled process and the control equipment.

The applications of fuzzy control reported until now point out two important aspects related to this control strategy:

DOI: 10.1201/9781003143444-7

- In certain situations (for example, the control of nonlinear processes that are subjected to difficult mathematical modeling or even the control of ill-defined processes), fuzzy control can be a **viable alternative** to classical, crisp control (conventional control).
- Compared to conventional control, fuzzy control can be strongly based and focused on the **experience** of a human operator, and a fuzzy controller can model more accurately this experience (in linguistic manner) versus a conventional controller.

The main features of fuzzy control are expressed as follows [Pre99a]:

- Fuzzy control employs the so-called **fuzzy controllers** or **fuzzy logic controllers** that ensure **nonlinear input-output static maps** that can be influenced/modified based on the designer's option.
- Fuzzy control can process several variables from the controlled process. Therefore, it is considered to belong to the class of multi input-multi output (MIMO) systems with interactions, and the fuzzy control can be viewed as a **multi input** controller (or even a multi output one as well), which is similar to state feedback controllers.
- Fuzzy controllers are essentially without dynamics. But the applications and performance of fuzzy controllers and **fuzzy control systems** can be enlarged significantly by inserting dynamics (i.e., derivative and/or integral components) to fuzzy controller structures resulting in the so-called **fuzzy controllers with dynamics** or **typical fuzzy controllers** because of the analogy to the typical conventional controllers, which are popular in industrial applications.
- **Fuzzy controllers are flexible with regard to the modification of the transfer features** (by input-output static maps), thus ensuring the possibility to develop a wide range of adaptive control system structures.

The approach based on human experience is acting in case of fuzzy controllers by expressing the control requirements and elaborating the control signal in terms of the "natural" IF-THEN **rules** that belong to the following set of rules:

$$\ldots$$

$$IF \ \ (antecedent) \ \ THEN \ \ (consequent), \qquad (7.1)$$

$$\ldots$$

where the **antecedent** (the **premise**) refers to the found-out situation concerning the controlled process evolution (usually compared to the desired evolution), and the **consequent** (the **conclusion**) refers to the measures which should be taken – under the form of the control signal u – in order to fulfill the desired evolution imposed by the performance specifications. The ensemble of these rules makes up the **rule base** of the fuzzy controller.

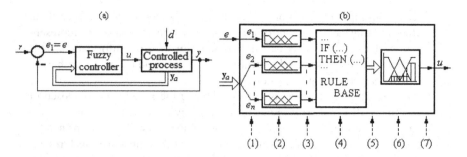

FIGURE 7.1 Fuzzy control system (a) and fuzzy controller (b) structure. (Adapted from Pre99a and Pre11.)

Proposition 7.1

The block diagram of the principle (considered as classical in the literature) of a fuzzy control system considered as a single-input system with respect to the **reference input** r and single-output system with respect to the **controlled output** y is presented in Figure 7.1. The reference input is also referred to as the set-point or the desired output y_d. The second input fed to the controlled process/the fuzzy control system is the disturbance input d.

Definition 7.1

The essential, already mentioned, particular feature of fuzzy control systems is the multiple cross-couplings regarded from the plant to the controller by the auxiliary variables $\mathbf{y}_a$, gathered in the vector $\mathbf{e}$

$$\mathbf{e} = \begin{bmatrix} e & \mathbf{y}_a^T \end{bmatrix} = \begin{bmatrix} e_1 & e_2 & \dots & e_n \end{bmatrix}^T, \tag{7.2}$$

which are direct or indirect inputs to the fuzzy controller. No matter how many inputs to the FC, the FC should possess at least one input, with the notation e_1, corresponding to the **control error** e:

$$e_1 = e = r - y. \tag{7.3}$$

Proposition 7.2

Figure 7.1b also highlights the operation principle of a fuzzy controller in its classical version, characterizing **Mamdani's fuzzy controllers**, with the following **variables and modules**: (i) the crisp inputs, (ii) the fuzzification module, (iii) the fuzzified inputs, (iv) the inference module, (v) the fuzzy conclusions, (vi) the defuzzification module and (vii) the crisp output. In this regard, the operation principle of a Mamdani fuzzy controller involves the **sequence of operations** (a), (b) and (c):

    a. The crisp input information – the measured variables, the reference input (the set-point) and the control error – is converted into a fuzzy representation. This operation is called **fuzzification** of crisp information. This operation makes use of **scheduling variables** instead of input variables in Takagi-Sugeno-Kang fuzzy controllers.

    b. The fuzzified information is processed using the **rule base**, composed of the fuzzy IF-THEN rules referred to as fuzzy control rules of type (10.1) that must be well defined in order to control the given plant. The principles to evaluate and process the rule base represent the **inference mechanism/ engine**, and the result is the "fuzzy" form of the control signal u, the **fuzzy control signal**.

    c. The fuzzy control signal must be converted into a crisp formulation, with well-specified physical nature, directly understandable and usable by the actuator in order to be capable of controlling the plant. This operation is known under the name of **defuzzification**.

Remark 7.1

The three operations described briefly above characterize the three modules in the structure of a fuzzy controller given in Figure 7.1b, namely **the fuzzification module** (2), **the inference module** (4) and **the defuzzification module** (6), all three being assisted by an adequate database. This section will be dedicated to treating basics of fuzzy set theory and presenting the operation mechanisms of these three modules.

7.1.1   FUZZY SETS, SET-THEORETIC OPERATORS, FUZZY RELATIONS

The essence of the fuzzy (vague) representation of information is the introduction of a measure that characterizes the membership of an element to a set. Generally, three approaches are used to define a set, (i–iii): (i) the enumeration of all elements of the set, (ii) the usage of a predicate $P(x)$, meaning that each element x has the property P, and (iii) the usage of the characteristic function. The generalization of the characteristic function specific to classical (crisp) sets leads to the notion of membership function specific to fuzzy sets.

Definition 7.2

Let $X \subseteq \Re$ be a **basic set** (**basic domain, universe, universe of discourse**) having the elements $x \in X$ and F a fuzzy set on X. The function μ_F, defined as

$$\mu_F : X \rightarrow [0,1], \tag{7.4}$$

is called **membership function** of the **fuzzy set** F, by which for each element $x \in X$ a value $\mu_F(x) \in [0,1]$ is mapped, which characterizes the **membership degree** of x to X. A **fuzzy set** F (defined) on X is completely defined by the set of pairs

$$F = \{(x, \mu_F(x)) \mid x \in X\}. \tag{7.5}$$

If the universe is a **discrete or countable set**, the following notation is used for the fuzzy set F:

$$F = \sum_{x \in X} \mu_F(x)/x, \tag{7.6}$$

where the symbol $\mu_F(x)/x = (x, \mu_F(x))$ represents the pair of discrete values "membership degree"/"crisp value belonging to the universe" and is named **singleton**. Equation (10.4) is often referred to as the sum of all these pairs, but the symbol $\sum$ (representing the union of these pairs on X) does not point out summation. In contrast, in case of a **noncountable or continuous universe**, the symbol $\sum$ is usually replaced by the symbol $\int$, which stands for union instead of integration [Pal97]

$$F = \int_X \mu_F(x)/x. \tag{7.7}$$

Proposition 7.3

The following ways are employed usually to represent a fuzzy set by means of its membership function, (i–iii): (i) the parametric representation under the form of an analytical function corresponding to the membership function, (ii) the direct graphical representation by means of the membership function graphics and (iii) the discrete representation by singletons for fuzzy sets with countable or discrete universes. The way (i) will be described as follows, and details on the other two ways are presented in [Pre97b]. The following membership functions are widely used in fuzzy control:

- the trapezoidal membership function:

$$\mu_F(x) = \begin{cases} 0, & \text{if} \quad x < \alpha, \\ (x-\alpha)/(\beta-\alpha), & \text{if} \quad x \in [\alpha, \beta], \\ 1, & \text{if} \quad x \in [\beta, \gamma] \\ (\delta-x)/(\delta-\gamma), & \text{if} \quad x \in [\gamma, \delta], \\ 0, & \text{if} \quad x > \delta, \end{cases} \quad , \; x \in \Re, \; \alpha < \beta \leq \gamma < \delta, \tag{7.8}$$

- the triangular membership function, with the particular parameters $\beta = \gamma$ in (7.8),
- the singleton membership function:

$$\mu_F(x) = \begin{cases} 1, & \text{if} \quad x = \alpha \\ 0, & \text{otherwise} \end{cases} \quad , \; x \in \Re, \tag{7.9}$$

- the Gaussian membership function:

$$\mu_F(x) = e^{-\frac{(x-\bar{x})^2}{2\sigma^2}}, \; x \in \Re, \tag{7.10}$$

with the parameters $\bar{x}$ – the center and $\sigma \neq 0$ – the width,
- the generalized bell-shaped membership function:

$$\mu(x) = \frac{1}{1 + \left(\dfrac{x - x_0}{a}\right)^{2b}}, \quad x \in \Re, \tag{7.11}$$

where $a > 0$ is the width of the bell, x_0 is the center, $b > 0$, with the feature $\mu(x_0) = 1$.

Remark 7.2

The parameter α associated with the singleton in (7.9) is referred to also as **modal value** [Gal95]. A singleton can be considered as a representation of a crisp value that is equal to the modal value.

Definition 7.3

The following **descriptors** are associated with the analytical characterization of a fuzzy set, and they are defined as follows for the fuzzy set $F = \{(x, \mu_F(x)) \mid x \in X\}$:
- **the support of a fuzzy set**, with the notation $\text{Supp}(F)$ and the definition

$$\text{Supp}(F) = \{x \mid x \in X, \ \mu_F(x) > 0\} \subset X, \tag{7.12}$$

- **the core of a fuzzy set**, marked by $K(F)$ and defined as

$$K(F) = \{x \mid x \in X, \ \mu_F(x) = 1\} \subset X, \tag{7.13}$$

- **the height of a fuzz set**, denoted by $\text{hgt}(F)$, defined in terms of

$$\text{hgt}(F) = \max\{x \mid x \in X\} \in [0,1]. \tag{7.14}$$

Definition 7.4

A fuzzy set $F = \{(x, \mu_F(x)) \mid x \in X\}$ is **normal** if $\text{hgt}(F) = 1$ and **subnormal** if $\text{hgt}(F) < 1$. A fuzzy set F is **(identical) null** if $\mu_F(x) = 0$, $\forall x \in X$ and **universal** if $\mu_F(x) = 1$, $\forall x \in X$.

Great interest in the automatic control applications is given to the feature of **fuzzy congruence** of fuzzy sets. For the sake of presenting the contents of this feature, the notion of α height cut in a fuzzy set will be presented firstly.

Definition 7.5

Let $F = \{(x, \mu_F(x)) \mid x \in X\}$ be a fuzzy set and α a real number, $\alpha \in (0,1]$. The α **height cut** in the fuzzy set F, with the notation F_α, is the fuzzy set

$$F_\alpha = \left\{(x, \mu_{F_\alpha}(x)) \mid x \in X\right\}, \quad \mu_{F_\alpha}(x) = \begin{cases} 1, & \text{if} \quad \mu_F(x) \geq \alpha \\ 0, & \text{otherwise} \end{cases}, x \in X, \ (7.15)$$

and it can be viewed also as a crisp set because μ_{F_α} can be considered as a characteristic function.

Definition 7.6

Two fuzzy sets $F_1 = \{(x, \mu_{F1}(x)) \mid x \in X\}$ and $F_2 = \{(x, \mu_{F2}(x)) \mid x \in X\}$ are **fuzzy congruent** if for any height α, $\alpha \in (0,1]$, there exist two real numbers α_1 and α_2 with the feature $0 < \alpha_1, \alpha_2 \leq 1$ such that

$$\text{Supp}((\alpha_1 F_1)_\alpha) \subseteq \text{Supp}((F_2)_\alpha), \ \text{Supp}((\alpha_2 F_2)_\alpha) \subseteq \text{Supp}((F_1)_\alpha), \quad (7.16)$$

where the multiplication stands for the multiplication of the membership functions. It is shown in [Pre97b] that the fuzzy congruence is equivalent to the equality of supports. A constraint in the fuzzy congruence of two fuzzy sets is brought by conditioning the same support and the same core for these fuzzy sets, and this leads to the **strictly fuzzy congruence**.

Remark 7.3

The fuzzy congruence and the strictly fuzzy congruence have two **consequences**, which are important from the point of view of fuzzy control, 1 and 2:

1. The monotonous increase/decrease of the flanks of the membership functions represent no requirement for the strictly fuzzy congruence of fuzzy sets. Therefore, in practical control applications, where the fuzzy information must be processed as quickly as possible, the flanks of membership functions are built from straight line segments, i.e., the classical shapes of membership functions will be used at least initially: triangular, rectangular, trapezoidal or singleton.
2. Major modifications in the characterization of fuzzy information are first obtained by modifying the support and/or core of fuzzy sets.

The equality and inclusions in case of fuzzy sets are derived on the basis of crisp set theory. The following definition is given in this context:

Definition 7.7

Two fuzzy sets $A = \{(x, \mu_A(x)) \mid x \in X\}$ and $B = \{(x, \mu_B(x)) \mid x \in X\}$ are **equal**, with the nomenclature $A = B$, if each element of the universe has the same membership degree in both sets, i.e.,

$$\mu_A(x) = \mu_B(x), \quad \forall x \in X. \quad (7.17)$$

Definition 7.8

A fuzzy set $A = \{(x, \mu_A(x)) \mid x \in X\}$ is named fuzzy subset of the fuzzy set $B = \{(x, \mu_B(x)) \mid x \in X\}$, with the notation $A \subseteq B$, if

$$\mu_A(x) \le \mu_B(x), \quad \forall x \in X. \tag{7.18}$$

Remark 7.4

In order to connect the fuzzy propositions using the logical operators AND, OR and NO, similar to the case of crisp sets, the **set-theoretic operators** (or **operators on fuzzy sets**) intersection, union and complement, respectively, are employed. The following operators are frequently used on the basis of accepting as above the two fuzzy sets $A = \{(x, \mu_A(x)) \mid x \in X\}$ and $B = \{(x, \mu_B(x)) \mid x \in X\}$:

- for **intersection**, denoted by $A \cap B$, the **MIN** operator:

$$A \cap B = \{(x, \mu_{A \cap B}(x)) \mid x \in X\}, \quad \mu_{A \cap B}(x) = \min(\mu_A(x), \mu_B(x)), \quad \forall x \in X, \tag{7.19}$$

- for **union**, marked by $A \cup B$, the **MAX** operator:

$$A \cup B = \{(x, \mu_{A \cup B}(x)) \mid x \in X\}, \quad \mu_{A \cup B}(x) = \max(\mu_A(x), \mu_B(x)), \quad \forall x \in X, \tag{7.20}$$

- for **complement**, with the notation A^c, **Mamdani's fuzzy complement** operator:

$$A^c = \{(x, \mu_{Ac}(x)) \mid x \in X\}, \quad \mu_{Ac}(x) = 1 - \mu_A(x), \quad \forall x \in X. \tag{7.21}$$

Definition 7.9

As a general rule, the **triangular norms** (or **t-norms**), **triangular conorms** (also called **s-norms** or **t-conorms**) and **c-norms** [Kle00] are defined to represent the intersection, union and complement, respectively. A widely used example of t-norm is the **PROD** operator:

$$\mathrm{PROD}(A, B) = \{(x, \mu_{\mathrm{PROD}(A,B)}(x)) \mid x \in X\}, \quad \mu_{\mathrm{PROD}(A,B)}(x) = \mu_A(x)\mu_B(x), \quad \forall x \in X, \tag{7.22}$$

and a widely used example of s-norm is the **SUM** operator:

$$\mathrm{SUM}(A, B) = \{(x, \mu_{\mathrm{SUM}(A,B)}(x)) \mid x \in X\}, \quad \mu_{\mathrm{SUM}(A,B)}(x) = \frac{\mu_A(x) + \mu_B(x)}{2}, \quad \forall x \in X, \tag{7.23}$$

where the arithmetic mean must be taken into account in case of more fuzzy sets in order to avoid membership degrees greater than 1.

Definition 7.10

The parameterized operators fuzzy AND and fuzzy OR result in the fuzzy set $F = \{(x, \mu_F(x)) \mid x \in X\}$, whose membership function is computed in terms of

$$\mu_F(x) = \gamma \min(\mu_A(x), \mu_B(x)) + (1 - \gamma) \frac{\mu_A(x) + \mu_B(x)}{2}, \quad \forall x \in X \qquad (7.24)$$

for fuzzy AND, and

$$\mu_F(x) = \gamma \max(\mu_A(x), \mu_B(x)) + (1 - \gamma) \frac{\mu_A(x) + \mu_B(x)}{2} / 2, \quad \forall x \in X \qquad (7.25)$$

for fuzzy OR, with the parameter γ, $\gamma \in [0,1]$, ensuring the possibility of applying these operators to represent the intersection or union. A similar possibility is also ensured using the **MIN-MAX** operator:

$$\mu_F(x) = \gamma \min(\mu_A(x), \mu_B(x)) + (1 - \gamma) \max(\mu_A(x), \mu_B(x)), \quad \forall x \in X. \qquad (7.26)$$

Remark 7.5

Well-acknowledged t-norms due to Hamacher, Frank, Yager and Dubois-Prade, and their corresponding s-norms due to Sugeno and c-norms due to Sugeno and Yager are available in the literature [Dri93, Zim01]. Relaxing the conditions in the definitions of t-norms and s-norms leads to other categories of operators with effects on the inference mechanisms of fuzzy controllers; such operators are the uninorms, nullnorms [Fod95, Cal01] and the entropy- and distance-based operators [Rud98, Rud06].

Remark 7.6

The **modification operators of fuzzy sets** represent operators based on arithmetic computations on the membership functions of fuzzy sets and are mainly meant for modeling linguistic hedges as "more/less…," "relatively more/less…," "very …". The particular feature of these operators is the fact that they do not affect the support and core of the fuzzy sets, and hence, they ensure the strict congruence of fuzzy sets.

Definition 7.11

The following frequently used modification operators are defined in the context of applying them to the initial fuzzy set $A = \{(x, \mu_A(x)) \mid x \in X\}$:

- the **concentration** operator, denoted by CON(A), with the result in a "denser" fuzzy set than the initial one:

$$\text{CON}(A) = \{(x, \mu_{\text{CON}(A)}(x)) \mid x \in X\}, \quad \mu_{\text{CON}(A)}(x) = (\mu_A(x))^n, \quad n = 2,3,\ldots, \quad \forall x \in X,$$

$$(7.27)$$

- the **dilation** operator, marked by DIL(A), with the result in a "less dense" fuzzy set in comparison with the initial one:

$$DIL(A) = \{(x, \mu_{DIL(A)}(x)) \mid x \in X\}, \quad \mu_{DIL(A)}(x) = \sqrt[n]{\mu_A(x)}, \quad n = 2,3,\dots, \quad \forall x \in X,$$

(7.28)

- the **contrast intensification** operator, with the notation INT(A), with the result in a fuzzy set "with intensified contrast" with respect to the initial fuzzy set:

$$INT(A) = \{(x, \mu_{INT(A)}(x)) \mid x \in X\},$$

$$\mu_{INT(A)}(x) = \begin{cases} 2(\mu_A(x))^2 & \text{if} \quad \mu_A(x) < 1/2 \\ 1 - 2(1 - (\mu_A(x))^2) & \text{otherwise} \end{cases}, \quad \forall x \in X.$$

(7.29)

Fuzzy relations represent the basis for the implications in fuzzy logic. Their definition is given as follows.

Definition 7.12

Accepting the universes $X_1, X_2, \dots, X_n$, the n-ary fuzzy relation is a fuzzy set R on the Cartesian product $X_1 \times X_2 \times \dots \times X_n$

$$R = \{((x_1, x_2, \dots, x_n), \mu_R(x_1, x_2, \dots, x_n)) \mid (x_1, x_2, \dots, x_n) \in X_1 \times X_2 \times \dots \times X_n\},$$

$$\mu_R : X_1 \times X_2 \times \dots \times X_n \to [0,1].$$

(7.30)

Since fuzzy relations are fuzzy sets, the operators presented in the previous section are valid for fuzzy relations too. In addition, it is defined also by the composition, denoted by ∘, that connects fuzzy sets and fuzzy relations.

Definition 7.13

Let us consider the fuzzy set $A = \{(x, \mu_A(x)) \mid x \in X\}$ and the fuzzy relation $R = \{((x,y), \mu_R(x,y)) \mid (x,y) \in X \times Y\}$. The **MAX-MIN composition** (**product**) leads to the fuzzy set B

$$B = A \circ R = \{(y, \mu_B(y)) \mid y \in Y\}, \mu_B(y) = \max_{x \in X} \min(\mu_A(x), \mu_R(x,y)), \quad \forall y \in Y. \quad (7.31)$$

Remark 7.7

In case of fuzzy relations defined on countable or discrete universes, the calculations in (7.31) corresponding to the MAX-MIN composition are done on the basis of expressing the operands in matrix forms and applying the matrix computation rules, where the classical product of elements is done using the MIN operator and the sum of products in the classical case (here it is the result of applying the MIN operator) is done using the MAX operator. Other compositions, which are often used in fuzzy control, are

- the **MAX-PROD composition**, where the classical product of elements is done using the PROD operator and the sum of products is done using the MAX operator and
- the **SUM-PROD composition**, where the classical product of elements is done using the PROD operator and the sum of products is done using the SUM operator (i.e., the arithmetic mean).

7.1.2 Information Processing in Fuzzy Controllers

This paragraph is dedicated to the presentation of the operating mechanisms of the modules in the fuzzy controller structure. The presentation is related to the block diagram given in Figure 7.1.

7.1.2.1 The Fuzzification Module
Proposition 7.4

The fundamental knowledge representation unit in fuzzy information processing is the notion of **linguistic variable**, associated with the quadruple (N_X, T_X, D_X, M_X), which indicates a structure, where [Pre99a]

- N_X – the symbolic name of the linguistic variable, with the examples N_X = speed, temperature, distance, control error, derivative/increment of control error, control signal, etc.
- T_X – the set of **linguistic terms** or **linguistic values**, which represent the linguistic values that can take N_X. A linguistic term associated with a linguistic value, denoted by LT_X as an element that belongs to the set T_X, is a symbol for a particular property of N_X. In order to define a linguistic variable and the corresponding linguistic terms as essential parts of the fuzzification module in a fuzzy controller, Figure 7.2 illustrates the

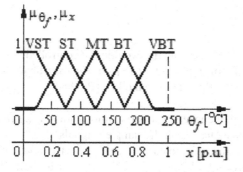

FIGURE 7.2 Example to illustrate the fuzzification of an input variable. (Adapted from Pre99a.)

transformation of the crisp value of furnace temperature θ_f (linguistic variable) into a fuzzy representation. This example deals with the fuzzification of an input variable. The following linguistic terms may be defined in the example for the temperature linguistic variable $(N_X = \theta_f)$: $T_X = \{$VST (Very Small Temperature), ST (Small Temperature), MT (Medium Temperature), BT (Big Temperature) and VBT (Very Big Temperature)$\}$.

- D_X – the domain of crisp values of the linguistic variable N_X. In case of the linguistic variable θ_f, it is obtained that $D_{\theta f} = [0, 250]\,^\circ$C. D_X can also be the universe associated to a fuzzy set. The example presented in Figure 7.2 highlights both the need to define the linguistic terms employed in the fuzzy characterization of the crisp information and normalizing (transforming into normalized/p.u. values) the universe by dividing to the maximum value, $\theta_{f\max} = 250\,^\circ$C, in terms of $x = \theta_c / \theta_{c\max}$, expressed in per unit [p.u.]. The result is $D_x = [0, 1]$ p.u. for the normalized domain.
- M_X – semantic function, which gives a meaning of a linguistic term in terms of the elements of D_X

$$M_X : LT_X \rightarrow \overline{LT}_X, \tag{7.32}$$

where $\overline{LT}_X$ is a fuzzy set on D_X

$$\overline{LT}_X = \{(x, \mu_{\overline{LT}_X}(x)) \mid x \in D_X\}, \ \mu_{\overline{LT}_X} : D_X \rightarrow [0, 1]. \tag{7.33}$$

In other words, the function M_X maps a linguistic term/a symbol onto an interpretation expressed as a fuzzy set. Proceeding this way, one may be able to make the difference (depending on the application) of linguistic terms regarded as either symbols or fuzzy sets.

Remark 7.8

In order to simplify the presentation in the sequel, the linguistic variables will be denoted identically to the physical variables they correspond, the meaning resulting from the context.

Remark 7.9

The crisp information concerning the evolution of controlled plant must be subject to the following **transforms** representing **steps of the fuzzification module** in order to be further processed by the inference module [Pre99a]:

- Analog-to-digital conversion of crisp information, i.e., sampling the analog signals and then quantizing the sampled signals.
- Digital processing of sampled and quantized crisp information, the processing of measured signals being accounted for here as digital filtering and also possibly as digital differentiation and integration as well.

- Transforming the crisp information into a fuzzy expression in terms of linguistic variables and corresponding linguistic terms. Accepting that the fuzzy controller inputs have well-stated values, and for the fuzzy characterization of crisp information, it is necessary to define the number of linguistic terms and their membership functions assigned to each input or scheduling linguistic variable.

Remark 7.10

As pointed out in [Pre99a and Pre11], the parameters in the fuzzification module to be chosen by the designer are the membership functions of the linguistic terms corresponding to the input linguistic variables and often also the scaling factors to be mentioned as follows. The literature does not provide general-purpose exhaustive recommendations with this regard, the final solution representing designer's option. Some relatively general recommendations in this context will be presented here, on the basis of [Pre99a], for the sake of taking them into consideration in the design and tuning of fuzzy controllers. All these recommendations are applied to scheduling variables as well.

Proposition 7.5

Recommendations to choose the number of linguistic terms corresponding to a linguistic variable. Usually this is an odd number, 3, 5 or 7. The number of linguistic terms sets the resolution of further information processing in the fuzzy controller. On the basis of several case studies, the literature proves – excepting certain special applications – that an increase in the number of linguistic terms over 7 does not contribute significantly to an efficient increase of the resolution. However, once this number increases for each input linguistic variable, this will result in an increased number of fuzzy rules, and the formulation of the rule base becomes more and more difficult. The linguistic terms are generally called to reflect an as general as possible contents, and they always depend on the variable involved.

Proposition 7.6

Recommendations to define and use the universe of input (linguistic) variables. It must be emphasized that the universe of input variables is predefined by the variation domain of sensors and interfaces (adaptation and conversion). Covering by linguistic terms, this domain will determine (in correlation with actuator's properties) the gain of the fuzzy controller. The existence of several inputs determines the possibility to define around a steady-state operating point more gains, one for each input. The reasoning and corresponding mathematical characterization are the same as in case of conventional control and are similar to the definition of the **proportional band** of a conventional controller.

Remark 7.11

The universe can be defined in several ways accounting for the nature of variables involved, the most frequent ways to express it being the following:

- in natural units,
- in p.u., in terms of division by a (nominal or maximum) value belonging to the universe (shown in Figure 7.2),
- in increments with respect to a reference value, expressed in natural units or p.u.

Remark 7.12

The definition of the universe is also referred to as **scaling**, and in case of using p.u., it is called **input normalization**. However, this operation must be seen in correlation with the universe of output (linguistic) variables, which requires the **output denormalization**. Both operations are necessary in case of discrete and continuous universes as well. As shown in [Pre99a], besides the normalization and denormalization in terms of multiplying/dividing by **scaling factors** that involve nominal or maximum values belonging to the universe, other values of scaling factors can also be used in either linear or nonlinear normalization and/or denormalization. The scaling factors represent key parameters of the fuzzy controllers. Therefore, the choice of their values is important because they affect the gains of the fuzzy controller and, further on, the dynamic performance indices of fuzzy control systems resulted after design and tuning. Furthermore, they may represent sources of instability and oscillations [Dri93, Pas98]. The authors will not carry out the scaling in this book as this operation is nonlinear, brings additional nonlinearities in the fuzzy control systems, which are essentially nonlinear, and creates additional abovementioned structural problems; the controller design and tuning will account for the possible domains of variations of the input and scheduling variables in important operating and realistic regimes.

Proposition 7.7

Recommendations for the initial choice of the membership functions of linguistic terms corresponding to the input linguistic variables. In cases of no available application-oriented experience in defining the linguistic terms and membership functions in case of input linguistic variables (experience gained by case studies and implementations), the following recommendations should be fulfilled in the first phase of design [Pre99a]:

- The membership functions of the linguistic terms corresponding to the input linguistic variables are chosen of triangular or trapezoidal type. The membership functions have (if possible) symmetrical shape excepting the extremity membership functions.
- Those allocations of membership functions are preferred that ensure the total overlap/covering of the universe such that each crisp value should

simultaneously fire two linguistic terms (finally, two rules). The overlap of the universe by a single linguistic term could cause discontinuities in the input-output static map of the fuzzy controller.

- The intersection point of the membership functions for two adjacent (with overlap) linguistic terms is recommended to have the ordinate greater than 0.4 (0.45), excepting the extremity zones of the universe. However, in case of linguistic terms corresponding to an output linguistic variable, this recommendation has to be correlated with the used defuzzification method to be presented later.
- If the linguistic variable involved is varying with ± values around zero, as it is for instance the situation of the control error, the symmetrical allocation with respect to zero is preferred.
- Zones of the universe that have no overlap by linguistic terms are not accepted. No overlap creates uncertainties in the crisp control signal u, having as effect, for example, $u = 0$ (this is the Matlab implementation to handle such situations).
- The definition of linguistic terms simultaneously having the core equal to one on their universe is not accepted.
- The too-rough quantization of the membership functions results in possible deformations of membership functions' support/core with effects on the inference module.

7.1.2.2 The Inference Module
Proposition 7.8

According to the structure presented in Figure 7.1, the fundamental subsystems of the inference module are the **rule base** and the **inference mechanism** (or the **inference engine**), assisted by an adequate rule base. The operation of a fuzzy controller is based on the set of rules (7.1) representing the **rule base**, which should ensure in terms of a linguistic characterization the controller operation on the universes of input and output variables. The information in the antecedent (premise) and consequent (conclusion) is expressed and then aggregated employing the operators defined in the previous section and the mechanism for consequent evaluation, namely **inference mechanism** (or the **inference engine**).

The rule base can be expressed either in a symbolic form of type (7.1) or as an inference matrix (or inference table or decision table or MacVicar-Whelan diagram). Both forms will be exemplified in this chapter.

Proposition 7.9

Considering the symbolic description of the rule base where a rule is expressed in terms of

$$\text{IF } (E = A) \text{ THEN } (U = B), \tag{7.34}$$

where E and U are input and output linguistic variables, respectively, A and B represent one linguistic term of E and U, respectively ($A \in T_E, B \in T_U$), the rule interpretation is given by a fuzzy relation on $D_E \times D_U$, with D_E and D_U – the universes of the linguistic variables E and U, respectively. Specifying again that the input linguistic variable E can also be used as a scheduling linguistic variable, the **construction of this fuzzy relation** is carried out according to the following steps [Pre99a]:

1. The interpretation of the fuzzy proposition ($E = A$), referred to as **rule antecedent (premise)**, is the fuzzy set $\bar{A}$

$$\bar{A} = \{(e, \mu_{\bar{A}}(x)) \mid e \in D_E\}, \quad \mu_{\bar{A}} : D_E \to [0,1] . \tag{7.35}$$

2. The interpretation of the fuzzy proposition ($U = B$), called **rule consequent (conclusion)**, is the fuzzy set $\bar{B}$

$$\bar{B} = \{(u, \mu_{\bar{B}}(y)) \mid u \in D_U\}, \quad \mu_{\bar{B}} : D_U \to [0,1] . \tag{7.36}$$

3. The interpretation of the rule is given by the fuzzy relation R that ensures the **implication**

$$R = \{((e,u), \mu_R(e,u)) \mid (e,u) \in D_E \times D_U\}, \quad \mu_R : D_E \times D_U \to [0,1] ,$$
$$\mu_R(e,u) = \mu_A(e) * \mu_B(u), \quad \forall (e,u) \in D_E \times D_U, \tag{7.37}$$

where $*$ can be either the Cartesian product or any other **fuzzy implication** operator. For example, this operator is the MIN one in case of Mamdani's implication.

Remark 7.13

Fuzzy propositions as those presented in (7.34) are simple **fuzzy propositions**. However, in many fuzzy control applications, the rule antecedents or consequents are composed fuzzy propositions, i.e., simple fuzzy proposition connected by means of intersection, union and/or complement operators. In this case, before step 1 and often also before step 2, the inference mechanism should determine the membership functions corresponding to each such proposition using adequate operators mentioned in the previous section.

Proposition 7.10

Subsequently, in the transition from fuzzy information processing using a single rule to the case of more rules, the **Mamdani fuzzy rule bases** consist of the rules $R^{(k)}$ according to

$$R^{(k)} : \text{IF } (E = A^{(k)}) \text{ THEN } (U = B^{(k)}), \quad k = 1 \ldots n, \tag{7.38}$$

where $A^{(k)}$ is a linguistic term corresponding to the input linguistic variable E, $A^{(k)} \in T_E$, the premise interpretation being the fuzzy set $\overline{A}^{(k)}$

$$\overline{A}^{(k)} = \{(e, \mu_{\overline{A}(k)}(e)) \mid e \in D_E\}, \quad \mu_{\overline{A}(k)} : D_E \to [0,1], \quad k = 1 \ldots n, \tag{7.39}$$

$B^{(k)}$ is a linguistic term corresponding the output linguistic variable U, $B^{(k)} \in T_U$, the conclusion interpretation being the fuzzy set $\overline{B}^{(k)}$

$$\overline{B}^{(k)} = \{(u, \mu_{\overline{B}(k)}(u)) \mid u \in D_U\}, \quad \mu_{\overline{B}(k)} : D_U \to [0,1], \quad k = 1 \ldots n. \tag{7.40}$$

Proposition 7.11

Mamdani's implication (characterized by the usage of MIN operator in implication) is interpreted in terms of the fuzzy relation $\overline{R}^{(k)}$

$$\overline{R}^{(k)} = \{((e,u), \mu_{\overline{R}(k)}(e,u)) \mid (e,u) \in D_E \times D_U\}, \quad \mu_{\overline{R}(k)} : D_E \times D_U \to [0,1] ,$$
$$\mu_{\overline{R}(k)}(e,u) = \min(\mu_A(e), \mu_B(u)), \quad \forall (e,u) \in D_E \times D_U, \quad k = 1 \ldots n , \tag{7.41}$$

where $\mu_{\overline{R}(k)}(e,u)$ indicates the **degree of fulfillment** (or the **firing strength**) of the rule $R^{(k)}$.

Remark 7.14

The interpretation of the rule base is done by the fuzzy relation $\overline{R}$, the result of the aggregation of all rules in terms of the union of fuzzy relations $\overline{R}^{(k)}$ afferent to the rules $R^{(k)}$

$$\overline{R} = \bigcup_{k=1}^{n} \overline{R}^{(k)}. \tag{7.42}$$

Remark 7.15

Employing the MAX operator for the union used in **rule aggregation** (with the note that the "aggregation" term is used also in connecting simple fuzzy propositions as part of composed fuzzy propositions), (7.41) and (7.42) lead to

$$\overline{R} = \{((e,u), \mu_{\overline{R}}(e,u)) \mid (e,u) \in D_E \times D_U\}, \quad \mu_{\overline{R}} : D_E \times D_U \to [0,1] ,$$
$$\mu_{\overline{R}}(e,u) = \max_{k=1 \ldots n} \min(\mu_{A(k)}(e), \mu_{B(k)}(u)), \quad \forall (e,u) \in D_E \times D_U. \tag{7.43}$$

Proposition 7.12

In the final part of the inference module, accepting the input variable e to take the crisp value e^*, $e = e^*$, the result of applying the rule base (7.38) is interpreted as the fuzzy set $\overline{U}$ that represents the **fuzzy conclusion** (or the **fuzzy control signal**)

$$\bar{U} = \{(u, \mu_{\bar{U}}(u)) \mid u \in D_U\}, \ \mu_{\bar{U}} : D_U \to [0,1],$$

$$\mu_{\bar{U}}(u) = \max_{k=1...n} \min(\mu_{A(k)}(e^*), \mu_{B(k)}(u)), \ \forall u \in D_U. \tag{7.44}$$

Remark 7.16

The interpretation of the rule base presented before corresponds to **Mamdani's MAX-MIN composition** (also referred to as the **MAX-MIN inference mechanism** or **Mamdani's MAX-MIN compositional rule of inference**) and characterizes **Mamdani's fuzzy controllers** or **Mamdani's fuzzy inference systems**. Concluding, this inference mechanism is characterized by treatment of AND linguistic connectors in the premise (the intersection of simple fuzzy propositions in the composed fuzzy proposition as part of the premise) by the MIN operator; treatment of OR linguistic connectors in the premise (the union of simple fuzzy propositions in the composed fuzzy proposition as part of the premise) by the MAX operator; implication using the MIN operator; rule aggregation in terms of the MAX operator.

Remark 7.17

Two other inference mechanisms are often used in fuzzy control, the MAX-PROD and SUM-PROD inference mechanisms. The **MAX-PROD inference mechanism** is characterized by the following operation mode [Pre99a]: treatment of AND linguistic connectors in the premise by the MIN operator; treatment of OR linguistic connectors in the premise by the MAX operator; implication using the PROD operator; rule aggregation in terms of the MAX operator. The **SUM-PROD inference mechanism** is characterized by the following operation mode [Pre99a]: treatment of AND linguistic connectors in the premise by the PROD operator; treatment of OR linguistic connectors in the premise by the SUM operator; implication using the PROD operator; rule aggregation in terms of the SUM operator.

An example concerning the automatic braking of a train that approaches red lighted traffic lights, which finally implies the train stopping, is presented as follows and adapted from [Pre97b] and [Pre99a] in order to illustrate the operation mode of the MAX-MIN inference mechanism.

Example 7.1

The information available for decision-making in braking is organized in terms of the two input variables d – the distance between the train and the traffic lights, constrained to $d \le 1,000$ m and v – the train velocity (speed), constrained to $v \le 100$ km/h. It is assumed that

- the railroad and the railway engine have sensors, which can provide the input variables with sufficient accuracy,

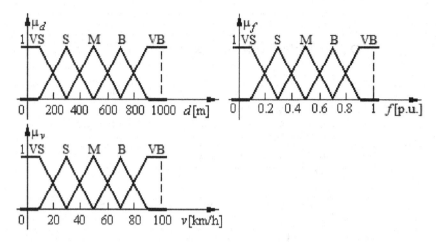

FIGURE 7.3 Input and output membership functions in train braking example. (Adapted from Pre99a.)

- the introduction of additional conditionings could be taken into consideration [Pre97b] by the definition of additional linguistic variables (the train mass could belong, for example, to this category) or the suitable adaptation of the membership functions of the linguistic terms corresponding to the input and output linguistic variables,
- the situations where d and v do not belong to the universes are not analyzed,
- the braking system of the train can be accounted for in setting the defuzzification method.

Five linguistic terms are assigned to each input linguistic variable, with the membership functions defined in accordance with Figure 7.3, for d : VS, S, M, B, VB, for v : VS, S, M, B, VB, and the nomenclature employed VS – Very Small, S – Small, M – Medium, B – Big and VB – Very Big. The output considered in this example is the braking force, f, expressed as degree of train braking in normalized values, $f[p.u.]$, $f \leq f_{max} = 1$, with five linguistic terms having the same symbols as those used for the inputs.

The chosen shape for the membership functions of different linguistic terms is trapezoidal for the extremity linguistic terms and symmetrical triangular for the middle linguistic terms. The entire universes of distance and speed are covered and overlapped by linguistic terms, with the usual situation characterized by firing two linguistic terms (for both d and v) and, finally, four rules. The rule base is expressed as the decision table (or the inference table) presented in Table 7.1.

The firing of one or another one of the rules in the rule base depends on the current (crisp) values of the input variables, d^* and v^*. The operation mode of the MAX-MIN inference mechanism is illustrated in Figure 7.4 considering the crisp values of the inputs $d^* = 350$ m and $v^* = 92$ km/h. Figure 7.4 highlights that two linguistic terms are fired in this case for the linguistic variable d, namely S and M, and one linguistic term for the linguistic variable v, i.e., VB. Therefore, only the following two rules in the rule base are fired, with the superscripts resulted from the notations defined in Table 7.1:

TABLE 7.1

Decision Table for Train Braking Example

		V					
F		VS	S	M	B	VB	
D	VS	M	M	B	VB	VB	1
	S	S	M	B	B	VB	2
	M	VS	S	M	M	B	3
	B	VS	S	S	M	B	4
	VB	VS	VS	VS	S	M	5
		1	2	3	4	5	

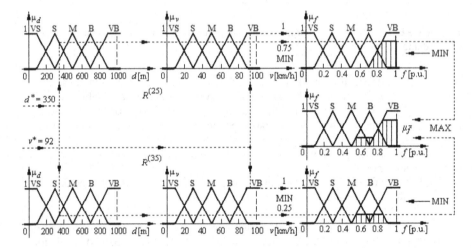

FIGURE 7.4 Illustration of MAX-MIN inference mechanism in train braking example. (Adapted from Pre99a.)

$$R^{(25)}:\text{IF } (d = \text{S AND } v = \text{VB}) \text{ THEN } (f = \text{VB}) \text{ OR}$$

$$(7.45)$$

$$R^{(35)}:\text{IF } (d = \text{M AND } v = \text{VB}) \text{ THEN } (f = \text{B}).$$

The value of v^* was deliberately chosen as $v^* > 90$ km/h in order to decrease the number of fired rules to enable a relatively easily understandable presentation of the inference engine. Nevertheless, the number of fired rules would be four if v^* would be chosen within $0 < v^* < 90$ km/h. Finally, the fuzzy control signal is expressed under the form of a fuzzy set with the membership function $\mu_{\bar{f}}$ according to Figure 7.4.

Remark 7.18

The inference mechanisms together with the rules presented above employ Mamdani fuzzy control rules, applicable mainly to fuzzy control with Mamdani fuzzy controllers. The **Takagi-Sugeno fuzzy rules** [Tak85, Sug88] are applicable in both fuzzy control – on the basis of the **Takagi-Sugeno fuzzy controllers**, referred to also as the **Takagi-Sugeno-Kang fuzzy controllers** – and system modeling of nonlinear processes with behaviors subjected to characterization in terms of a set of operating regimes (defined in the vicinity of a set of operating points). The most general form of these rule bases is

$$R^{(k)}\text{:IF } (E = A^{(k)}) \text{ THEN } (U = f_k(E)), \quad k = 1...n, \tag{7.46}$$

where the premise part is identical to Mamdani's case and the difference appears in the conclusion by the fact that $f_k(E)$ is a nonlinear or linear function that describes the dynamics of the process/the fuzzy controller or even the dynamics of the fuzzy control system for the particular value of the linguistic term $A^{(k)}$ corresponding to the scheduling and also input linguistic variable E. In other words, the function $f_k(E)$ represents a local model of the fuzzy controller or the process or the fuzzy control system.

Remark 7.19

As specified above, the premise part of the rules can involve scheduling variables that are different to the input ones. However, the consequent part of the rules remains the same as that given in (7.46), i.e., it involves a nonlinear or linear function that maps the inputs to the outputs.

Remark 7.20

The **inference mechanisms** used in the **Takagi-Sugeno-Kang rule bases** are similar to those in Mamdani ones, with the obvious difference in the rule aggregation part due to the different expressions in rule conclusions. Accepting an input variable e taking the crisp value e^*, i.e., $e = e^*$, this special form of the conclusion results in the fuzzy conclusion expressed as the (fuzzy) set of singletons

$$\{(u_{(k)}, \mu_{(k)} = \mu_{\bar{R}(k)}(e^*, u_{(k)})) \mid (e^*, u_{(k)}) \in D_E \times D_U\}, \quad k = 1...n, \tag{7.47}$$

where $u_{(k)}$ is the modal value afferent to each singleton (i.e., to each fired rule), and $\mu_{(k)} = \mu_{\bar{R}(k)}(e^*, u_{(k)})$ is the firing strength of the rule $R^{(k)}$.

Remark 7.21

If the linguistic variable E corresponds to the state vector, then $A^{(k)}$ defines a fuzzy subset of the fuzzy set on the state space corresponding to a particular operating regime, and $f_k(E)$ describes the dynamics of the process or the controller or the system in this operating regime. This type of fuzzy rules (associated with a certain inference mechanism) can be regarded as an interpolating mechanism that weights more or less certain local models/controllers afferent to different operating regimes depending on the current operating point.

Proposition 7.13

The parameters in the inference module to be set by the designer are the rule base and the inference mechanism (or the inference engine). The **rule base** should be defined correctly to ensure the fulfillment of the following three important **properties** that ensure a good operation of the fuzzy controller:

1. A **rule base** must be **complete**, i.e., any combination of input values leads to a certain output value. In relation with (7.4), this means that

$$\text{hgt}(\bar{U}) > 0, \quad \forall e^* \in D_E. \tag{7.48}$$

   The number of rules as part of a complete rule base (illustrated in the train braking example) equals the product of numbers of linguistic terms corresponding to the input linguistic variables.
2. A rule base must be **consistent**, i.e., it does not contain contradiction. In other words, it should not contain rules with the same rule antecedent but having mutually exclusive rule consequents (the same premise leads always to the same conclusion).
3. A rule base must be **continuous**, i.e., it does not contain neighboring rules with output fuzzy sets that have empty intersection, with the note that two rules are neighboring if the intersection of the fuzzy sets obtained by their premises (of type (7.40)) are fuzzy sets with nonzero heights [Dri01].

Proposition 7.14

The **definition of rule base** becomes more and more difficult if the number of inputs is increasing. General recommendations to define the rule base cannot be given, the role of expert (who knows the evolution of the plant) being of exquisite importance. The following **approaches** are recommended in [Pre99a] as suitable for the definition or rule bases:

- the use of an expert's knowledge in controlling the given process, with attention paid to preparing the interview,
- heuristic methods, based on engineering-like analyses of the possible evolutions of the process variables, done in cooperation with the process engineer,
- the use of previous experimental results, real-world or simulated ones, if a more or less detailed mathematical model of controlled process is available,
- special methods employing knowledge gained from conventional controllers and identification techniques that involve clustering, (neural) network-based learning and nature-inspired optimization algorithms.

Remark 7.22

As far as the **choice of inference mechanism** is concerned, the usual option is one of the three mechanisms discussed above, namely MAX-MIN, MAX-PROD and

SUM-PROD. Case studies investigated by several authors show that the inference mechanism has less significant effect on the shape of the input-output map of the fuzzy controller. Therefore, it is preferred to work with the methods that ensure good efficiency in information processing.

7.1.2.3 The Defuzzification Module

The defuzzification is the conversion of the fuzzy control signal, which is a fuzzy set produced by the inference module, into a crisp value. It is obvious that the crisp value calculated by defuzzification should belong to the universe of the fuzzy control signal.

Remark 7.23

Several defuzzification methods are used in **Mamdani fuzzy controllers**. The two most often used defuzzification methods for these fuzzy controllers are the mean of the maxima method and the center of gravity method in several versions including the specific situation of singleton-type membership functions of the linguistic terms afferent to the output linguistic variables.

Proposition 7.15

The **mean of maxima method** calculates the average of crisp values of output (control signal) that correspond to the conclusions with maximum firing strength. Accepting that the result of inference module is obtained under the form of the fuzzy set $\bar{u}$ on the universe D_u, which represents the fuzzy control signal, the crisp control signal u^* is obtained in terms of

$$u^* = 0.5\left[\inf_{u \in D_u}\left\{u \in D_u \mid \mu_{\bar{u}}(u) = \mathrm{hgt}(\bar{u})\right\} + \sup_{u \in D_u}\left\{u \in D_u \mid \mu_{\bar{u}}(u) = \mathrm{hgt}(\bar{u})\right\} \right] \quad (7.49)$$

and exemplified in Figure 7.5a.

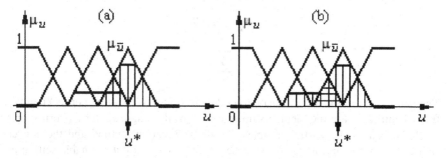

FIGURE 7.5 Illustration of MAX-MIN inference mechanism in train braking example. (Adapted from Pre99a.)

The **center of gravity** determines the crisp value of output taking into consideration, in a weighted manner, all influences obtained from the rules fired by the particular state of the inputs at a certain moment. The formulae giving the crisp control signal are adapted from mechanics, and they are specific to calculating the abscissa of center of gravity:

- in the continuous case:

$$u^* = \frac{\int_{D_u} u \cdot \mu_{\bar{u}}(u)\,du}{\int_{D_u} \mu_{\bar{u}}(u)\,du}, \tag{7.50}$$

- - in the discrete case:

$$u^* = \frac{\sum_{i=1}^{m} u_i \cdot \mu_{\bar{u}}(u_i)}{\sum_{i=1}^{m} \mu_{\bar{u}}(u_i)}, \quad m = \mathrm{card}(D_u). \tag{7.51}$$

Remark 7.24

The operation mode of the center of the gravity method is sketched in Figure 7.5b. This illustration shows that the overlapping area is not reflected in the above calculations.

Proposition 7.16

The most widely used defuzzification method in case of the **Takagi-Sugeno-Kang fuzzy controllers** or **Takagi-Sugeno-Kang fuzzy systems** is the weighted area method. Accepting the fuzzy control signal in (7.47), the following relationship is used to calculate the crisp control signal:

$$u^* = \frac{\sum_{k=1}^{n} u_{(k)} \cdot \mu_{(k)}}{\sum_{k=1}^{n} \mu_{(k)}}. \tag{7.52}$$

The parameters in the defuzzification module to be set by the designer in Mamdani fuzzy controllers are the membership functions of the linguistic terms corresponding to the output linguistic variables, the defuzzification method and the conversion of the crisp control signal. Analyses must be carried out to enable setting all these parameter values, and they should account for the following: the characteristics of the actuators (without dynamics or with negligible dynamics, possibly absorbed by the controlled process); the performance indices imposed to the fuzzy control system; the hardware and software implementation of the fuzzy controller. Aspects concerning the fuzzy controller design from the point of view of setting the parameters in the defuzzification module will be presented as follows.

Proposition 7.17

Aspects concerning the choice of linguistic terms and membership functions corresponding to the output/the control signal linguistic variable. The main aspects of interest in this case are the following:

- **The number of chosen linguistic terms** is usually odd (3, 5, or possibly 7). A larger number of linguistic terms does not bring spectacular results in the shape of the input-output static map of the fuzzy controller.
- **The existence of zones in the universe with no covering by membership functions** does not represent a serious problem. The covering of universe by continuous or discrete crisp values of the control signal is solved in terms of the convenient choice of the defuzzification method.
- **The scaling/definition of the universe**. The output denormalization specified in relation with the fuzzification module should be taken into consideration. In addition, the universe has to be always scaled/defined in such a way that it should fulfill the control requirements of the actuator. This means that the crisp control signal is not permitted to exceed the extreme values accepted by the actuator, and this exceed is related to the safe turning off/on of the actuator and the dynamic forcing to obtain an as reduced as possible settling time; the variation domain of the signal fed to the actuator and, accordingly, the variation domain of the actuating signal must be sufficiently well overlapped by the variation domain of the control signal in order to guarantee that the operating points imposed to the control system are actually reached.

Remark 7.25

This last remark can also become a specific aspect known as **the extremity problem** of the fuzzy controller, which is characterized by no overlap of the necessary domain of the control signal by the obtained crisp control signal due to the wrong definition of the membership functions of the output linguistic terms in correlation with an inadequate chosen defuzzification method (for example, in case of the center of gravity method). This problem can be avoided in terms of taking several measures, and the following ones are viable [Pre99a]:

- the symmetrical extension of the extremity membership functions,
- the modification of the shapes of membership functions to singleton ones,
- the choice of another defuzzification method, overall or just in the extremity zones.

- **The shapes of the membership functions** will be chosen to ensure – in relation with the defuzzification method – maximum efficiency in information processing (usually reflected in an as reduced as possible computation time). As mentioned in the previous section, singleton, rectangular, triangular and trapezoidal membership functions are recommended to be used for the linguistic terms of the output linguistic variables, accounting for the following remarks [Pre97b]:
- the singleton membership functions are the easiest ones to be processed,
- the rectangular membership functions modify significantly the amount of computations, but, by varying the width of the rectangle (the support), additional modifications of fuzzy controller features are obtained; nevertheless, the overlap problem can be avoided in comparison with the situation of triangular membership functions (the computation of the center of gravity becomes heavier),
- on the basis of all aspects presented above, the triangular membership functions seem to be the most unfavorable ones.

Proposition 7.18

Aspects concerning the choice of the defuzzification method. The following criteria have to be accounted for when choosing the defuzzification method [Dri93, Pre97b]:

- The type of actuator: in case of actuators with a finite number of discrete states, the mean of the maxima method is recommended along with other methods with a similar operating mechanism; in case of actuators with compact variation domain/universe, the center of gravity method is preferred.
- The continuity of the input-output static of the fuzzy controller has to be ensured. This means that a small change in the fuzzy control signal should not result in a large change in the crisp control signal. The center of gravity method satisfies this criterion, and the mean of maxima method does not.
- The disambiguity must be offered. This is characterized by avoiding situations when two relatively large areas in the membership functions of the fuzzy control signal are covered by two areas in the membership functions of the fuzzy sets as a result of implication. Both center of gravity and mean of maxima methods satisfy this criterion.
- The plausibility is necessary. This is characterized by placing the crisp control signal approximately in the middle of the support of the fuzzy control signal. The center of gravity method does not satisfy this criterion, and the mean of maxima method satisfies this criterion only if it is combined with the MAX-PROD inference mechanism.
- The computational complexity is particularly important in practical applications of fuzzy controllers. The mean of the maxima method is a computationally fast method, whereas the center of gravity method is much slower. Although the implementation of the center of gravity method is

rather difficult, choosing particular shapes of membership functions (i.e., singleton, rectangular and even triangular or trapezoidal) along with well-acknowledged inference methods determines significantly smaller computational load and fuzzy control signals with membership functions having convenient shapes that enable relatively easy analytical calculations.

Proposition 7.19

Conversion of crisp control signal. Depending on the defuzzification method and the type of actuator, the crisp control signal will be (in case of digital control):

- the current crisp value of the control signal u_k or
- the increment of the current crisp value with respect to the previous crisp value of the control signal, $\Delta u_k = u_k - u_{k-1}$.

Remark 7.26

In both above situations, the resulted value is converted into analog form by a digital-to-analog converter, excepting the situations when the actuator directly accepts the binary form of the crisp control signal from the case of continuous process control as an input. The problems and results related to information quantization are the same as in the case of analog-to-digital conversion.

Remark 7.27

Concluding, the relatively simple analysis of the operation of fuzzy controllers presented in this section highlights that the number of parameters that allow the modification of the transfer properties of fuzzy controllers – which are mainly represented by the input-output maps – is large and related to all three information processing modules in the fuzzy controller structure. The effects of modifying the values of these parameters should be known in order to compensate for the process nonlinearities by the fuzzy controller in the desired way aiming to fulfill the performance specifications imposed to the control system. This compensation can be carried out on the basis of the indications given in this section and also the measures to modify the shapes of the input-output maps of fuzzy controllers given in [Pre97b] and [Pre99a].

7.1.3 FUZZY CONTROLLERS AND DESIGN APPROACHES

The fuzzy controller represents a nonlinear controller with one or more inputs and one or more outputs. The shape of the nonlinear input-output static map attached to the fuzzy controller can be modeled such that to achieve a large variety of forms by the adequate setting of the parameters in the modules of the fuzzy controller. As shown in [Pre99a], the additional dynamic processing of some of system variables (by differentiation and/or integration) can provide dynamic features that yield fuzzy controllers with dynamics. As a result, the entire range of controllers known in their conventional versions can be generated but with parameters subjected to modifications because of the desired nonlinear character of the input-output relationships.

7.1.3.1 Fuzzy Controllers without Dynamics

The following members of this family of fuzzy controllers will be briefly exemplified as follows: the single input-single output (SISO) nonlinear proportional fuzzy controllers, the SISO two-positional and three-positional fuzzy controllers and the multiinput-single output (MISO) nonlinear proportional fuzzy controllers.

Proposition 7.20

The SISO nonlinear proportional fuzzy controller. If the universes of input and output variables are covered by a very small number of linguistic terms, a conventional controller with limitation (or saturation) is obtained. This controller has more or less nonlinear input-output static map depending on the features of the informational modules of the fuzzy controller [Pre99a]:

- **The fuzzy controller has a single rule.** Trapezoidal membership functions are used for the linguistic term of the input linguistic variable and the linguistic term of the output linguistic variable, having symmetric universes with respect to 0 in the regions where the membership degrees are less than one. The defuzzification is carried out by the center of gravity method. The modification of the slope of the input-output static map can be achieved by
  - scaling the input and output universes and
  - modifying the shapes of the membership functions of the input and output linguistic terms.
- **The fuzzy controller has two rules.** Trapezoidal membership functions are used for the two linguistic terms of the input linguistic variable, having symmetric universes with respect to 0 in the regions where the membership degrees are less than one. Singleton membership functions are used for the two linguistic terms of the output linguistic variable. The defuzzification is carried out again by the center of gravity method. The modification of the slope of the input-output static map of the fuzzy controller can be achieved as in the case of the fuzzy controller with one rule.

The further increase of the number of input and output linguistic terms leads to convenient nonlinear input-output static maps with shapes that can be of interest for the control system designer to compensate for the nonlinearities of the controlled process [Pre97b].

Proposition 7.21

SISO two-positional and three-positional fuzzy controllers. The description of these controllers is given in [Pre97b and Pre99a]. The two-positional fuzzy controller is also using a reduced number of input and output linguistic terms, namely two ones for the input and output linguistic variables, and two rules are defined in this regard. Triangular membership functions are used for the two linguistic terms of the input and output linguistic variables, having symmetric universes with respect to 0 and no overlap. Singleton membership functions are also accepted for the two linguistic terms of the output linguistic variable, with the modal values that are equal to the modal values of the triangular membership functions. To create a **hysteresis zone** (which mandatory in relation with two-positional controllers), a zone with no overlap must be ensured in the vicinity of the value 0 that belongs to the input universe.

The three-positional fuzzy controller makes use of three rules. Triangular membership functions are employed for the three linguistic terms of the input and output linguistic variables, having symmetric universes with respect to 0 and no overlap. The other aspects concerning the two-positional controller are valid here as well.

This approach to define the membership functions of the linguistic terms of the input linguistic variable can also be applied to increase the number of linguistic terms in order to build multipositional fuzzy controllers. All these fuzzy controllers operate on the basis of the largest of maxima method referred to also as the height method or defuzzification method [Pre99a].

Proposition 7.22

The MISO nonlinear proportional fuzzy controller. The structure of principle of MISO nonlinear output proportional fuzzy controller is actually the general one given in Figure 7.1a. The construction of such a controller assumes that the input variables obtained from the controlled process and grouped in the vector **e** are signals that are separated by blocks with dynamics inside the controlled process.

A widely used case in practical applications occurs if the vector **e** is the state vector of the controlled process. This allows the design of a **state feedback fuzzy controller** without or with control error correction, with measured or estimated states. The nonlinearity introduced by this fuzzy controller can ensure the improvement of the dynamic behavior of the fuzzy control system referred to as the **state feedback fuzzy control system**.

Example 7.2

An electrohydraulic servo system (EHS) is considered as a representative example of state feedback fuzzy control system, with the structure presented in Figure 7.6, where: EHC – the electrohydraulic converter, SD – the slide-valve distributor, MSM – the main servo motor, NL 1... NL 5 – the nonlinear blocks, M 1 and M 2 – measuring blocks (position sensors). The characteristic variables of the EHS are: u (V) – the control signal applied to EHC, x_1 (m) (the position of SD) and x_2 (m) (the position of MSM) – the state variables, x_{1M} (V) and x_{2M} (V) – the measured state variables and y (m) – the controlled output. The values of the parameters of the controlled process are [Pre81] $u_l = 10$ V, $g_0 = 0.0625$ mm/V, $\varepsilon_2 = 0.02$ mm, $\varepsilon_4 = 0.2$ mm, $x_{1l} = 21.8$ mm, $y_l = 210$ mm, $T_{i1} = 0.001872$ s, $T_{i2} = 0.0756$ s, $k_{M1} = 0.2$ V/mm and $k_{M2} = 0.032$ V/mm.

The presence of two integrators recommends the design of a state feedback control system for EHS. The presence of nonlinearities and the possibility of some parameters of EHS to modify their value justifies the design of a state feedback fuzzy control system. State feedback Mamdani fuzzy controllers for this system have been designed in [Pre97b, Pre99c and Pre05]. The design of a **state feedback Takagi-Sugeno-Kang fuzzy controller** will be presented as follows.

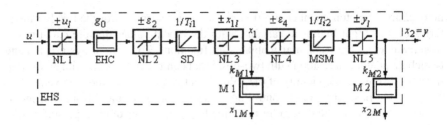

FIGURE 7.6 Structure of EHS. (Adapted from Pre81 and Pre20.)

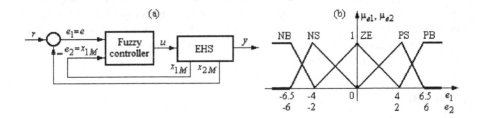

FIGURE 7.7 State feedback fuzzy control system (a) and membership functions of the linguistic terms of the input linguistic variables (b). (Adapted from Pre97b and Pre20.)

Several versions of state feedback fuzzy controller can be designed. Figure 7.7a illustrates the structure of a state feedback fuzzy control system, with the input variables $e_1 = e$ – the control error and e_2 (r indicates the reference input). The fuzzy controller is designed starting with the separate design of three linear state feedback controllers that stabilize the linearized mathematical model of EHS. A Takagi-Sugeno-Kang fuzzy controller is used because of the already mentioned feature of interpolator of three local controllers that correspond to different operating regimes depending on the input values. In this regard, the membership functions of the linguistic terms of the input linguistic variables (which are the scheduling variables as well) are set in accordance with Figure 7.7b.

Input scaling blocks are not inserted in Figure 7.7a for the sake of simplicity. A simplified state feedback Takagi-Sugeno-Kang fuzzy controller with three input linguistic terms instead of five has been designed in [Pre20].

Dropping out the nonlinearities in Figure 7.6 for a convenient design of the linear state feedback controllers, the state-space model of EHS viewed as a controlled process is obtained in terms of

$$\begin{bmatrix} \dot{x}_1 \\ \dot{x}_2 \end{bmatrix} = \begin{bmatrix} 0 & 0 \\ a & 0 \end{bmatrix} \cdot \begin{bmatrix} x_1 \\ x_2 \end{bmatrix} + \begin{bmatrix} b^* \\ 0 \end{bmatrix} \cdot u,$$

$$y = \begin{bmatrix} 0 & 1 \end{bmatrix} \cdot \begin{bmatrix} x_1 \\ x_2 \end{bmatrix},$$

(7.53)

where the parameter definitions and values are

$$a = \frac{1}{T_{i2}} = 14.05 , \quad b^* = \frac{g_0}{T_{i1}} = 26.04.$$

(7.54)

Considering that the three linear state feedback controllers are characterized by the control laws

$$u = -k_1^{(j)} x_{1M} + k_2^{(j)} (r - x_{2M}), \quad j = 1...3, \tag{7.55}$$

with the superscript corresponding to one of the three linear controllers and accounting for the expression of the characteristic polynomial of the linear state feedback control system in the three situations

$$\Delta^{(j)}(s) = s(s + k_{M1} k_1^{(j)}) + a \, b^* k_{M2} k_2^{(j)}, \quad j = 1...3, \tag{7.56}$$

imposing the linear closed-loop system poles $p_{1,2}^{(1)} = -3$ for small absolute values of the two inputs, $p_{1,2}^{(2)} = -5 \pm 2j$ for medium (average) absolute values of the two inputs and $p_{1,2}^{(3)} = -10 \pm 10j$ for big absolute values of the two inputs leads to the following values of the parameters that will be used in the rule consequents:

$$k_1^{(1)} = 30, \ k_2^{(1)} = 0.79, \ k_1^{(2)} = 50, \ k_2^{(2)} = 2.57, \ k_1^{(3)} = 100, \ k_2^{(3)} = 17.76. \tag{7.57}$$

The MAX-PROD mechanism is used in the inference engine, and the rule base is expressed in Table 7.2, where the rule consequents are

$$u^{(j)} = k_1^{(j)} e_1 - k_2^{(j)} e_2, \quad j = 1...3. \tag{7.58}$$

The weighted average method is employed in the defuzzification module.

The validation of this state feedback Takagi-Sugeno-Kang fuzzy controller is carried out in terms of its numerical implementation in the angular speed control of a nonlinear laboratory direct current (DC) drive (AMIRA DR300). As specified in [Pre06], the DC motor is loaded using a current controlled DC generator, mounted on the same shaft, and the drive has built-in analog current controllers for both DC machines having rated speed equal to 3,000 rpm, rated power equal to 30 W and rated current equal to 2 A. The speed control of the DC motor is digitally implemented using an A/D-D/A converter card. The speed sensors are a tachogenerator and an additional incremental rotary encoder mounted at the free drive shaft. A sample of experimental results is presented in Figure 7.8 for a sinusoidal reference input (or set-point) that is specific to servo systems.

TABLE 7.2

Decision Table of State Feedback Takagi-Sugeno-Kang Fuzzy Controller for EHS

		e_2				
U		NB	NS	ZE	PS	PS
e_1	PB	$u^{(3)}$	$u^{(3)}$	$u^{(3)}$	$u^{(3)}$	$u^{(3)}$
	PS	$u^{(3)}$	$u^{(2)}$	$u^{(2)}$	$u^{(2)}$	$u^{(3)}$
	ZE	$u^{(3)}$	$u^{(2)}$	$u^{(1)}$	$u^{(2)}$	$u^{(3)}$
	NS	$u^{(3)}$	$u^{(2)}$	$u^{(2)}$	$u^{(2)}$	$u^{(3)}$
	NB	$u^{(3)}$	$u^{(3)}$	$u^{(3)}$	$u^{(3)}$	$u^{(3)}$

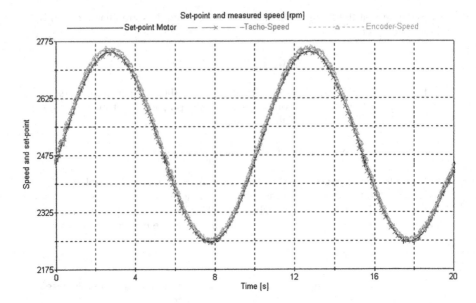

FIGURE 7.8 Reference input (set-point) and angular speed versus time for the state feedback Takagi-Sugeno-Kang fuzzy control system.

Proposition 7.23

Example 7.2 also proves an **approach to the design of Takagi-Sugeno-Kang fuzzy control systems**, which consists of the following steps:

Step 1. The mathematical modeling of the controlled process. Both simplified and fuzzy models [Bab98, Abo03] can be used in this regard.

Step 2. The derivation of local mathematical models of the process as a function of the particular operating features of the process and the performance specifications imposed to the control system.

Step 3. The development of local linear or fuzzy controllers that control the simplified local models or the local fuzzy models, respectively.

Step 4. Setting the structure of the fuzzy controller in order to ensure the bumpless transfer from one local controller to another one and fulfill the control system performance specifications.

The above steps can be carried out iteratively until the control system performance specifications are fulfilled.

This design approach is known in the literature under various names, and the most frequently used one for Takagi-Sugeno-Kang fuzzy controllers is the **parallel distributed compensation** (PDC) [Wan96]. The **modal equivalence principle** [Gal95] is applied to Mamdani fuzzy controllers; according to this approach, the fuzzy controller and the linear one (which is used in starting the fuzzy controller design) have the same values of their outputs for the modal values of the inputs. The modal equivalence principle ensures a relatively simple design of Takagi-Sugeno-Kang fuzzy controllers because it avoids the use formulation and solving of linear matrix inequalities (LMIs).

7.1.3.2 Fuzzy Controllers with Dynamics

It is generally acknowledged that the features of linear conventional control systems can be enhanced by the introduction of some dynamic components in the controller structure. The effects of these components can be reflected in

- steady-state regimes, by the rejection or, from one case to another, alleviation of the control error,
- dynamic regimes, by improving the phase margin (of linear control systems)/reducing the overshoot, reducing the settling time and/or improving the stability conditions.

The same general usefulness could be given by introducing the dynamic components in the case of fuzzy controllers as well.

Remark 7.28

As a matter of principle, the dynamic processing of an input signal – some additional integral (I) or derivative (D) components – creates additional signals on the fuzzy controller input. In this context, if the FC input, so, if $e_1 = e$ (the control error), then e_2 and e_3 could, for instance, become $e_2 = \dot{e}$, $e_3 = \ddot{e}$ or $e_3 = \int_0^{} e(\tau)d\tau$. As shown in [Pre97b], the principles for the implementation of dynamic components are quite different, and lead to several structures of fuzzy controllers with dynamics.

Proposition 7.24

Figure 7.9b–h gives approaches to the dynamic processing (of D or I type) of different variables of a fuzzy controller. The controller structures illustrated in Figure 7.9b–g outline that both the input signals (before defuzzification) and the crisp (defuzzified) output of the fuzzy controller can be subjected to dynamic processing; i.e., the dynamic processing takes place outside the strictly speaking fuzzy controller, which remains, in essence, nonlinear and without dynamics.

Remark 7.29

Above, the D and I components can be implemented in either analog or digital version. The symbols of different processing types employed in Figure 7.9 are well known. The importance of the analog implementation versions of the D components (the real-world processing is of DT1 (i.e., derivative with first-order lag) type) and the I components is rather reduced in fuzzy control. However, the digital versions of

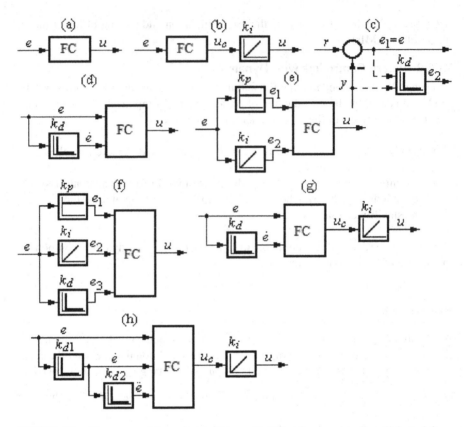

FIGURE 7.9 Structures of fuzzy controllers, with FC – fuzzy controller. (Adapted from Pre97b.)

D (DT1) and I components create a quasi-continuous equivalent of the analog D and I components. There are several **approaches to build quasi-continuous D and I components,** but just few of them are pointed out as follows.

Lemma 7.1

For the D component, using the notation e for its input and d for its output, the usual computation relation is

$$d(k) = \frac{e(k) - e(k-1)}{T_s},\tag{7.59}$$

where T_s is the sampling period, and k is the discrete time index. If the input variable e has very rapid variation which could be harmful to the implementation of the D component, then either e is prefiltered in terms of a PT1 (i.e., proportional to first-order lag or low-pass filter) law, or the D component is created on the basis of the actual sample $e(k)$ and an old sample, $e(k-m)$:

$$d(k) = \frac{e(k) - e(k - m)}{m\, T_s}, \quad m \in N, \; m \geq 2, \tag{7.60}$$

and the efficiency of (7.60) has to be checked depending on the application involved.

Lemma 7.2

For the D component, using the notation e for its input and σ for its output, one version of computation relation is

$$\sigma(k) = \sum_{i=0}^{k} e(i) = e(k) + \sum_{i=0}^{k-1} e(i), \tag{7.61}$$

which is equivalent to

$$\sigma(k) = \sigma(k - 1) + e(k), \quad \sigma(k) = x(k) + e(k), \quad x(k) = \sum_{i=0}^{k-1} e(i). \tag{7.62}$$

Such a characterization will also permit a relatively easy quasi-continuous equivalence of the digital case. Using the first-order Padé approximation

$$e^{-s\, T_s} \approx \frac{1 - s\, T_s/2}{1 + s\, T_s/2}, \tag{7.63}$$

the following relations for the two components with D and I dynamics are obtained [Pre99a]:

$$d(s) \approx \frac{s}{1 + s\, T_s/2} e(s), \quad \sigma(s) \approx \frac{1 + s\, T_s/2}{s\, T_s} e(s), \tag{7.64}$$

which will enable the definition of a pseudo-transfer function attached to the fuzzy controller with dynamics.

Remark 7.30

Since the fuzzy controllers are essentially a nonlinear controller, it is obviously wrong to speak about their transfer functions. However, accepting that the fuzzy controllers ensure continuous input-output static maps, then in the vicinity of a steady-state operating point (e.g., the origin of the strictly speaking fuzzy controller with an I component on the fuzzy controller output in Figure 7.9b), a quasi-continuous pseudo-transfer function can be associated with the fuzzy controllers with dynamics, in the conditions of this assumption. In this context, the controller structures presented in Figure 7.9 are referred to as proportional (P) fuzzy controller (Figure 7.9a), I fuzzy controller (Figure 7.9b), proportional-derivative (PD) fuzzy controller (Figure 7.9d), position Proportional-Integral (PI) fuzzy controller or PI fuzzy controller with input/control error integration (Figure 7.9e), position Proportional-Integral-Derivative (PID) fuzzy controller (Figure 7.9f), incremental PI fuzzy controller or PI fuzzy controller with output/control signal integration (Figure 7.9g) and PID fuzzy controller with output/control signal integration (Figure 7.9h). The computation of the expressions of these pseudo-transfer functions should account for the presence of the scaling factors that correspond to the fuzzification and defuzzification modules [Pre97b].

Proposition 7.25

Figure 7.9c suggests the possibility to build **two-degree-of-freedom (2-DOF) fuzzy controllers**. The concept of 2-DOF fuzzy control has been proposed by Precup and Preitl in 1999 and 2003 [Pre99b, Pre03] as fuzzy control with nonhomogenous dynamics with respect to the input channels (i.e., the reference input channel and the controlled output channel). This concept was further developed in other papers applied to servo systems and electrical drives [Pre09, Pre12].

Proposition 7.26

The usefulness of PI fuzzy controllers concerns their systematic design and tuning starting with the features (that are known and accepted as suitable) of a basic linear PI controller, which is employed (along with the PID controllers) in the majority of industrial control applications [Åst95]. In addition, knowing the PI fuzzy controllers with output/control signal integration and getting experience on them creates the premises to design and tune other controller structures synthesized in Figure 7.9 including PD fuzzy controllers and PID fuzzy controllers and other complex fuzzy controllers as well. Those are the reasons why **PI fuzzy controllers with output/ control signal integration** are discussed as follows. The introduction of dynamics in these controllers is carried out in terms of (Figure 7.10a)

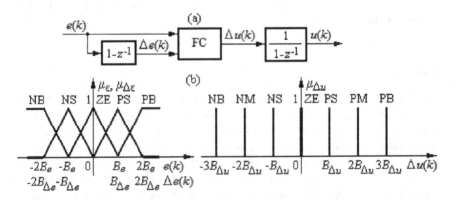

FIGURE 7.10 Structure of PI fuzzy controller with output integration (a) and membership functions of the linguistic terms of the input and output linguistic variables, with FC – fuzzy controller. (Adapted from Pre97b.)

- the numerical differentiation of the control error $e(k)$ under the form of the increment of control error, $\Delta e(k)$

$$\Delta e(k) = e(k) - e(k-1), \tag{7.65}$$

- the numerical integration of the increment of control signal, $\Delta u(k)$.

Theorem 7.1

The design and tuning of the controller proposed in *Proposition 7.26* starts from the expression of the recurrent equation of the incremental (velocity-type) quasi-continuous PI digital control algorithm

$$\Delta u(k) = K_P \Delta e(k) + K_I e(k) = K_P[\Delta e(k) + \alpha\ e(k)], \tag{7.66}$$

where the parameters K_P, K_I and α, with

$$K_P \alpha = K_I \tag{7.67}$$

are functions of the parameters of the continuous-time linear PI controller with the transfer function

$$H_C(s) = k_C \left(1 + \frac{1}{s\ T_i} \right), \tag{7.68}$$

with k_C – the gain and T_i – the integral time constant. The parameters K_P, K_I and α get the following expressions if Tustin's method is applied to discretize the continuous-time PI controller:

$$K_P = k_C \left(1 - \frac{T_s}{2T_i} \right),\ K_I = \frac{k_C T_s}{T_i},\ \alpha = \frac{K_I}{K_P} = \frac{2T_s}{2T_i - T_s}. \tag{7.69}$$

On the basis of (7.66) and the representation of the increment of control signal $\Delta u(k)$ in the phase plane $< e(k), \Delta e(k) >$, the features that allow the design and tuning of this fuzzy controller including the formulation of its rule are highlighted [Pre97b]:

- there exists the so-called zero-control signal line, $\Delta u(k) = 0$, with the equation

$$\Delta e(k) + \alpha\ e(k) = 0, \tag{7.70}$$

which divides the phase plane in two half-planes,

- with respect to this line, the increment of control signal takes the values $\Delta u(k) > 0$ in the upper half-plane and $\Delta u(k) < 0$ in the lower half-plane,
- the distance from any point in the phase plane to the zero-control signal line corresponds to the absolute value of the increment of control signal, $|\Delta u(k)|$.

The fuzzification is carried out as follows for a Mamdani fuzzy controller in terms of Figure 7.10b:

- for the input variables $e(k)$ and $\Delta e(k)$: five linguistic terms are chosen with regularly distributed triangular membership functions having an overlap of 1,
- for the output variable $\Delta u(k)$: seven linguistic terms are chosen with regularly distributed singleton membership functions,
- setting the scaling factors (for the sake of simplicity and avoiding additional nonlinearities, they are dropped out in Figure 7.10b).

Remark 7.31

The tuning parameters of the PI fuzzy controller structure, proposed in *Proposition 7.26* and *Theorem 7.1*, are B_e, $B_{\Delta e}$ and $B_{\Delta u}$. They are correlated to the shapes of the membership functions of the linguistic terms corresponding to the input and output linguistic variables. This reduced number of parameters ensures a low-cost implementation, design and tuning of the fuzzy controller. A larger number of parameters can contribute to an improved compensation of the process nonlinearities.

Remark 7.32

Continuing *Remark 7.31*, the rule base is defined accounting for the abovementioned features. If the number of input linguistic terms is five for the two input variables, the complete rule base is synthesized in the decision table presented in Table 7.3.

Proposition 7.27

Applying the modal equivalence principle [Gal95], the following steps are proceeded to tune the parameters B_e, $B_{\Delta e}$ and $B_{\Delta u}$ [Pre93]:

Step A. The following relation is expressed for the zero-control signal line:

$$\alpha = -\frac{\Delta e_k}{e_k} = \frac{B_{\Delta e}}{B_e}. \tag{7.71}$$

TABLE 7.3

Decision Table of PI Fuzzy Controller with Output Integration

Δu_k		e_k				
		NB	NS	ZE	PS	PB
Δe_k	PB	ZE	PS	PM	PB	PB
	PS	NS	ZE	PS	PM	PB
	ZE	NM	NS	ZE	PS	PM
	NS	NB	NM	NS	ZE	PS
	NB	NB	NB	NM	NS	ZE

Step B. The following condition is fulfilled for the constant control signal line, $\Delta u(k) = B_{\Delta u}$:

$$B_{\Delta u} = \Delta u_k = K_P[\Delta e(k) + \alpha\, e(k)],$$

$$\Delta u_k = K_P B_{\Delta e}. \tag{7.72}$$

Step C. Next, (7.71) is equivalent to

$$B_{\Delta u} = K_P B_{\Delta e}. \tag{7.73}$$

Step D. The relation (7.70) and next its substitution in (7.73) using (7.67) leads to the tuning relations

$$B_{\Delta e} = \alpha\, B_e, \quad B_{\Delta u} = K_P \alpha\, B_e = K_I B_e. \tag{7.74}$$

Step E. One of the parameters, for example, B_e, is chosen, and the other two parameters, $B_{\Delta e}$ and $B_{\Delta u}$, result in terms of (7.74).

Remark 7.33

The choice of the inference method and the defuzzification method represents the designer's options. Mamdani's MAX-MIN inference mechanism and the center of defuzzification method are the most widely used in this regard.

Remark 7.34

The obtained control signal in its incremental form $\Delta u(k)$ can be further used in the control system as follows:

- directly, if the actuator is of integral type or it contains a pure integral component,
- by calculating the effective value of the control signal $u(k)$

$$u_k = u_{k-1} + \Delta u_k, \tag{7.75}$$

where (7.75) corresponds to the numerical integration block placed on the fuzzy controller output illustrated in Figure 7.10b.

Proposition 7.28

The results presented above allow an approach to design and tune control systems with Mamdani PI fuzzy controllers, which consists of the following steps:

Step 1. The mathematical modeling of the controlled process.
Step 2. The design and tuning of a basic PI controller with the transfer function given in (7.68) using an approach that is specific to conventional linear controllers.

Step 3. The choice of the value of the settling time, the discretization of the continuous-time PI controller and the derivation of the recurrent equation of the incremental quasi-continuous PI digital control algorithm given in (7.66).

Step 4. The choice of the fuzzy controller structure.

Step 5. The choice of the tuning parameter B_e and the application of the tuning equations (7.74), which is the result of the application of the modal equivalence principle [Gal95], to obtain the values of the other two tuning parameters, $B_{\Delta e}$ and $B_{\Delta u}$.

Remark 7.35

This tuning approach can also be applied to Takagi-Sugeno-Kang PI fuzzy controllers, where only the first relation in (7.74) is valid. The Takagi-Sugeno-Kang versions of the PI fuzzy controllers prove to be advantageous in certain situations as they exploit the advantage of these fuzzy controllers of being bumpless interpolators between several local controllers, namely PI ones in this case [Bab96, Pre04, Pre15].

Remark 7.36

The choice of the value of the parameter B_e in step 5 is important as it influences the behavior of the fuzzy control system and thus the performance indices. The experience of the control system designer can be used in this regard, and the possible domains of variations of the reference input can be used in this regard. However, one way to set systematically the value of the parameter B_e is to carry out the stability analysis of the control system. The stability conditions represent useful information in setting the values of B_e. The most successful stability analysis approaches applied to fuzzy control systems will be briefly discussed as follows on the basis of the discussion presented in [Pre20].

Remark 7.37

The general approach to deal with the stability analysis in model-based fuzzy control, treated in the seminal publications [Tan92, Wan96, Tan01], is to make use of Takagi-Sugeno-Kang fuzzy models of the process and express the stability analysis conditions as LMIs in terms of PDC, which states that the dynamics of each local subsystem in the rule consequents of the Takagi-Sugeno-Kang fuzzy models of the process are controlled using the eigenvalue analysis [Wan96, Tan01]. Some recent results on the LMI-based stability analysis include the relaxation of stability conditions [Fre19, Gun19, Moo19, Wan19, Jia20, Liu20, Sak20, Sha20a], the negative absolute eigenvalue approach [Gan20] and Lyapunov-Krasovskii functionals [Xia20a].

Remark 7.38

The idea behind the PDC-based approach to the stability analysis and stable design of Takagi-Sugeno-Kang fuzzy control systems based on LMIs is the extensive use of quadratic Lyapunov function candidates. The effects of various parameters of the fuzzy models are considered resulting in nonquadratic Lyapunov function-based approaches exemplified as follows: the membership-function-dependent analysis [Lam18, Yan19], nonquadratic stabilization of uncertain systems, exponential stability with guaranteed cost control [Pan16], piecewise continuous and smooth functions, piecewise continuous exact fuzzy models, general polynomial approaches [Li20a, Xia20b], sum-of-squares-based polynomial membership functions [Yu18, Zha19, Li20b, Pan20], superstability conditions, integral structure-based Lyapunov functions [Yon17], the subspace-based improved sector nonlinearity approach [Rob17], the fractional intelligent approach and interpolation function-based approaches [Med18].

Remark 7.39

As shown in [Gue01 and Pre11], the classical approach based on PDC to stabilize fuzzy control systems and the use of LMIs in the stability analysis may introduce computational burden, complexity and coupling of subsystems. Therefore, different approaches to LMI-based ones are justified. The most successful approaches in this category are [Pre20]: Bilinear Matrix Inequalities [Wu18, Wu21], Popov's hyperstability theory [Bin95, Böh95, Pre97a Kum16], the limit cycle-based approach [Pre08a, Pre08b, Rad09], the circle criterion [Ara89, Opi93, Hab03, Ban07], the harmonic balance method [Bin95, Kie93, Bol94, Cue99, Ros19], the center manifold theory [Pre99c, Pre20], LaSalle's invariant set principle [Tom07] and Barbashin-Krasovskii theorem [Pre07]. Many of this non-LMI-based approaches are applied to control systems with Mamdani fuzzy controllers and not with Takagi-Sugeno-Kang ones.

Remark 7.40

Another way to set systematically the value of the parameter B_e is to carry out the optimal tuning of fuzzy controllers. Optimization problems must be defined in this regard, where the variables are the tuning parameters of the fuzzy controllers. As specified in [Pre21a], metaheuristic algorithms are successfully applied because they ensure higher performance and require lower computing capacity and time versus deterministic algorithms in several optimization problems. The following metaheuristic algorithms are selected in [Pre21a] as they have been applied most recently to the optimal tuning of fuzzy controllers: adaptive weight Genetic Algorithm (GA) for gear shifting control [Eck19], GA-based multiobjective optimization for electric vehicle powertrain control [Eck20], GA for hybrid power system control [Aba20], engine control [Mas20], energy management in hybrid vehicles [Fu20], wellhead back pressure control systems [Lia20], micro-unmanned helicopter control [Hu20], Particle Swarm Optimization (PSO) algorithm with compensating coefficient of inertia weight factor for filter time constant adaptation in hybrid energy storage systems control [Wu19], set-based PSO algorithm with adaptive weights for optimal path planning of unmanned aerial vehicles [Wai19], PSO algorithm for zinc production [Xie19], inverted pendulum control [Gir21] and servo system control [Pre19], hybrid PSO-Artificial Bee Colony (ABC) algorithm for frequency regulation in microgrids

[Moh20a], Imperialist Competitive Algorithm for human immunodeficiency control [Rei19], Grey Wolf Optimizer algorithms for sun-tracker systems [Tri20] and servo system control [Pre19], PSO, Cuckoo Search and Differential Evolution (DE) for gantry crane systems position control [Sol20], Whale Optimization Algorithm (WOA) for vibration control of steel structures [Azi19], Grasshopper Optimization Algorithm for load frequency control [Nos19], DE for EHS control [Don19], Gravitational Search, Charged System Search [Pre19] and Slime Mould algorithms for servo system control [Pre21a].

Example 7.3

The design and tuning of a Mamdani PI fuzzy controller is exemplified as follows focusing on the angular speed control of the nonlinear laboratory DC drive described in this section, where the controlled process is well approximated by the transfer function $P(s)$ [Pre06]

$$P(s) = \frac{k_P}{s(1 + T_\Sigma s)}, \tag{7.76}$$

where k_P is the process gain, and T_Σ is the sum of small (or parasitic) time constants, with the values $k_P = 4,900$ and $T_\Sigma = 0.035$ s. One solution to control this class of processes is built around PI controllers [Åst95], and simple and efficient approach to tune the parameters of the PI controller is the extended symmetrical optimum (ESO) method proposed and described in [Pre96 and Pre99d]. The ESO method is characterized by a single tuning parameter, β, whose choice within the domain $1 < \beta < 20$ leads to modifying in the desired way the empirical performance indices of the control system with respect to the step modification of the reference input. The ESO method ensures a tradeoff to these performance indices in terms of using the performance diagrams illustrated in Figure 7.11, where: σ_1 – the overshoot, $\hat{t}_r = t_r / T_\Sigma$ – the 10% to 90% rise time in normed values, $\hat{t}_s = t_s / T_\Sigma$ – the 2% settling time in normed values (defined in the controlled output with respect to the unit step modification of the reference input) and ϕ_m – the phase margin.

The control system performance indices can be improved by the additional filtering of the reference input, which leads to a 2-DOF control system structure. The tuning relations specific to the ESO method are

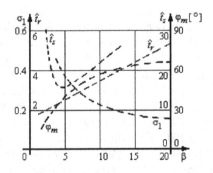

FIGURE 7.11   Performance indices versus tuning parameter β. (Adapted from Pre99d.)

$$k_C = \frac{1}{\sqrt{\beta T_\Sigma k_P}}, \quad T_i = \beta T_\Sigma. \tag{7.77}$$

The ESO method is applied in the design and tuning of the basic PI controller that is needed in the design and tuning of the PI fuzzy controller. Setting the value of the tuning parameter to $\beta = 6$, the parameters of the continuous-time PI controller result from (7.77), $k_C = 0.0024$ and $T_i = 0.21$ s. Choosing the value of the sampling period $T_s = 0.01$ s, which accommodates for quasi-continuous digital control, the application of (7.69) leads to the parameters of the incremental quasi-continuous digital PI controller $K_P = 0.0023$, $K_I = 0.00011$ and $\alpha = 0.0488$. Finally, setting $B_e = 3$, the tuning relations (7.74) result in $B_{\Delta e} = 0.146$ and $B_{\Delta u} = 0.00034$. The behavior of the fuzzy control system is illustrated in Figure 7.12a with respect to the modification of the reference input (or the speed set-point) and Figure 7.12b with respect to the ±10 % of the nominal value step type modification of the load-type disturbance input.

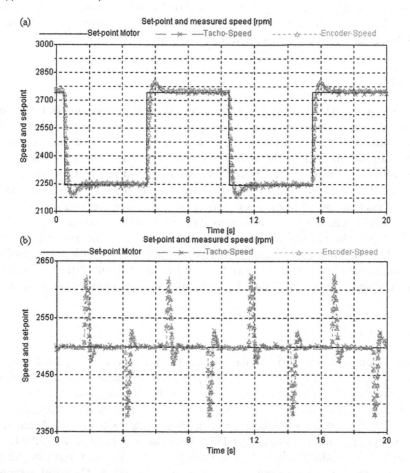

FIGURE 7.12 Reference input (or set-point) and controlled output (angular speed) versus time with respect to reference input modification (a) and load-type disturbance input modification. (Adapted from Pre06.)

Remark 7.41

The PI fuzzy controllers in the versions presented in this section are successful in many applications and allow the relatively easy design and tuning of other fuzzy controller structures presented in Figure 7.9. The existence of several specific features in the dynamics of the controlled process may require modification in the controller structures, which deal with [Pre97b, Pas98]

   a. modifications of the rule base and
   b. modifications of the fuzzification and defuzzification modules.

The category (a) of modifications is imposed by the actual behavior of the process as an effect of certain properties as, for example, those specific to nonminimum phased systems. The category (b) has to be correlated to the requirements concerning the modifications of the operating point of the process and can lead to complex fuzzy controllers [Sal05, Pre11, Pre15], whose design and tuning requires systematic analyses that depend on the processes and controllers involved.

Remark 7.42

As emphasized in [Pre21b], type-2 fuzzy logic and control can ensure the further performance improvement of fuzzy control systems with respect to type-1 fuzzy logic and control described in this section. Interval type-2 fuzzy controllers are especially successful in this regard. The main differences between the interval type-2 fuzzy controllers and the type-1 fuzzy controllers pointed out in [Wu12] (changing the embedded type-1 fuzzy sets in computing the bounds of the type-reduced interval with respect to input changes plus the upper and lower membership functions of the same interval type-2 fuzzy set may be used simultaneously in computing each bound of the type-reduced interval) lead to the main advantage of interval type-2 fuzzy controllers, highlighted in [Lam08], i.e., they are able to handle the uncertainties in nonlinear control systems directly as, for example, parameter uncertainties, mismeasurement uncertainties, observation uncertainties and communication uncertainties.

Remark 7.43

Several metaheuristic algorithms have been applied recently to the optimal tuning of interval type-2 fuzzy controllers. Some representative examples extracted from the list given in [Pre21b] are as follows: multitasking GA for couple-tank water level in both interval type-2 and type-1 fuzzy control, genetic programming for anesthesia control [Wei20], WOA, black-hole optimization algorithm with Lévy flight and sine-cosine algorithm for frequency control in power systems [Ghe20, Kho20], PSO algorithm for zinc production in both type-2 and type-1 fuzzy control versions [Xie19], PSO algorithm for robot control and chaotic systems [Le18], ABC algorithm for rectifiers [Ack20], hybrid PSO-ABC for frequency control [Moh20a], biogeography-based optimization algorithm for glucose level regulation [Moh20b], Crow Search

Algorithm for blood pressure control [Sha20b], elitist Invasive Weed Optimization for maximum power point tracking [Bay20], firefly algorithm and galactic swarm optimization [Ber21], GA, cuckoo search, PSO, DE, ABC and combined PSO-DE algorithms for servo processes [DeM20].

7.2 HYBRID MODEL-FREE FUZZY CONTROLLERS

This section describes two combinations of model-free controllers and fuzzy controllers in terms of adding fuzzy logic terms for performance improvement. The presentation will be focused on the SISO controllers, and two combinations will be presented in the next two sections, the first-order discrete-time intelligent Proportional-Integral (iPI) controllers with Takagi-Sugeno-Kang PD fuzzy terms and the second-order discrete-time iPI controllers with Takagi-Sugeno-Kang PD fuzzy terms.

7.2.1 FIRST-ORDER DISCRETE-TIME iPI CONTROLLERS WITH TAKAGI-SUGENO-KANG PD FUZZY TERMS

Theorem 7.2

The design and tuning of these controllers start with the expression of the discrete-time MIMO iPI controller given in (3.76). A particular form of (3.76) is the control law of the discrete-time SISO iPI controller

$$u(k) = \alpha^{-1}(-\hat{F}(k) + y_d(k+1) - y_d(k) - K_1 e(k) - K_2 e(k-1)), \qquad (7.78)$$

where, as specified in Section 3.2.5.2 for the discrete-time MIMO iPI controller, $u(k) \in \Re$ is the control signal, $\alpha \in \Re$ is a parameter chosen by the designer such that $\Delta y(k+1) = y(k+1) - y(k)$ and $\alpha u(k)$ have the same order of magnitude, $\hat{F}(k) \in \Re$ is the estimate of $F(k) \in \Re$ in terms of the following particular form of (3.70):

$$\hat{F}(k) = y(k) - y(k-1) - \alpha u(k-1), \qquad (7.79)$$

$y_d(k) \in \Re$ is the reference input (or the set-point), which can be obtained as the output of the reference model, the variable $F(k) \in \Re$ plays a disturbance role in the first-order local model of the process expressed as a particular form of (3.68)

$$y(k+1) = y(k) + \alpha u(k) + F(k), \qquad (7.80)$$

with $y(k) \in \Re$ – the controlled output, $e(k)$ is the control error

$$e(k) = y_d(k) - y(k), \qquad (7.81)$$

and $K_1 \in \Re$ and $K_2 \in \Re$ are the gains of the iPI controller.

The estimate $\hat{F}(k) \in \Re$ defined in (7.79) is next substituted in (7.78), leading to the control law of the discrete-time SISO iPI controller

$$u(k) = u(k-1) + \alpha^{-1}(y_d(k+1) - y_d(k) - y(k) + y(k-1) - K_1 e(k) - K_2 e(k-1)), \quad (7.82)$$

and it illustrates the presence of the I term.

Using the increment of control error $\Delta e(k)$

$$\Delta e(k) = e(k) - e(k-1) \tag{7.83}$$

and inserting $e(k-1) = e(k) - \Delta e(k)$ resulted from (7.83) in (7.82), the expression of the control law of the discrete-time MIMO iPI controller becomes

$$u(k) = u(k-1) + \alpha^{-1}(y_d(k+1) - y_d(k) - y(k) + y(k-1) - M_1\Delta e(k) - M_2e(k)), \tag{7.84}$$

where $M_1 \in \Re$ is the proportional gain and $M_2 \in \Re$ is the integral gain of the iPI controller

$$M_1 = -K_2, \quad M_2 = K_1 + K_2. \tag{7.85}$$

Inserting the following notation $\psi_{PD}(k)$ for the PD term in (7.84):

$$\psi_{PD}(k) = M_1\Delta e(k) + M_2e(k), \tag{7.86}$$

the expression of the control law of the discrete-time MIMO iPI controller is

$$u(k) = u(k-1) + \alpha^{-1}(y_d(k+1) - y_d(k) - y(k) + y(k-1) - \psi_{PD}(k)). \tag{7.87}$$

Remark 7.44

The idea behind the first-order discrete-time iPI controllers with Takagi-Sugeno-Kang PD fuzzy terms suggested in [Rom17] is to replace the PD term $\psi_{PD}(k)$ in (7.87) with the Takagi-Sugeno-Kang PD fuzzy term $\phi(k)$ resulting in the general expression of the control law of these hybrid model-free fuzzy controllers

$$u(k) = u(k-1) + \alpha^{-1}(y_d(k+1) - y_d(k) - y(k) + y(k-1) - \phi(k)). \tag{7.88}$$

Proposition 7.29

The discrete-time Takagi-Sugeno-Kang PD fuzzy term is built around the two-input-single-output fuzzy controller (TISO-FC) in terms of the structure given in Figure 7.13a. Figure 7.13b points out the input (and also scheduling) membership functions and their parameters, B_{e1} and $B_{\Delta e}$.

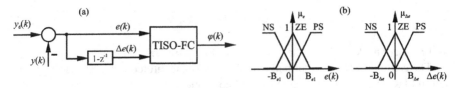

FIGURE 7.13 Discrete-time Takagi-Sugeno-Kang PD fuzzy term (a) and input membership functions (b). (Adapted from Rom17, Rom18, Rom19a and Rom19b.)

TABLE 7.4

Rule Base of Hybrid Model-Free Fuzzy Controller for Cart Control, Arm Angular Position Control and Payload Position Control of Tower Crane Systems

$\dfrac{e(k)}{\Delta e(k)}$	NS	ZE	PS
PS	$\varphi_2(k)$	$\varphi_3(k)$	$\varphi_1(k)$
ZE	$\varphi_2(k)$	$\varphi_3(k)$	$\varphi_3(k)$
NS	$\varphi_2(k)$	$\varphi_2(k)$	$\varphi_2(k)$

Source: Adapted from [Rom17] and [Rom19b].

Remark 7.45

The weighted average method is employed in the defuzzification module. The SUM and PROD operators are used in the inference engine of the discrete-time Takagi-Sugeno-Kang PD fuzzy term included in these hybrid model-free fuzzy controllers. The inference engine is associated with nine rules. An example of rule base is given in Table 7.4 for cart control, arm angular position control and payload position control of tower crane systems.

The following notations are employed in the rule consequents:

$$\phi_i(k) = \chi_i \ \psi_{PD}(k), \ i = 1...3, \tag{7.89}$$

where $\chi_i > 0$, $i = 1...3$, are the parameters inserted in order to alleviate the overshoot of the control system.

Remark 7.46

Table 7.4, (7.86) and (7.89) lead to the conclusion that this hybrid model-free fuzzy controller is a nonlinear combination of four discrete-time linear PD controllers placed in the rule consequents. Figure 7.13b and (7.89) indicate that this hybrid model-free fuzzy controller has eight tuning parameters, namely

- the parameters of the iPI controller: α, K_1 and K_2,
- the parameters of the Takagi-Sugeno-Kang PD fuzzy term: B_{e1}, $B_{\Delta e}$ and $\chi_i > 0$, $i = 1...3$.

Lemma 7.3

As shown in [Rom17], the parameter B_{e1} is set such that to fire all fuzzy rules in the operating regimes of the control system, and their values depend on the reference input domain. The value of the parameter $B_{\Delta e}$ is obtained in terms of the modal equivalence principle

$$B_{\Delta e} = -(M_2 / M_1) \cdot B_{e1}. \tag{7.90}$$

Therefore, the parameter vector of this hybrid model-free fuzzy controller is

$$\chi = [\alpha \ K_1 \ K_2 \ B_{e1} \ \chi_1 \ \chi_2 \ \chi_3]^T. \tag{7.91}$$

A part of these parameters can be set according to Chapter 3 and Section 7.1. The values of the remaining parameters can also be set or optimally tuned using meta-heuristic algorithms as, for example, Gravitational Search Algorithm in [Rom17].

7.2.2   SECOND-ORDER DISCRETE-TIME iPI CONTROLLERS WITH TAKAGI-SUGENO-KANG PD FUZZY TERMS

Theorem 7.3

The design and tuning of these controllers start with the expression of the second-order local model of the process expressed as a particular form of (3.82) valid for MIMO controllers but expressed as follows for SISO controllers:

$$y(k+1) = 2y(k) - y(k-1) + \alpha \ u(k) + F(k). \tag{7.92}$$

A particular form of the discrete-time MIMO iPI controller given in (3.86) is the control law of the discrete-time SISO iPI controller

$$u(k) = \alpha^{-1}(-\hat{F}(k) + y_d(k+1) - 2y_d(k) + y_d(k-1) - K_1 e(k) - K_2 e(k-1)), \tag{7.93}$$

which obtains the following expression using (7.83) and (7.85):

$$u(k) = \alpha^{-1}(-\hat{F}(k) + y_d(k+1) - 2y_d(k) + y_d(k-1) - M_1 \Delta e(k) - M_2 e(k)). \tag{7.94}$$

The estimate $\hat{F}(k) \in \Re$ of $F(k) \in \Re$ is obtained as the following particular form of (3.84):

$$\hat{F}(k) = y(k-2) - 2y(k-1) + y(k) - \alpha \ u(k-1), \tag{7.95}$$

which is next substituted in (7.94), leading to the control law of the discrete-time SISO iPI controller

$$u(k) = u(k-1) + \alpha^{-1}(y_d(k+1) - 2y_d(k) + y_d(k-1)$$
$$-y(k) + 2y(k-1) - y(k-2) - M_1 \Delta e(k) - M_2 e(k)). \tag{7.96}$$

Using in (7.96) the notation $\psi_{PD}(k)$ for the PD term defined in (7.86), the expression of the control law of the discrete-time MIMO iPI controller is

$$u(k) = u(k-1) + \alpha^{-1}(y_d(k+1) - 2y_d(k) + y_d(k-1)$$
$$-y(k) + 2y(k-1) - y(k-2) - \psi_{PD}(k)). \tag{7.97}$$

Remark 7.47

The idea that supports the design of the second-order discrete-time iPI controllers with Takagi-Sugeno-Kang PD fuzzy terms suggested in [Rom18] is to replace the PD term $\psi_{PD}(k)$ in (7.97) with the Takagi-Sugeno-Kang PD fuzzy term $\phi(k)$ resulting in the general expression of the control law of these hybrid model-free fuzzy controllers

$$u(k) = u(k-1) + \alpha^{-1}(y_d(k+1) - 2y_d(k) + y_d(k-1)$$
$$-y(k) + 2y(k-1) - y(k-2) - \varphi(k)). \tag{7.98}$$

Proposition 7.30

The discrete-time Takagi-Sugeno-Kang PD fuzzy term is built around TISO-FC in terms of the structure illustrated in Figure 7.13. The weighted average method is employed in the defuzzification module of these fuzzy controllers as well. The SUM and PROD operators are used in the inference engine of the discrete-time Takagi-Sugeno-Kang PD fuzzy term included in these hybrid model-free fuzzy controllers. The inference engine is associated with nine rules, and an example of rule base is given in Table 7.4 adapted from [Rom18] for cart control, arm angular position control and payload position control of tower crane systems.

Remark 7.48

The modal equivalence principle is applied to the fuzzy term in order to reduce the number of parameters. Several parameter vectors characterize these hybrid model-free fuzzy controllers. One such parameter vector is exemplified in (7.91) in relation with the rule base given in Table 7.4.

7.3 HYBRID MODEL-FREE ADAPTIVE FUZZY CONTROLLERS

This section gives a combination of discrete-time model-free adaptive controllers and fuzzy controllers in terms of adding fuzzy logic terms for performance improvement. The presentation is focused on SISO controllers.

Theorem 7.4

The controller design and tuning start with the expression of the discrete-time MIMO Model-Free Adaptive Control-Compact Form Dynamic Linearization (MFAC-CFDL) algorithms presented in (5.7). A particular form of (5.7) is the control law of the discrete-time SISO MFAC-CFDL controllers

$$u(k) = u(k-1) + \frac{\rho\,\hat{\Phi}(k)[y_d(k+1) - y(k)]}{\lambda + (\hat{\Phi}(k))^2}, \tag{7.99}$$

where $\hat{\Phi}(k)$ is the estimate of the scalar that corresponds to the PPD matrix $\Phi(k)$ for MIMO controllers, $\lambda = \text{const} > 0$ is the weighting parameter involved in the objective function defined in (5.3) for MIMO controllers, and $\rho > 0$ is the step size. Additional details on discrete-time MIMO MFAC-CFDL controllers are given in Section 5.2.1.1.

These hybrid model-free adaptive fuzzy controllers are built in terms of adding, as suggested in [Rom19a and Rom19b], the Takagi-Sugeno-Kang PD fuzzy term $\phi(k)$ that appears in (7.88) and (7.98) to the right-hand term in (7.99). This leads to the general expression of the control law of these hybrid model-free adaptive fuzzy controllers

$$u(k) = u(k-1) + \frac{\rho \, \hat{\Phi}(k)[y_d(k+1) - y(k)]}{\lambda + (\hat{\Phi}(k))^2} + \phi(k). \tag{7.100}$$

The discrete-time Takagi-Sugeno-Kang PD fuzzy term is built around TISO-FC in terms of the structure given in Figure 7.13a. Figure 7.13b highlights the input (and also scheduling) membership functions and their parameters B_{e1} and $B_{\Delta e}$.

Remark 7.49

The weighted average method is employed in the defuzzification module. The SUM and PROD operators are used in the inference engine of the discrete-time Takagi-Sugeno-Kang PD fuzzy term included in these hybrid model-free fuzzy controllers. The inference engine is associated with nine rules. An example of rule base is given in Table 7.4 for cart control, arm angular position control and payload position control of tower crane systems. The notations defined in (7.89) are employed in the rule consequents, where $\chi_i > 0$, $i = 1...3$ are the parameters inserted, as in the previous section, in order to alleviate the overshoot of the control system.

Proposition 7.31

Applying the modal equivalence principle in (7.90) to reduce the number of tuning parameters, Figure 7.13b and (7.89) indicate that this hybrid model-free adaptive fuzzy controller has 11 tuning parameters, namely

- the parameters of the MFAC-CFDL control law described in Chapter 5: η, λ, μ, ρ and $\hat{\phi}(1)$,
- the parameters of the discrete-time linear PD term with the output $\psi_{PD}(k)$ computed in accordance with (7.86): M_1 and M_2,
- the parameters of the fuzzy term: B_{e1} and $\chi_i > 0$, $i = 1...3$.

Therefore, the parameter vector of this hybrid model-free adaptive fuzzy controller is

$$\chi = [\eta \ \lambda \ \mu \ \rho \ \hat{\phi}(1) \ M_1 \ M_2 \ B_{e1} \ \chi_1 \ \chi_2 \ \chi_3]^T. \tag{7.101}$$

A part of these parameters can be set according to Chapter 5 and Section 7.1. The values of the remaining parameters can also be set or optimally tuned using meta-heuristic algorithms as, for example, Grey Wolf Optimizer in [Rom19a and Rom19b].

7.4 EXAMPLE AND APPLICATION

7.4.1 SISO CONTROL SYSTEMS

The objective function used to measure the performance of each SISO control system structure with first-order discrete-time iPI controllers with Takagi-Sugeno-Kang PD fuzzy terms, second-order discrete-time iPI controllers with Takagi-Sugeno-Kang PD fuzzy terms and hybrid model-free adaptive fuzzy controllers presented in Section 7.2 is

$$J_{(\psi)}(\chi_{(\psi)}) = \frac{1}{2} E \left\{ \sum_{k=1}^{N} [e_{(\psi)}(k, \chi_{(\psi)})]^2 \right\},$$
(7.102)

where $\chi_{(\psi)}$ is the general notation for the parameter vector of each controller, the subscript $\psi \in \{c, a, p\}$ indicates both the controlled output and its corresponding controller, namely c – the cart position (i.e., $y_{(c)} = y_1$), a – the arm angular position (i.e., $y_{(a)} = y_2$) and p – the payload position (i.e., $y_{(p)} = y_3$), and $e_{(\psi)} = y_{d(\psi)} - y_{(\psi)}$ is the control error. The mathematical expectation $E\{\Xi\}$ is taken with respect to the stochastic probability distribution of the disturbance inputs applied to the process and thus affects the control system behavior. The sampling period is $T_s = 0.01$ s, and the time horizon is 70 s, thus leading to a number of $N = 70 / 0.01 = 7,000$ samples.

The first-order discrete-time iPI controllers with Takagi-Sugeno-Kang PD fuzzy terms, the second-order discrete-time iPI controllers with Takagi-Sugeno-Kang PD fuzzy terms and the hybrid model-free adaptive fuzzy controllers are tuned separately, one for each controlled output of the process, in terms of the steps given in Section 7.2. The description of the steps is given in relation with the Matlab & Simulink programs and schemes.

The reference inputs involved in this section are obtained by first generating the following step signals:

$$y_c^*(k) = 0.15 \text{ if } k \in [0, 2,000], 0.1 \text{ if } k \in (2,000, 3,500], -0.05 \text{ if } k \in (3,500, 5,000],$$

$$0 \text{ if } k \in (5,000, 7,000]$$
(7.103)

for cart position control,

$$y_a^*(k) = 0 \text{ if } k \in [0, 500], 0.15 \text{ if } k \in (500, 2,500], -0.15 \text{ if } k \in (2,500, 4,000],$$

$$0 \text{ if } k \in (4,000, 7,000]$$
(7.104)

for arm angular position control and

$$y_p^*(k) = 0 \text{ if } k \in [0,1,500], 0.1 \text{ if } k \in (1,500,3,000), -0.05 \text{ if } k \in (3,000,4,500),$$

$$0 \text{ if } k \in (4,500,7,000] \tag{7.105}$$

for payload position control. They are next filtered using the filters with the transfer functions

$$H_{(c)}(s) = \frac{1}{1+0.2s} \tag{7.106}$$

for cart position control,

$$H_{(a)}(s) = \frac{1}{1+0.2s} \tag{7.107}$$

for arm angular position control and

$$H_{(p)}(s) = \frac{1}{1+0.2s} \tag{7.108}$$

for payload position control, with the corresponding discrete transfer functions:

$$H_{(c)}(z^{-1}) = \frac{0.0488}{1-0.9512z^{-1}}, \tag{7.109}$$

$$H_{(a)}(z^{-1}) = \frac{0.0465}{1-0.9535z^{-1}}, \tag{7.110}$$

$$H_{(p)}(z^{-1}) = \frac{0.0328}{1-0.9672z^{-1}}. \tag{7.111}$$

Finally, the reference inputs $y_{d(c)}$, $y_{d(a)}$ and $y_{d(p)}$ applied to the three SISO control loops are obtained as the outputs of the above filters. Although this section gives only a sample of simulation results for first-order discrete-time iPI controllers, second-order discrete-time iPI controllers and MFAC with Takagi-Sugeno-Kang PD fuzzy terms, the above algorithms can be hybridized with other fuzzy terms.

7.4.1.1 The First-Order Discrete-Time iPI Controllers with Takagi-Sugeno-Kang PD Fuzzy Terms

Three first-order discrete-time iPI controllers with Takagi-Sugeno-Kang PD fuzzy terms are used separately to control the three SISO control system structures, where the parameter vectors of the first-order discrete-time iPI controllers with Takagi-Sugeno-Kang PD fuzzy terms controller are

$$\chi_{(\psi)} = [\alpha \ \ K_1 \ \ K_2 \ \ B_{el} \ \ \chi_1 \ \ \chi_2 \ \ \chi_3]^T, \tag{7.112}$$

with $\psi \in \{c, a, p\}$, i.e., one parameter vector for each controller.

The controllers are nonlinear since saturation-type nonlinearities are inserted on the three controller outputs in order to match the power electronics of the actuators, which operate on the basis of the pulse width modulation (PWM) principle. As shown in Chapter 1, the three control signals (or control inputs) are within −1 and 1, ensuring the necessary connection to the tower crane system subsystem, which models the controlled process in the control system structures developed in this chapter and the next chapters and also includes three zero-order hold blocks to enable digital control. The tower crane system subsystem belongs to the Process.mdl Simulink diagram, which is included in the accompanying Matlab & Simulink programs given in Chapter 1.

The parameter $\alpha = 0.0015$ is first chosen according to step 1 in Section 3.2.5.2, and then the parameters K_1 and K_2 are established to ensure the condition (3.78) according to step 2 in Section 3.2.5.2, and the parameters B_{e1}, $B_{\Delta e}$, χ_1, χ_2 and χ_3 of the fuzzy term are next set:

$$\chi_{(c)} = [0.0015 \ -0.75 \ 0.33 \ 0.7 \ 1 \ 3 \ 1.1]^T, \tag{7.113}$$

$$\chi_{(a)} = [0.0015 \ -0.72 \ 0.4 \ 0.7 \ 1.1 \ 0.9 \ 1]^T, \tag{7.114}$$

$$\chi_{(p)} = [0.0015 \ -0.85 \ 0.53 \ 0.7 \ 1 \ 1.5 \ 2]^T. \tag{7.115}$$

The details related to the first-order discrete-time iPI controllers with Takagi-Sugeno-Kang PD fuzzy terms and the programs of the above steps are implemented in the MFC _ FUZZY _ iPI _ 1st _ order _ c.m Matlab program for cart position control, the MFC _ FUZZY _ iPI _ 1st _ order _ a.m Matlab program for arm angular position control and the MFC _ FUZZY _ iPI _ 1st _ order _ p.m Matlab program for payload position control, which are included in the accompanying Matlab & Simulink programs. The simulation of the control systems with the three first-order discrete-time iPI controllers with Takagi-Sugeno-Kang PD fuzzy terms are conducted using the same Simulink diagrams, i.e., the CS _ iPI _ FUZZY _ SISO _ c.mdl diagram for cart position control, the CS _ iPI _ FUZZY _ SISO _ a.mdl diagram for arm angular position control and the CS _ iPI _ FUZZY _ SISO _ p.mdl diagram for payload position control.

The simulation results obtained for the first-order discrete-time iPI controllers with Takagi-Sugeno-Kang PD fuzzy terms with parameters in (7.113–7.115) are illustrated in Figure 7.14 for cart position control, Figure 7.15 for arm angular position control and Figure 7.16 for payload position control.

The experimental results obtained for the control systems with first-order discrete-time iPI controllers with Takagi-Sugeno-Kang PD fuzzy terms with parameters in (7.113–7.115) are illustrated in Figure 7.17 for cart position control, Figure 7.18 for arm angular position control and Figure 7.19 for payload position control.

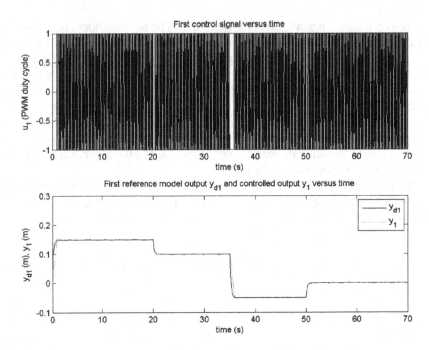

FIGURE 7.14 Simulation results of cart position control system with first-order discrete-time iPI controllers with Takagi-Sugeno-Kang PD fuzzy terms.

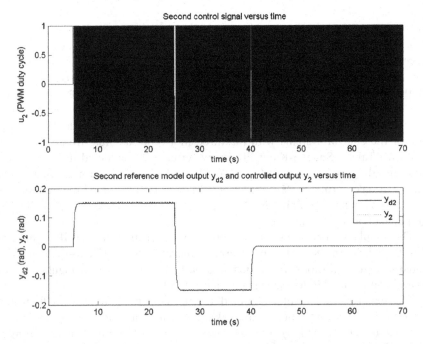

FIGURE 7.15 Simulation results of arm angular position control system with first-order discrete-time iPI controllers with Takagi-Sugeno-Kang PD fuzzy terms.

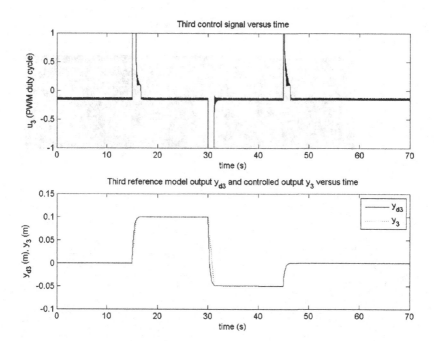

FIGURE 7.16 Simulation results of payload position control system with first-order discrete-time iPI controllers with Takagi-Sugeno-Kang PD fuzzy terms.

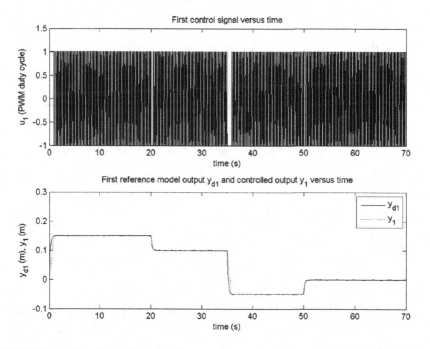

FIGURE 7.17 Experimental results of cart position control system with first-order discrete-time iPI controllers with Takagi-Sugeno-Kang PD fuzzy terms.

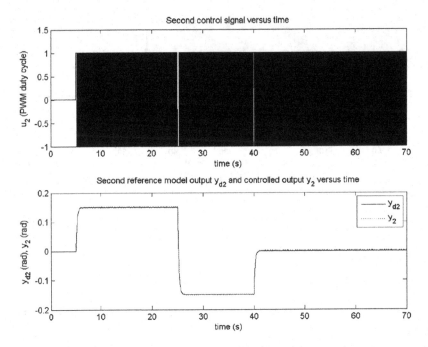

FIGURE 7.18 Experimental results of arm angular position control system with first-order discrete-time iPI controllers with Takagi-Sugeno-Kang PD fuzzy terms.

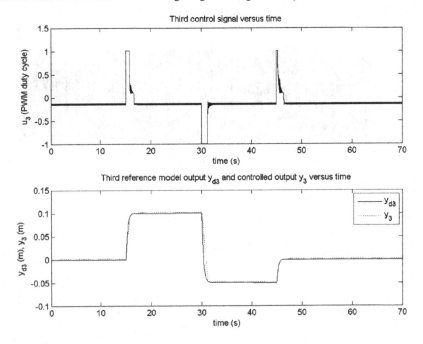

FIGURE 7.19 Experimental results of payload position control system with first-order discrete-time iPI controllers with Takagi-Sugeno-Kang PD fuzzy terms.

7.4.1.2 The Second-Order Discrete-Time iPI Controllers with Takagi-Sugeno-Kang PD Fuzzy Terms

Three second-order discrete-time iPI controllers with Takagi-Sugeno-Kang PD fuzzy terms are tuned separately for the three SISO control system structures, and the parameter vector of the second-order discrete-time iPI controllers with Takagi-Sugeno-Kang PD fuzzy terms controller are

$$\chi_{(\psi)} = [\alpha \ K_1 \ K_2 \ B_{e1} \ \chi_1 \ \chi_2 \ \chi_3]^T, \quad (7.116)$$

with $\psi \in \{c, a, p\}$, i.e. one parameter for each controller.

The parameter $\alpha = 0.0015$ is first chosen according to step 1 in Section 3.2.6.2, and then the parameters K_1 and K_2 are set to ensure the condition (3.89) according to step 2 in Section 3.2.6.2, and finally the parameters B_{e1}, $B_{\Delta e}$, χ_1, χ_2 and χ_3 of the fuzzy term are set as follows:

$$\chi_{(c)} = [0.0015 \ -0.75 \ 0.7 \ 0.7 \ 1 \ 0.8 \ 3]^T, \quad (7.117)$$

$$\chi_{(a)} = [0.0015 \ -0.9 \ 0.85 \ 0.7 \ 1 \ 0.1 \ 1.5]^T, \quad (7.118)$$

$$\chi_{(p)} = [0.0015 \ -0.85 \ 0.5 \ 0.7 \ 1 \ 0.9 \ 1.4]^T. \quad (7.119)$$

The details related to second-order discrete-time iPI controllers with Takagi-Sugeno-Kang PD fuzzy terms and the programs of the above steps are implemented in the MFC _ FUZZY _ iPI _ 2nd _ order _ c.m Matlab program for cart position control, the MFC _ FUZZY _ iPI _ 2nd _ order _ a.m Matlab program for arm angular position control and the MFC _ FUZZY _ iPI _ 2nd _ order _ p.m Matlab program for payload position control, which are included in the accompanying Matlab & Simulink programs. The simulation of the control system structures with the three second-order discrete-time iPI controllers with Takagi-Sugeno-Kang PD fuzzy terms are conducted using the same Simulink diagrams, i.e., the CS _ iPI _ 2nd _ order _ FUZZY _ SISO _ c.mdl diagram for cart position control, the CS _ iPI _ 2nd _ order _ FUZZY _ SISO _ a.mdl diagram for arm angular position control and the CS _ iPI _ 2nd _ order _ FUZZY _ SISO _ p.mdl diagram for payload position control.

The simulation results obtained for the control systems with second-order discrete-time iPI controllers with Takagi-Sugeno-Kang PD fuzzy terms with parameters in (7.117–7.119) are illustrated in Figure 7.20 for cart position control, Figure 7.21 for arm angular position control and Figure 7.22 for payload position control.

The experimental results obtained for the control systems with second discrete-time iPI controllers with Takagi-Sugeno-Kang PD fuzzy terms with the parameters in (7.117–7.119) are illustrated in Figure 7.23 for cart position control, Figure 7.24 for arm angular position control and Figure 7.25 for payload position control.

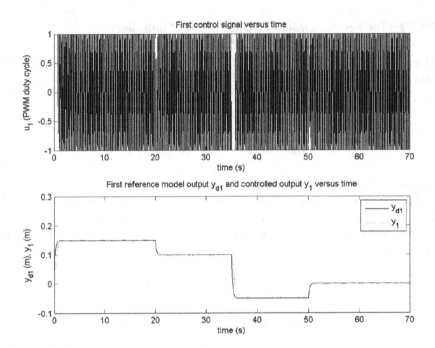

FIGURE 7.20 Simulation results of cart position control system with second-order discrete-time iPI controllers with Takagi-Sugeno-Kang PD fuzzy terms.

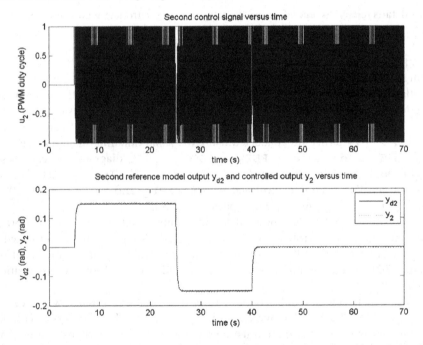

FIGURE 7.21 Simulation results of arm angular position control system with second-order discrete-time iPI controllers with Takagi-Sugeno-Kang PD fuzzy terms.

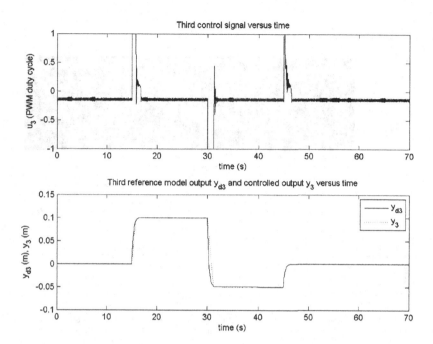

FIGURE 7.22 Simulation results of payload position control system with second-order discrete-time iPI controllers with Takagi-Sugeno-Kang PD fuzzy terms.

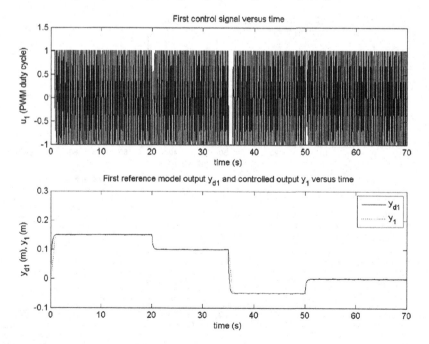

FIGURE 7.23 Experimental results of cart position control system with second-order discrete-time iPI controllers with Takagi-Sugeno-Kang PD fuzzy terms.

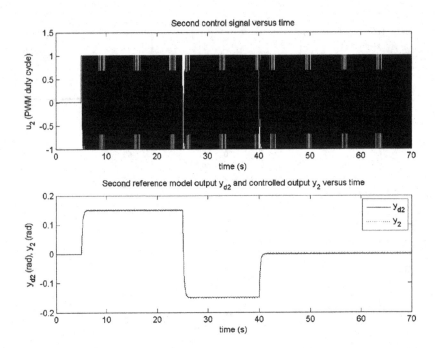

FIGURE 7.24 Experimental results of arm angular position control system with second-order discrete-time iPI controllers with Takagi-Sugeno-Kang PD fuzzy terms.

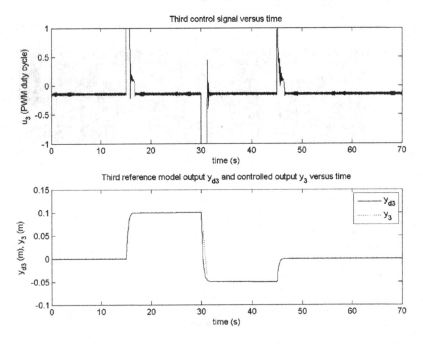

FIGURE 7.25 Experimental results of payload position control system with second-order discrete-time iPI controllers with Takagi-Sugeno-Kang PD fuzzy terms.

7.4.1.3 The Hybrid Model-Free Adaptive Fuzzy Controllers

Three hybrid model-free adaptive fuzzy controllers are used separately in the three SISO control system structures, where the parameter vector of hybrid model-free adaptive fuzzy controllers is

$$\chi = [\eta \ \lambda \ \mu \ \rho \ \hat{\varphi}(1) \ M_1 \ M_2 \ B_{e1} \ \chi_1 \ \chi_2 \ \chi_3]^T, \qquad (7.120)$$

with $\psi \in \{c, a, p\}$, i.e., one parameter for each controller.

The value of the initial value of the estimate of the PPD matrix $\hat{\varphi}(1)$ is first chosen according to step 1 in Section 5.2.1.1, the parameters are next considered as $1,000\hat{\varphi}_{(\psi)}(1)$ for the upper limit and $0.001\hat{\varphi}_{(\psi)}(1)$ for the lower limit and are imposed according to step 2 in Section 5.2.1.1 to ensure the conditions (5.4) and (5.6). Next, the remaining parameters $\eta \in (0,1)$, $\lambda > 0$, $\mu > 0$ and $\rho > 0$ are chosen according to step 3 in Section 5.2.1.1, and finally, the parameters B_{e1}, $B_{\Delta e}$, χ_1, χ_2 and χ_3 of the fuzzy term are set as follows:

$$\chi_{(c)} = [0.6 \ 3.5 \ 28 \ 45 \ 3.5 \ 80 \ 3.99 \ 0.7 \ 1 \ 0.8 \ 3]^T, \qquad (7.121)$$

$$\chi_{(a)} = [0.9 \ 0.01 \ 1 \ 990 \ 12 \ 190 \ 50 \ 0.7 \ 1 \ 0.1 \ 1.5]^T, \qquad (7.122)$$

$$\chi_{(p)} = [0.81 \ 0.71 \ 14.9 \ 25.2 \ 0.89 \ 165 \ 85 \ 0.7 \ 1 \ 0.9 \ 1.4]^T. \qquad (7.123)$$

The details related to hybrid model-free adaptive fuzzy controllers and the programs of the above steps are implemented in the MFAC _ CFDL _ FUZZY _ c.m Matlab program for cart position control, the MFAC _ CFDL _ FUZZY _ a.m Matlab program for arm angular position control and the MFAC_CFDL _ FUZZY _ p.m Matlab program for payload position control, which are included in the accompanying Matlab & Simulink programs. The digital simulations of the control systems with the three hybrid model-free adaptive fuzzy controllers are conducted using the same Simulink diagrams, i.e., the CS _ MFAC _ CFDL _ FUZZY _ SISO _ c.mdl diagram for cart position control, the CS _ MFAC _ CFDL _ FUZZY _ SISO _ a.mdl diagram for arm angular position control and the CS _ MFAC _ CFDL _ FUZZY _ SISO _ p.mdl diagram for payload position control.

The simulation results obtained for the control systems with hybrid model-free adaptive fuzzy controllers with the parameters in (7.121–7.123) are illustrated in Figure 7.26 for cart position control, Figure 7.27 for arm angular position control and Figure 7.28 for payload position control.

The experimental results obtained for the control systems with hybrid model-free adaptive fuzzy controllers with the parameters in (7.121–7.123) are illustrated in Figure 7.29 for cart position control, Figure 7.30 for arm angular position control and Figure 7.31 for payload position control.

7.4.2 MIMO Control Systems

In this section, the exemplification of the MIMO loops is given considering three SISO loops that are running in parallel. These loops consist of the first-order

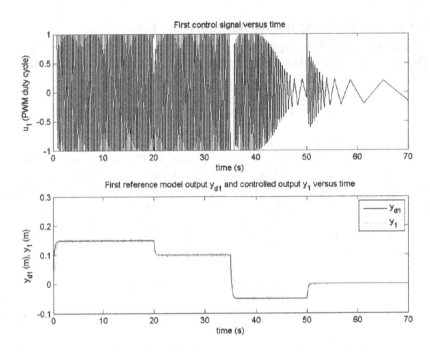

FIGURE 7.26 Simulation results of cart position control system with hybrid model-free adaptive fuzzy controllers.

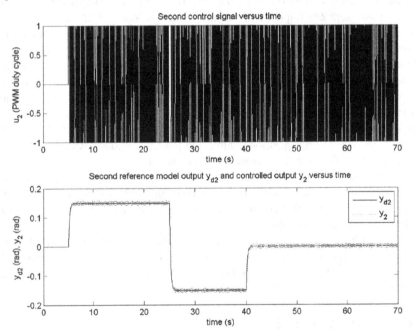

FIGURE 7.27 Simulation results of arm angular position control system with hybrid model-free adaptive fuzzy controllers.

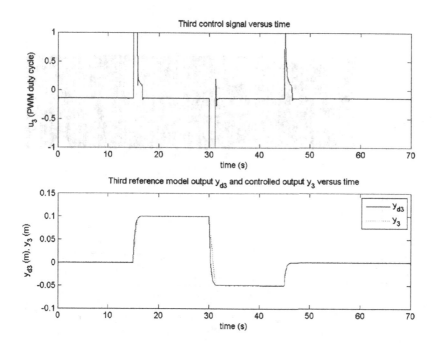

FIGURE 7.28 Simulation results of payload position control system with hybrid model-free adaptive fuzzy controllers.

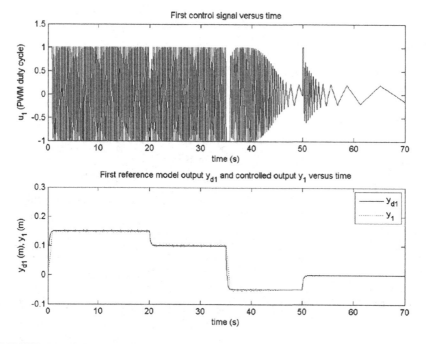

FIGURE 7.29 Experimental results of cart position control system with hybrid model-free adaptive fuzzy controllers.

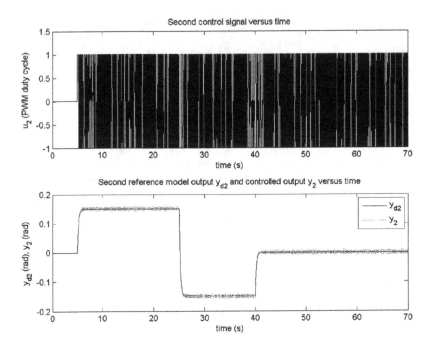

FIGURE 7.30 Experimental results of arm angular position control system with hybrid model-free adaptive fuzzy controllers.

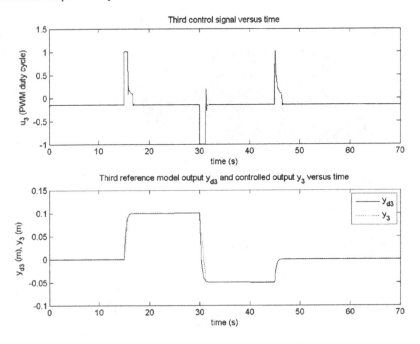

FIGURE 7.31 Experimental results of payload position control system with hybrid model-free adaptive fuzzy controllers.

discrete-time iPI controllers with Takagi-Sugeno-Kang PD fuzzy terms, the second-order discrete-time iPI controllers with Takagi-Sugeno-Kang PD fuzzy terms and the hybrid model-free adaptive fuzzy controllers, applied to the tower crane system described in Chapter 1. The performance is measured using the following objective function:

$$J(\chi) = \frac{1}{2} E \left\{ \sum_{k=1}^{N} [\mathbf{e}(k, \chi)]^2 \right\}, \tag{7.124}$$

where χ is the parameter vector of the controller, i.e., one of the three controllers specified above, $\mathbf{e}$ is the control error vector

$$\mathbf{e} = \mathbf{y}^* - \mathbf{y} = [e_1 \ e_2 \ e_3]^T = [y_{d1} - y_1 \ y_{d2} - y_2 \ y_{d3} - y_3]^T, \tag{7.125}$$

and the scalar control errors e_1, e_2 and e_3 are related to cart position, arm angular position and payload position, respectively, the controlled outputs y_1, y_2 and y_3 are also related to cart position, arm angular position and payload position, respectively, and the expressions of the reference inputs y_{d1}, y_{d2} and y_{d3} are given in relation with (7.103–7.111). The mathematical expectation $E\{\Xi\}$, the sampling period, the time horizon and the number of samples are the same as in the SISO case.

7.4.2.1 The First-Order Discrete-Time iPI Controllers with Takagi-Sugeno-Kang PD Fuzzy Terms

A MIMO first-order discrete-time iPI controller with Takagi-Sugeno-Kang PD fuzzy terms consisting of three SISO controllers is tuned, and the parameter vector of this MIMO first-order discrete-time iPI controller with Takagi-Sugeno-Kang PD fuzzy terms is

$$\chi = [\chi_{(c)}^T \ \chi_{(a)}^T \ \chi_{(p)}^T]^T$$
$$= [\alpha_{(c)} \ K_{(c)1} \ K_{(c)2} \ B_{(c)e1} \ \chi_{(c)1} \ \chi_{(c)2} \ \chi_{(c)3} \ \alpha_{(a)} \ K_{(a)1} \ K_{(a)2} \tag{7.126}$$
$$B_{(a)e1} \ \chi_{(a)1} \ \chi_{(a)2} \ \chi_{(a)3} \ \alpha_{(p)} \ K_{(p)1} \ K_{(p)2} \ B_{(p)e1} \ \chi_{(p)1} \ \chi_{(p)2} \ \chi_{(p)3}]^T,$$

which includes three parameter vectors, namely one for each SISO controller.

The parameters $\alpha_{(c)} = 0.0015$, $\alpha_{(a)} = 0.0015$ and $\alpha_{(p)} = 0.0015$ are first chosen according to step 1 in Section 3.2.5.2, then the parameters $K_{(c)1}$, $K_{(c)2}$, $K_{(a)1}$, $K_{(a)2}$, $K_{(p)1}$ and $K_{(p)2}$ are set to ensure the condition (3.78) according to step 2 in Section 3.2.5.2, next the parameters $B_{(c)e1}$, $B_{(c)\Delta e}$, $\chi_{(c)1}$, $\chi_{(c)2}$, $\chi_{(c)3}$, $B_{(a)e1}$, $B_{(a)\Delta e}$, $\chi_{(a)1}$, $\chi_{(a)2}$, $\chi_{(a)3}$, $B_{(p)e1}$, $B_{(p)\Delta e}$, $\chi_{(p)1}$, $\chi_{(p)2}$ and $\chi_{(p)3}$ of the fuzzy term are set:

$$\chi = [0.0015 \ -0.75 \ 0.33 \ 0.7 \ 1 \ 3 \ 1.1 \ 0.0015 \ -0.72 \ 0.4 \ 0.7 \ 1.1 \ 0.9 \ 1$$
$$0.0015 \ -0.85 \ 0.53 \ 0.7 \ 1 \ 1.5 \ 2]^T. \tag{7.127}$$

The details related to MIMO first-order discrete-time iPI controllers with Takagi-Sugeno-Kang PD fuzzy terms and the programs of the above steps are implemented in MFC _ FUZZY _ iPI _ 1st _ order.m Matlab program for tower crane system control, which is included in the accompanying Matlab & Simulink programs. The simulation of the control system with MIMO first-order discrete-time iPI controller with Takagi-Sugeno-Kang PD fuzzy terms is conducted using the same Simulink diagram, i.e., the CS _ iPI _ FUZZY _ SISO _ MIMO.mdl diagram for tower crane system control.

The simulation results obtained for the control system with MIMO first-order discrete-time iPI controller with Takagi-Sugeno-Kang PD fuzzy terms with the parameters in (7.127) are illustrated in Figure 7.32 for cart position control, Figure 7.33 for arm angular position control and Figure 7.34 for payload position control.

The experimental results obtained for the control system with MIMO first-order discrete-time iPI controller with Takagi-Sugeno-Kang PD fuzzy terms with the parameters in (7.127) are illustrated in Figure 7.35 for cart position control, Figure 7.36 for arm angular position control and Figure 7.37 for payload position control.

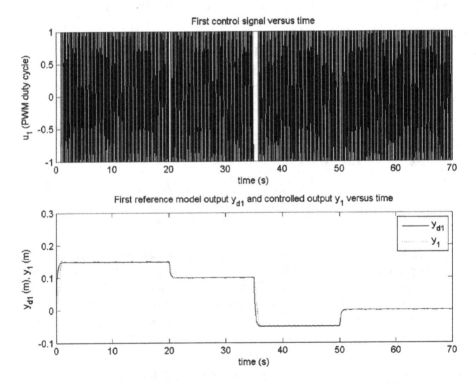

FIGURE 7.32 Simulation results of cart position control system with MIMO first-order discrete-time iPI controller with Takagi-Sugeno-Kang PD fuzzy terms.

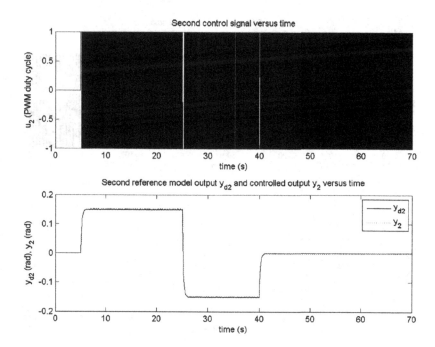

FIGURE 7.33 Simulation results of arm angular position control system with MIMO first-order discrete-time iPI controller with Takagi-Sugeno-Kang PD fuzzy terms.

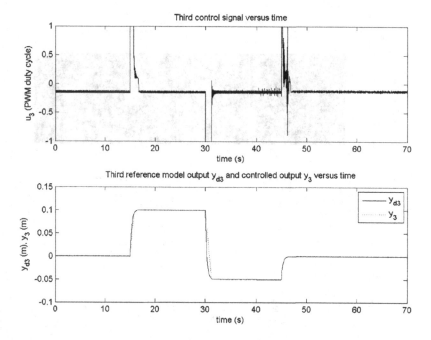

FIGURE 7.34 Simulation results of payload position control system with MIMO first-order discrete-time iPI controller with Takagi-Sugeno-Kang PD fuzzy terms.

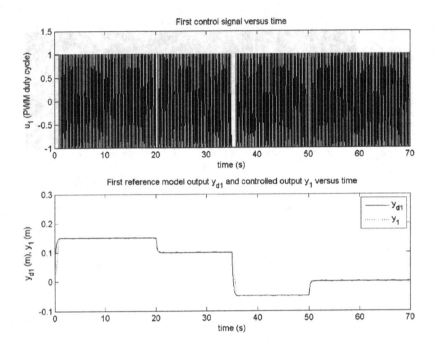

FIGURE 7.35   Experimental results of cart position control system with MIMO first-order discrete-time iPI controller with Takagi-Sugeno-Kang PD fuzzy terms.

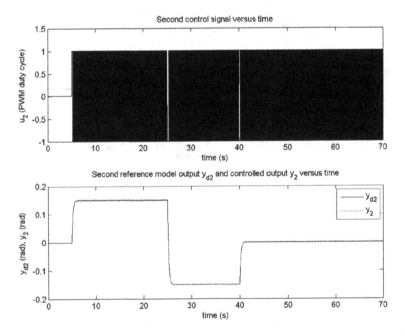

FIGURE 7.36   Experimental results of arm angular position control system with MIMO first-order discrete-time iPI controller with Takagi-Sugeno-Kang PD fuzzy terms.

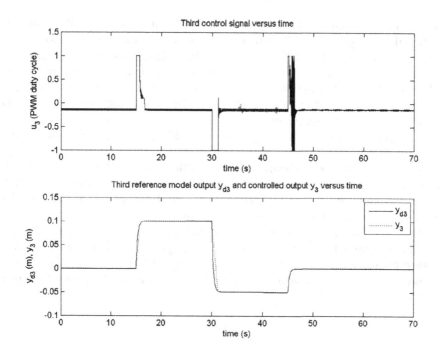

FIGURE 7.37 Experimental results of payload position control system with MIMO first-order discrete-time iPI controller with Takagi-Sugeno-Kang PD fuzzy terms.

7.4.2.2 The Second-Order Discrete-Time iPI Controllers with Takagi-Sugeno-Kang PD Fuzzy Terms

A MIMO second-order discrete-time iPI controller with Takagi-Sugeno-Kang PD fuzzy terms consisting of three SISO controllers running in parallel is implemented. The parameter vector of the MIMO second-order discrete-time iPI controller with Takagi-Sugeno-Kang PD fuzzy terms controller is

$$\chi = [\chi_{(c)}{}^T \ \chi_{(a)}{}^T \ \chi_{(p)}{}^T]^T = [\alpha_{(c)} \ K_{(c)1} \ K_{(c)2} \ B_{(c)e1} \ \chi_{(c)1} \ \chi_{(c)2} \ \chi_{(c)3} \ \alpha_{(a)} \ K_{(a)1} \ K_{(a)2}$$

$$B_{(a)e1} \ \chi_{(a)1} \ \chi_{(a)2} \ \chi_{(a)3} \ \alpha_{(p)} \ K_{(p)1} \ K_{(p)2} \ B_{(p)e1} \ \chi_{(p)1} \ \chi_{(p)2} \ \chi_{(p)3}]^T. \qquad (7.128)$$

The parameters $\alpha_{(c)} = 0.0015$, $\alpha_{(a)} = 0.0015$ and $\alpha_{(p)} = 0.0015$ are chosen according to step 1 in Section 3.2.6.2, then the parameters $K_{(c)1}$, $K_{(c)2}$, $K_{(a)1}$, $K_{(a)2}$, $K_{(p)1}$ and $K_{(p)2}$ are set to ensure the condition (3.89) according to step 2 in Section 3.2.6.2, and next the parameters $B_{(c)e1}$, $B_{(c)\Delta e}$, $\chi_{(c)1}$, $\chi_{(c)2}$, $\chi_{(c)3}$, $B_{(a)e1}$, $B_{(a)\Delta e}$, $\chi_{(a)1}$, $\chi_{(a)2}$, $\chi_{(a)3}$, $B_{(p)e1}$, $B_{(p)\Delta e}$, $\chi_{(p)1}$, $\chi_{(p)2}$ and $\chi_{(p)3}$ of the fuzzy term are set:

$$\chi = [0.0015 \ -0.75 \ 0.7 \ 0.7 \ 1 \ 0.8 \ 3 \ 0.0015 \ -0.9 \ 0.85 \ 0.7 \ 1 \ 0.1 \ 1.5$$

$$0.0015 \ -0.85 \ 0.5 \ 0.7 \ 1 \ 0.9 \ 1.4]^T. \qquad (7.129)$$

The details related to the MIMO second-order discrete-time iPI controller with Takagi-Sugeno-Kang PD fuzzy terms and the programs of the above steps are implemented in the MFC _ FUZZY _ iPI _ 2nd _ order.m Matlab program for tower crane system control, which is included in the accompanying Matlab & Simulink programs. The simulations of the control system with the MIMO second-order discrete-time iPI controller with Takagi-Sugeno-Kang PD fuzzy terms are conducted using the same Simulink diagram, i.e., the CS _ iPI _ 2nd _ order _ FUZZY _ SISO _ MIMO.mdl diagram for tower crane system control.

The simulation results obtained for the MIMO second-order discrete-time iPI controllers with Takagi-Sugeno-Kang PD fuzzy terms with the parameters in (7.129) are illustrated in Figure 7.38 for cart position control, Figure 7.39 for arm angular position control and Figure 7.40 for payload position control.

The experimental results obtained for the control system with the MIMO second-order discrete-time iPI controller with Takagi-Sugeno-Kang PD fuzzy terms with parameters in (7.129) are illustrated in Figure 7.41 for cart position control, Figure 7.42 for arm angular position control and Figure 7.43 for payload position control.

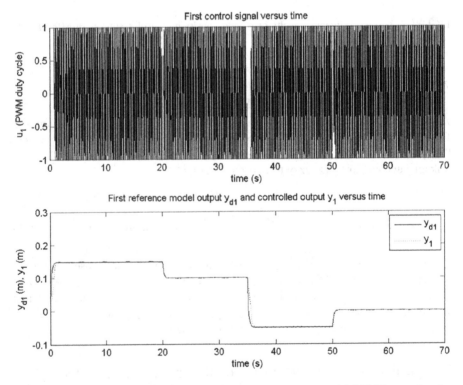

FIGURE 7.38 Simulation results of cart position control system with MIMO second-order discrete-time iPI controller with Takagi-Sugeno-Kang PD fuzzy terms.

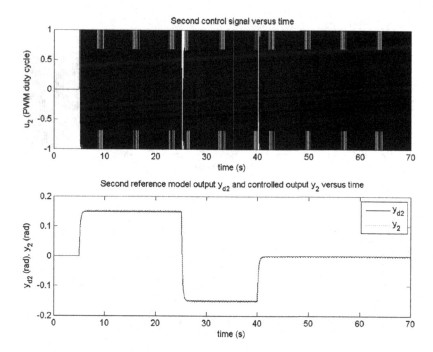

FIGURE 7.39 Simulation results of arm angular position control system with MIMO second-order discrete-time iPI controller with Takagi-Sugeno-Kang PD fuzzy terms.

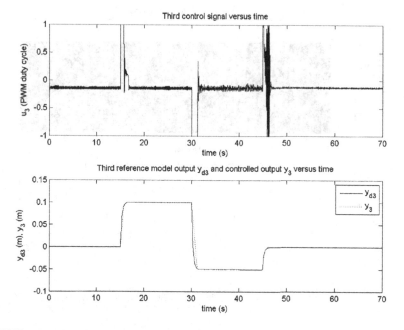

FIGURE 7.40 Simulation results of payload position control system with MIMO second-order discrete-time iPI controller with Takagi-Sugeno-Kang PD fuzzy terms.

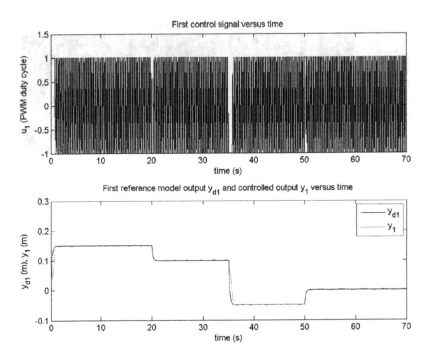

FIGURE 7.41 Experimental results of cart position control system with MIMO second-order discrete-time iPI controller with Takagi-Sugeno-Kang PD fuzzy terms.

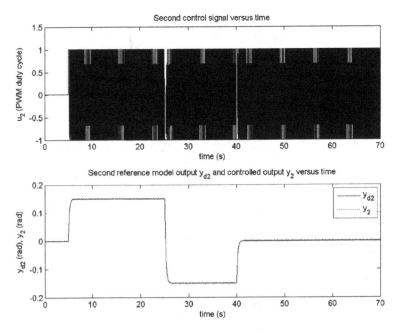

FIGURE 7.42 Experimental results of arm angular position control system with MIMO second-order discrete-time iPI controller with Takagi-Sugeno-Kang PD fuzzy terms.

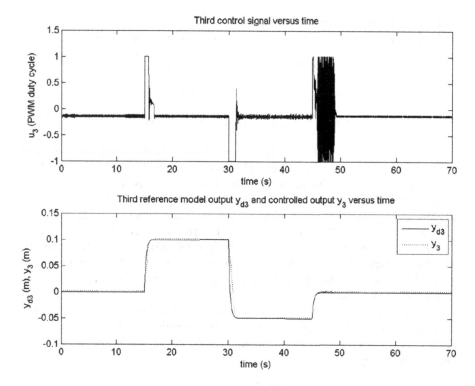

FIGURE 7.43 Experimental results of payload position control system with MIMO second-order discrete-time iPI controller with Takagi-Sugeno-Kang PD fuzzy terms.

7.4.2.3 The Hybrid Model-Free Adaptive Fuzzy Controllers

A MIMO hybrid model-free adaptive fuzzy controller consisting of three SISO controllers that are running in parallel is applied. The parameter vector of the MIMO hybrid model-free adaptive fuzzy controller is

$$\chi = [\chi_{(c)}^T \ \chi_{(a)}^T \ \chi_{(p)}^T]^T$$

$$= [\eta_{(c)} \ \lambda_{(c)} \ \mu_{(c)} \ \rho_{(c)} \ \hat{\varphi}_{(c)}(1) \ M_{(c)1} \ M_{(c)2} \ B_{(c)e1} \ \chi_{(c)1} \ \chi_{(c)2} \ \chi_{(c)3}$$

$$\eta_{(a)} \ \lambda_{(a)} \ \mu_{(a)} \ \rho_{(a)} \ \hat{\varphi}_{(a)}(1) \ M_{(a)1} \ M_{(a)2} \ B_{(a)e1} \ \chi_{(a)1} \ \chi_{(a)2} \ \chi_{(a)3} \quad (7.130)$$

$$\eta_{(p)} \ \lambda_{(p)} \ \mu_{(p)} \ \rho_{(p)} \ \hat{\varphi}_{(p)}(1) \ M_{(p)1} \ M_{(p)2} \ B_{(p)e1} \ \chi_{(p)1} \ \chi_{(p)2} \ \chi_{(p)3}]^T.$$

First, the initial values of the estimate of the PPD matrix $\hat{\varphi}_{(c)}(1)$, $\hat{\varphi}_{(a)}(1)$ and $\hat{\varphi}_{(p)}(1)$ are chosen according to step 1 in Section 5.2.1.1, and the parameters of the lower and upper limits are next imposed as $1{,}000\hat{\varphi}_{(\psi)}(1)$ and $0.001\hat{\varphi}_{(\psi)}(1)$, respectively, according to step 2 in Section 5.2.1.1 to ensure the conditions (5.4) and (5.6). Next, the parameters $\eta_{(c)} \in (0,1)$, $\lambda_{(c)} > 0$, $\mu_{(c)} > 0$, $\rho_{(c)} > 0$, $\eta_{(a)} \in (0,1)$, $\lambda_{(a)} > 0$, $\mu_{(a)} > 0$, $\rho_{(a)} > 0$, $\eta_{(p)} \in (0,1)$, $\lambda_{(p)} > 0$, $\mu_{(p)} > 0$ and $\rho_{(p)} > 0$ are chosen according to step 3 in

Section 5.2.1.1, finally parameters $B_{(c)e1}$, $B_{(c)\Delta e}$, $\chi_{(c)1}$, $\chi_{(c)2}$, $\chi_{(c)3}$, $B_{(a)e1}$, $B_{(a)\Delta e}$, $\chi_{(a)1}$, $\chi_{(a)2}$, $\chi_{(a)3}$, $B_{(p)e1}$, $B_{(p)\Delta e}$, $\chi_{(p)1}$, $\chi_{(p)2}$ and $\chi_{(p)3}$ of the fuzzy term are set:

$$\chi = [0.6 \ 3.5 \ 28 \ 45 \ 3.5 \ 80 \ 3.99 \ 0.7 \ 2.99 \ 0.8 \ 3$$

$$0.9 \ 0.01 \ 1 \ 990 \ 12 \ 190 \ 50 \ 1.49 \ 0.1 \ 1.5$$

$$0.81 \ 0.71 \ 14.9 \ 25.2 \ 0.89 \ 165 \ 85 \ 0.7 \ 1.41 \ 0.9 \ 1.4]^{T}. \qquad (7.131)$$

The details related to the MIMO hybrid model-free adaptive fuzzy controller and the programs of the above steps are implemented in the MFAC _ CFDL _ FUZZY.m Matlab program for tower crane system control, which is included in the accompanying Matlab & Simulink programs. The simulations of the MIMO hybrid model-free adaptive fuzzy controllers are conducted using the same Simulink diagram, i.e., the CS _ MFAC _ CFDL _ FUZZY _ MIMO.mdl diagram for tower crane system control.

The simulation results obtained for the control system with MIMO hybrid model-free adaptive fuzzy controller with the parameters in (7.131) are illustrated in Figure 7.44 for cart position control, Figure 7.45 for arm angular position control and Figure 7.46 for payload position control.

The experimental results obtained for the control system with MIMO hybrid model-free adaptive fuzzy controller with the parameters in (7.131) are illustrated in Figure 7.47 for cart position control, Figure 7.48 for arm angular position control and Figure 7.49 for payload position control.

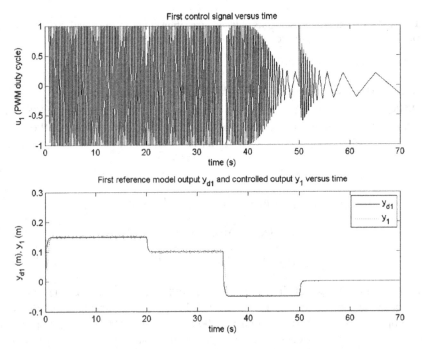

FIGURE 7.44 Simulation results of cart position control system with MIMO hybrid model-free adaptive fuzzy controller.

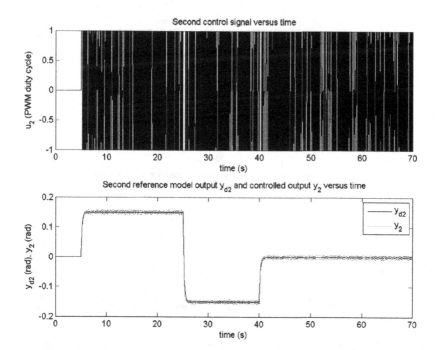

FIGURE 7.45 Simulation results of arm angular position control system with MIMO hybrid model-free adaptive fuzzy controller.

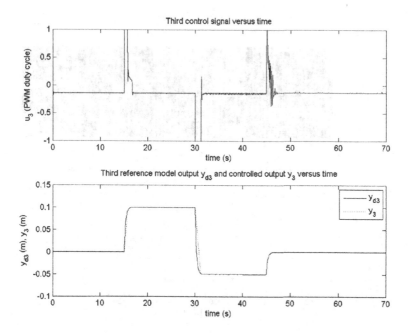

FIGURE 7.46 Simulation results of payload position control system with MIMO hybrid model-free adaptive fuzzy controller.

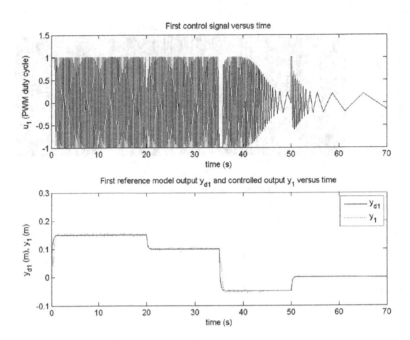

FIGURE 7.47 Experimental results of cart position control system with MIMO hybrid model-free adaptive fuzzy controller.

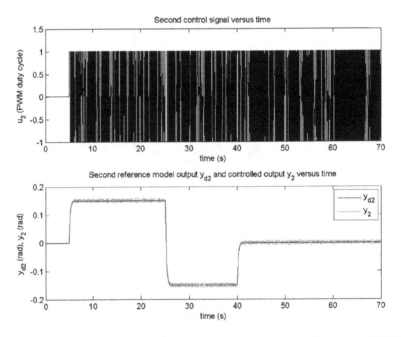

FIGURE 7.48 Experimental results of arm angular position control system with MIMO hybrid model-free adaptive fuzzy controller.

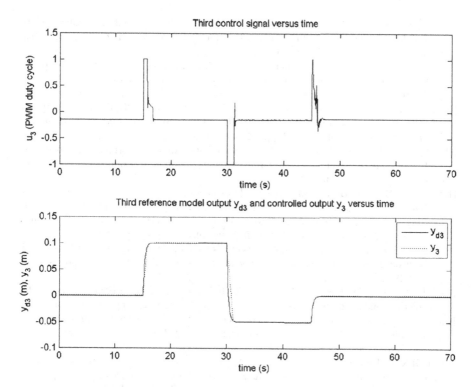

FIGURE 7.49 Experimental results of payload position control system with MIMO hybrid model-free adaptive fuzzy controller.

REFERENCES

[Aba20] I. Abadlia, L. Hassaine, A. Beddar, F. Abdoune and M. R. Bengourina, "Adaptive fuzzy control with an optimization by using genetic algorithms for grid connected a hybrid photovoltaic-hydrogen generation system," *International Journal of Hydrogen Energy*, vol. 45, no. 43, pp. 22589–22599, Sep. 2020.

[Abo03] J. Abonyi, *Fuzzy Model Identification for Control*, Birkhäuser, Boston, MA, 2003.

[Ack20] H. Acikgoz, C. Yildiz, R. Coteli and B. Dandil, "DC-link voltage control of three-phase PWM rectifier by using artificial bee colony based type-2 fuzzy neural network," *Microprocessors and Microsystems*, vol. 78, p. 103250, Oct. 2020.

[Ara89] J. Aracil, A. Ollero and A. Garcia-Cerezo, "Stability indices for the global analysis of expert control systems," *IEEE Transactions on Systems, Man, and Cybernetics*, vol. 19, no. 5, pp. 998–1007, Sep.–Oct. 1989.

[Azi19] M. Azizi, R. G. Ejlali, S. A. M. Ghasemi and S. Talatahari, "Upgraded whale optimization algorithm for fuzzy logic based vibration control of nonlinear steel structure," *Engineering Structures*, vol. 192, pp. 53–70, Aug. 2019.

[Åst95] K. J. Åström and T. Hägglund, *PID Controllers Theory: Design and Tuning*, Instrument Society of America, Research Triangle Park, NC, 1995.

[Bab96] R. Babuška, and H. B. Verbruggen, "An overview on fuzzy modeling for control," *Control Engineering Practice*, vol. 4, no. 11, pp. 1593–1606, Nov. 1996.

[Bab98] R. Babuška, *Fuzzy Modeling for Control*, Kluwer Academic Publishers, Boston, MA, 1998.

[Ban07] X.-J. Ban, X. Z. Gao, X.-L. Huang, and A. V. Vasilakos, "Stability analysis of the simplest Takagi-Sugeno fuzzy control system using circle criterion," *Information Sciences*, vol. 177, no. 20, pp. 4387–4409, Jun. 2007.

[Bay20] P. Bayat and A. Baghramian, "A novel self-tuning type-2 fuzzy maximum power point tracking technique for efficiency enhancement of fuel cell based battery chargers," *International Journal of Hydrogen Energy*, vol. 45, no. 43, pp. 23275–23293, Sep. 2020.

[Ber21] E. Bernal, M. L. Lagunes, O. Castillo, J. Soria and F. Valdez, "Optimization of type-2 fuzzy logic controller design using the GSO and FA algorithms," *International Journal of Fuzzy Systems*, vol. 23, no. 1, pp. 42–57, Feb. 2021.

[Bin95] T. Bindel and R. Mikut, "Entwurf, Stabilitätsanalyse und Erprobung von Fuzzy-Reglern am Beispiel einer Durchflussregelung," *Automatisierungstechnik*, vol. 43, no. 5, pp. 249–255, May 1995.

[Bol94] M. Boll, J. Bornemann and F. Dörrscheidt, "Anwendung der harmonischen Balance auf Regelkreise mit unsymmetrischen Fuzzy-Komponenten und konstante Eingangsgrössen," Workshop "Fuzzy Control" des GMA-UA 1.4.2, Dortmund, *Forshungsberichte der Fakultät für Elektrotechnik*, no. 0194, pp. 70–84, Jun. 1994.

[Böh95] R. Böhm and M. Bosch, "Stabilitätsanalyse von Fuzzy-Mehrgrössenregelungen mit Hilfe der Hyperstabilitätstheorie," *Automatisierungstechnik*, vol. 43, no. 4, pp. 181–186, Apr. 1995.

[Cal01] T. Calvo, B. De Baets and J. Fodor, "The functional equations of Frank and Alsina for uninorms and nullnorms," *Fuzzy Sets and Systems*, vol. 120, no. 3, pp. 385–394, Jun. 2001.

[Cue99] F. Cuesta, F. Gordillo, J. Aracil and A. Ollero, "Stability analysis of nonlinear multivariable Takagi-Sugeno fuzzy control systems," *IEEE Transactions on Fuzzy Systems*, vol. 7, no. 5, pp. 508–520, Oct. 1999.

[DeM20] R. R. De (Maity), R. K. Mudi and C. Dey, "Nature-inspired and hybrid optimization algorithms on interval type-2 fuzzy controller for servo processes: a comparative performance study," *SN Applied Sciences*, vol. 2, no. 7, p. 1292, Jul. 2020.

[Don19] M. Dong, X. Luan, B. Wu and J. Liang, "The fuzzy control of electro-hydraulic servo system based on DE algorithm," in *Proceedings of 2018 Chinese Intelligent Systems Conference*, Y. Jia, J. Du and W. Zhang, Eds., Springer, Singapore, Lecture Notes in Electrical Engineering, vol. 529, pp. 747–757, 2019.

[Dri93] D. Driankov, H. Hellendoorn and M. Reinfrank, *An Introduction to Fuzzy Control*, Springer-Verlag, Berlin, 1993.

[Dri01] D. Driankov, "A reminder on fuzzy logic," in *Fuzzy Logic Techniques for Autonomous Vehicle Navigation*, D. Driankov and A. Saffiotti, Eds., Physica-Verlag, Springer-Verlag, Heidelberg, pp. 25–47, 2001.

[Eck19] J. J. Eckert, F. M. Santiciolli, R. Y. Yamashita, F. C. Corrêa, L. C. A. Silva and F. G. Dedini, "Fuzzy gear shifting control optimisation to improve vehicle performance, fuel consumption and engine emissions," *IET Control Theory and Applications*, vol. 13, no. 16, pp. 2658–2669, Nov. 2019.

[Eck20] J. J. Eckert, L. Corrêa de Alkmin Silva, F. G. Dedini and F. C. Corrêa, "Electric vehicle powertrain and fuzzy control multi-objective optimization, considering dual hybrid energy storage systems," *IEEE Transactions on Vehicular Technology*, vol. 69, no. 4, pp. 3773–3782, Apr. 2020.

[Fod95] J. Fodor, "Contrapositive symmetry of fuzzy implications," *Fuzzy Sets and Systems*, vol. 69, no. 2, pp. 141–156, Jan. 1995.

[Fre19] L. Frezzatto, M. J. Lacerda, R. C. L. F. Oliveira, and P. L. D. Peres, "H_2 and H_∞ fuzzy filters with memory for Takagi-Sugeno discrete-time systems," *Fuzzy Sets and Systems*, vol. 371, pp. 78–95, Sep. 2019.

[Fu20] Z.-M. Fu, L.-L. Zhu, F.-Z. Tao, P.-J. Si and L.-F. Sun, "Optimization based energy management strategy for fuel cell/battery/ultracapacitor hybrid vehicle considering fuel economy and fuel cell lifespan," *International Journal of Hydrogen Energy*, vol. 45, no. 15, pp. 8875–8886, Mar. 2020.

[Gal95] S. Galichet and L. Foulloy, "Fuzzy controllers: synthesis and equivalences," *IEEE Transactions on Fuzzy Systems*, vol. 3, no. 2, pp. 140–148, May 1995.

[Gan20] R. V. Gandhi and D. M. Adhyaru, "Takagi-Sugeno fuzzy regulator design for nonlinear and unstable systems using negative absolute eigenvalue approach," *IEEE/CAA Journal of Automatica Sinica*, vol. 7, no. 2, pp. 482–493, Mar. 2020.

[Ghe20] M. Gheisarnejad, M. H. Khooban and T. Dragičević, "The future 5G network-based secondary load frequency control in shipboard microgrids," *IEEE Journal of Emerging and Selected Topics in Power Electronics*, vol. 8, no. 1, pp. 836–844, Mar. 2020.

[Gir21] M. E. Girgis and R. I. Badr, "Optimal fractional-order adaptive fuzzy control on inverted pendulum model," *International Journal of Dynamics and Control*, vol. 9, no. 1, pp. 288–298, Mar. 2021.

[Gue01] T. M. Guerra and L. Vermeiren, "Control laws for Takagi-Sugeno fuzzy models," *Fuzzy Sets and Systems*, vol. 120, no. 1, pp. 95–108, May 2001.

[Gun19] N. Gunasekaran and Y. H., "Stochastic sampled-data controller for T-S fuzzy chaotic systems and its applications," *IET Control Theory & Applications*, vol. 13, no. 12, pp.1834–1843, Aug. 2019.

[Hab03] R. E. Haber Guerra, G. Schmitt-Braess, R. Haber Haber, A. Alique and J. R. Alique, "Using circle criteria for verifying asymptotic stability in PI-like fuzzy control systems: application to the milling process," *IEE Proceedings - Control Theory and Applications*, vol. 150, no. 6, pp. 619–627, Nov. 2003.

[Hol82] L. P. Holmblad and J. J. Ostergaard, "Control of a cement kiln by fuzzy logic," in *Fuzzy Information and Decision Processes*, M. M. Gupta and E. Sanchez, Eds., North Holland, Amsterdam, pp. 389–399, 1982.

[Hu20] Y.-P. Hu, Y.-P. Yang, S. Li and Y.-M. Zhou, "Fuzzy controller design of micro-unmanned helicopter relying on improved genetic optimization algorithm," *Aerospace Science and Technology*, vol. 98, p. 105685, Mar. 2020.

[Jia20] B.-P. Jiang, H. R. Karimi, Y.-G. Kao and C.-C. Gao, "Takagi-Sugeno model based event-triggered fuzzy sliding-mode control of networked control systems with semi-Markovian switchings," *IEEE Transactions on Fuzzy Systems*, vol. 28, no. 4, pp. 673–683, Apr. 2020.

[Kho20] M. H. Khooban and M. Gheisarnejad, "Islanded microgrid frequency regulations concerning the integration of tidal power units: real-time implementation," *IEEE Transactions on Circuits and Systems II: Express Briefs*, vol. 67, no. 6, pp. 1099–1103, Jun. 2020.

[Kie93] H. Kiendl, "Harmonic balance for fuzzy control systems," in *Proceedings of First European Congress on Fuzzy and Intelligent Technologies*, Aachen, Germany, 1993, vol. 1, pp. 127–141.

[Kle00] E. P. Klement, R. Mesiar and E. Pap, *Triangular Norms*, Kluwer Academic Publishers, Dordrecht, 2000.

[Kum16] T. Kumbasar, "Robust stability analysis and systematic design of single-input interval type-2 fuzzy logic controllers," *IEEE Transactions on Fuzzy Systems*, vol. 24, no. 3, pp. 675–694, Jun. 2016.

[Lam08] H.-K. Lam and L. D. Seneviratne, "Stability analysis of interval type-2 fuzzy-model-based control systems," *IEEE Transactions on Systems, Man, and Cybernetics, Part B (Cybernetics)*, vol. 38, no. 3, pp. 617–628, Jun. 2008.

[Lam18] H.-K. Lam, "A review on stability analysis of continuous-time fuzzy-model-based control systems: From membership-function-independent to membership-function-dependent analysis," *Engineering Applications of Artificial Intelligence*, vol. 67, pp. 390–408, Jan. 2018.

[Le18] T.-L. Le, C.-M. Lin and T.-T. Huynh, "Self-evolving type-2 fuzzy brain emotional learning control design for chaotic systems using PSO," *Applied Soft Computing*, vol. 73, pp. 418–433, Dec. 2018.

[Li20a] G.-L. Li, C. Peng, M.-R. Fei and Y.-C. Tian, "Local stability conditions for T-S fuzzy time-delay systems using a homogeneous polynomial approach," *Fuzzy Sets and Systems*, vol. 385, pp. 111–126, Apr. 2020.

[Li20b] X.-M. Li, K. Mehran, H.-K. Lam, B. Xiao and Z.-Y. Bao, "Stability analysis of discrete-time positive polynomial-fuzzy-model-based control systems through fuzzy co-positive Lyapunov function with bounded control," *IET Control Theory & Applications*, vol. 14, no. 2, pp. 233–243, Jan. 2020.

[Lia20] H.-B. Liang, J.-L. Zou, K. Zuo and M. J. Khan, "An improved genetic algorithm optimization fuzzy controller applied to the wellhead back pressure control system," *Mechanical Systems and Signal Processing*, vol. 142, P. 106708, Aug. 2020.

[Liu20] D. Liu, G.-H. Yang and M. J. Er, "Event-triggered control for T-S fuzzy systems under asynchronous network communications," *IEEE Transactions on Fuzzy Systems*, vol. 28, no. 2, pp. 390–399, Feb. 2020.

[Mam74] E. H. Mamdani, "Applications of fuzzy control for control of simple dynamic plant," *Proceedings of the IEE*, vol. 121, no. 12, pp. 1585–1588, Dec. 1974.

[Mam75] E. H. Mamdani and S. Assilian, "An experiment in linguistic synthesis with a fuzzy logic controller," *International Journal of Man-Machine Studies*, vol. 7, no. 1, pp. 1–13, Jan. 1975.

[Mas20] A. P. Masoumi, A. R. Tavakolpour-Saleh and A. Rahideh, "Applying a genetic-fuzzy control scheme to an active free piston Stirling engine: Design and experiment," *Applied Energy*, vol. 268, P. 115045, Jun. 2020.

[Med18] J. A. Meda-Campaña, A. Grande-Meza, J. de Jesús Rubio, R. Tapia-Herrera, T. Hernández-Cortés, A. V. Curtidor-López, L. A. Páramo-Carranza and I. O. Cázares-Ramírez, "Design of stabilizers and observers for a class of multivariable TS fuzzy models on the basis of new interpolation function," *IEEE Transactions on Fuzzy Systems*, vol. 26, no. 5, pp. 2649–2662, Oct. 2018.

[Moh20a] A. Mohammadzadeh and E. Kayacan, "A novel fractional-order type-2 fuzzy control method for online frequency regulation in AC microgrid," *Engineering Applications of Artificial Intelligence*, vol. 90, P. 103483, Apr. 2020.

[Moh20b] A. Mohammadzadeh and T. Kumbasar, "A new fractional-order general type-2 fuzzy predictive control system and its application for glucose level regulation," *Applied Soft Computing*, vol. 91, p. 106241, Jun. 2020.

[Moo19] H. Moodi, M. Farrokhi, T.-M. Guerra and J. Lauber, "On stabilization conditions for T-S systems with nonlinear consequent parts," *International Journal of Fuzzy Systems*, vol. 21, no. 1, pp. pp. 84–94, 2019.

[Nos19] S. M. Nosratabadi, M. Bornapour and M. A. Gharaei, "Grasshopper optimization algorithm for optimal load frequency control considering predictive functional modified PID controller in restructured multi-resource multi-area power system with redox flow battery units," *Control Engineering Practice*, vol. 89, pp. 204–227, Aug. 2019.

[Opi93] H.-P. Opitz, "Fuzzy control and stability criteria," in *Proceedings of First European Congress on Fuzzy and Intelligent Technologies*, Aachen, Germany, 1993, vol. 1, pp. 130–136.

[Pal97] R. Palm, H. Hellendoorn and D. Driankov, *Model Based Fuzzy Control: Fuzzy Gain Schedulers and Sliding Mode Fuzzy Controllers*, Springer-Verlag, Berlin, 1997.

[Pan16] B. Pang, X. Liu, Q. Jin and W. Zhang, "Exponentially stable guaranteed cost control for continuous and discrete-time Takagi-Sugeno fuzzy systems," *Neurocomputing*, vol. 205, no. 1, pp. 210–221, Sep. 2016.

[Pan20] B. Pang and Q.-L. Zhang, "Interval observers design for polynomial fuzzy singular systems by utilizing sum-of-squares program," *IEEE Transactions on Systems, Man, and Cybernetics: Systems*, vol. 50, no. 6, pp. 1999–2006, Jun. 2020.

[Pas98] K. M. Passino and S. Yurkovich, *Fuzzy Control*, Addison-Wesley, Menlo Park, CA, 1998.

[Pre81] S. Preitl and D. Onea, "Analytical and experimental identification of electro-hydraulic speed controllers meant for hydro-generators," *Buletinul Stiintific si Tehnic al IPTVT, Electrical Engineering Series*, vol. 26 (40), pp. 83–93, Dec. 1981.

[Pre93] R.-E. Precup and S. Preitl, "Fuzzy control of an electro-hydraulic servo system under nonlinearity constraints," in *Proceedings of First EUFIT'93 European Congress*, Aachen, Germany, 1993, vol. 3, pp. 1524–1530.

[Pre96] S. Preitl and R.-E. Precup, "On the algorithmic design of a class of control systems based on providing the symmetry of open-loop Bode plots," *Scientific Bulletin of UPT, Transactions on Automatic Control and Computer Science*, vol. 41 (55), no. 2, pp. 47–55, Dec. 1996.

[Pre97a] R.-E. Precup and S. Preitl, "Popov-type stability analysis method for fuzzy control systems," in *Proceedings of Fifth European Congress on Intelligent Technologies and Soft Computing*, Aachen, Germany, 1997, vol. 2, pp. 1306–1310.

[Pre97b] S. Preitl and R.-E. Precup, *Introducere in conducerea fuzzy a proceselor* (in Romanian), Editura Tehnica, Bucharest, 1997.

[Pre99a] R.-E. Precup and S. Preitl, *Fuzzy Controllers*, Editura Orizonturi Universitare, Timisoara, 1999.

[Pre99b] R.-E. Precup and S. Preitl, "Development of some fuzzy controllers with non-homogenous dynamics with respect to the input channels meant for a class of systems," in *Proceedings of 1999 European Control Conference*, Karlsruhe, Germany, 1999, pp. 61–66.

[Pre99c] R.-E. Precup, S. Preitl and S. Solyom, "Center manifold theory approach to the stability analysis of fuzzy control systems," in *Computational Intelligence. Theory and Applications*, B. Reusch, Ed., Springer-Verlag, Berlin, Lecture Notes in Computer Science, vol. 1625, pp. 382–390, 1999.

[Pre99d] S. Preitl and R.-E. Precup, "An extension of tuning relations after symmetrical optimum method for PI and PID controllers," *Automatica*, vol. 35, no. 10, pp. 1731–1736, Oct. 1999.

[Pre03] R.-E. Precup and S. Preitl, "Development of fuzzy controllers with non-homogeneous dynamics for integral-type plants," *Electrical Engineering*, vol. 85, no. 3, pp. 155–168, Jul. 2003.

[Pre04] R.-E. Precup and S. Preitl, "Optimisation criteria in development of fuzzy controllers with dynamics," *Engineering Applications of Artificial Intelligence*, vol. 17, no. 6, pp. 661–674, Sep. 2004.[Pre05] R.-E. Precup and S. Preitl, "On the stability and sensitivity analysis of fuzzy control systems for servo-systems," in *Fuzzy Systems Engineering, Theory and Practice*, N. Nedjah and L. Macedo Mourelle, Eds., Springer-Verlag, Berlin, Studies in Fuzziness and Soft Computing, vol. 181, pp. 131–161, 2005.

[Pre06] R.-E. Precup and S. Preitl, "Development method for low cost fuzzy controlled servosystems," in *Proceedings of 2006 IEEE International Symposium on Intelligent Control ISIC*, München, Germany, 2006, pp. 2707–2712.

[Pre07] R.-E. Precup, M. L. Tomescu and S. Preitl, "Lorenz system stabilization using fuzzy controllers," *International Journal of Computers Communications and Control*, vol. 2, no. 3, pp. 279–287, Sep. 2007.

[Pre08a] R.-E. Precup, S. Preitl, P. A. Clep, I.-B. Ursache, J. K. Tar and J. Fodor, "Stable fuzzy control systems with iterative feedback tuning," in *Proceedings of 12th International Conference on Intelligent Engineering Systems*, Miami, FL, USA, 2008, pp. 287–292.

[Pre08b] S. Preitl, R.-E. Precup, M.-B. Radac, C.-A. Dragos, J. K. Tar and J. Fodor, "On the stable design of stable fuzzy control systems with iterative learning control," in *Proceedings of 9th International Symposium of Hungarian Researchers on Computational Intelligence and Informatics*, Budapest, Hungary, 2008, pp. 345–360.

[Pre09] R.-E. Precup, S. Preitl, E. M. Petriu, J. K. Tar, M. L. Tomescu and C. Pozna, "Generic two-degree-of-freedom linear and fuzzy controllers for integral processes," *Journal of The Franklin Institute*, vol. 346, no. 10, pp. 980–1003, Dec. 2009.

[Pre11] R.-E. Precup and H. Hellendoorn, "A survey on industrial applications of fuzzy control," *Computers in Industry*, vol. 62, no. 3, pp. 213–226, Apr. 2011.

[Pre12] S. Preitl, A.-I. Stinean, R.-E. Precup, Z. Preitl, E. M. Petriu, C.-A. Dragos and M.-B. Radac, "Controller design methods for driving systems based on extensions of symmetrical optimum method with DC and BLDC motor applications," *IFAC Proceedings Volumes*, vol. 45, no. 3, pp. 264–269, Mar. 2012.

[Pre15] R.-E. Precup, P. Angelov, B. S. J. Costa and M. Sayed-Mouchaweh, "An overview on fault diagnosis and nature-inspired optimal control of industrial process applications," *Computers in Industry*, vol. 74, pp. 75–94, Dec. 2015.

[Pre19] R.-E. Precup and R.-C. David, *Nature-Inspired Optimization Algorithms for Fuzzy Controlled Servo Systems*, Butterworth-Heinemann, Elsevier, Oxford, 2019.

[Pre20] R.-E. Precup, S. Preitl, E. M. Petriu, R.-C. Roman, C.-A. Bojan-Dragos, E.-L. Hedrea and A.-I. Szedlak-Stinean, "A center manifold theory-based approach to the stability analysis of state feedback Takagi-Sugeno-Kang fuzzy control systems," *Facta Universitatis, Series: Mechanical Engineering*, vol. 18, no. 2, pp. 189–204, Jun. 2020.

[Pre21a] R.-E. Precup, R.-C. David, R.-C. Roman, E. M. Petriu and A.-I. Szedlak-Stinean, "Slime mould algorithm-based tuning of cost-effective fuzzy controllers for servo systems," *International Journal of Computational Intelligent Systems*, vol. 14, no. 1, pp. 1042–1052, Mar. 2021.

[Pre21b] R.-E. Precup, R.-C. David, R.-C. Roman, A.-I. Szedlak-Stinean and E. M. Petriu, "Optimal tuning of interval type-2 fuzzy controllers for nonlinear servo systems using slime mould algorithm," *International Journal of Systems Science*, doi: 10.1080/00207721.2021.1927236, pp. 1–16, Jun. 2021.

[Rad09] M.-B. Radac, R.-E. Precup, S. Preitl, J. K. Tar and K. J. Burnham, "Tire slip fuzzy control of a laboratory anti-lock braking system," in *Proceedings of 2009 European Control Conference*, Budapest, Hungary, 2009, pp. 940–945.

[Rei19] N. A. Reisi, S. Hadipour Lakmesari, M. J. Mahmoodabadi and S. Hadipour, "Optimum fuzzy control of human immunodeficiency virus type1 using an imperialist competitive algorithm," *Informatics in Medicine Unlocked*, vol. 16, 100241, Dec. 2019.

[Rob17] R. Robles, A. Sala, M. Bernal and T. González, "Subspace-based Takagi-Sugeno modeling for improved LMI performance," *IEEE Transactions on Fuzzy Systems*, vol. 25, no. 4, pp. 754–767, Aug. 2017.

[Rom17] R.-C. Roman, R.-E. Precup and M.-B. Radac, "Model-free fuzzy control of twin rotor aerodynamic systems," in *Proceedings of 25th Mediterranean Conference on Control and Automation*, Valletta, Malta, 2017, pp. 559–564.

[Rom18] R.-C. Roman, R.-E. Precup and R.-C. David, "Second order intelligent proportional-integral fuzzy control of twin rotor aerodynamic systems," *Procedia Computer Science*, vol. 139, pp. 372–380, Oct. 2018.

[Rom19a] R.-C. Roman, R.-E. Precup, C.-A. Bojan-Dragos and A.-I. Szedlak-Stinean, "Combined model-free adaptive control with fuzzy component by virtual reference feedback tuning for tower crane systems," *Procedia Computer Science*, vol. 162, pp. 267–274, Oct. 2019.

[Rom19b] R.-C. Roman, R.-E. Precup, E. M. Petriu, E.-L. Hedrea, C.-A. Bojan-Dragos and M.-B. Radac, "Model-free adaptive control with fuzzy component for tower crane systems," in *Proceedings of 2019 IEEE International Conference on Systems, Man and Cybernetics*, Bari, Italy, 2019, pp. 1400–1405.

[Ros19] A. Rosales, L. Ibarra, P. Ponce and A. Molina, "Fuzzy sliding mode control design based on stability margins," *Journal of the Franklin Institute*, vol. 356, no. 10, pp. 5260–5273, Jul. 2019.

[Rud98] I. J. Rudas and O. Kaynak, "New types of generalized operations", in *Computational Intelligence, Soft Computing and Neuro-Fuzzy Integration with Applications*, O. Kaynak, B. Turksen and I. J. Rudas, Eds., Springer NATO ASI Series, Series F, Computer and Systems Sciences, vol. 192, Springer-Verlag, Berlin, Heidelberg, pp. 128–156, 1998.

[Rud06] I. J. Rudas and J. Fodor, "Information aggregation in intelligent systems using generalized operators," *International Journal of Computers, Communications & Control*, vol. I, no. 1, pp. 47–57, Feb. 2006.

[Sak20] R. Sakthivel, S. Mohanapriya, B. Kaviarasan, Y. Ren and S. M. Anthoni, "Non-fragile control design and state estimation for vehicle dynamics subject to input delay and actuator faults," *IET Control Theory & Applications*, vol. 14, no. 1, pp. 134–144, Jan. 2020.

[Sal05] A. Sala, T. M. Guerra and R. Babušk, "Perspectives of fuzzy systems and control," *Fuzzy Sets and Systems*, vol. 156, no. 2, pp. 432–444, Dec. 2005.

[Sha20a] N. F. Shamloo, A. A. Kalat and L. Chisci, "Indirect adaptive fuzzy control of nonlinear descriptor systems," *European Journal of Control*, vol. 51, pp. 30–38, Jan. 2020.

[Sha20b] R. Sharma, K. K. Deepak, P. Gaur and D. Joshi, "An optimal interval type-2 fuzzy logic control based closed-loop drug administration to regulate the mean arterial blood pressure," *Computer Methods and Programs in Biomedicine*, vol. 185, P. 105167, Mar. 2020.

[Sol20] M. I. Solihin, C. Y. Chuan and W. Astuti, "Optimization of fuzzy logic controller parameters using modern meta-heuristic algorithm for Gantry Crane System (GCS)," *Materials Today: Proceedings*, vol. 29, part 1, pp. 168–172, Jun. 2020.

[Sug88] M. Sugeno and G. T. Kang, "Structure identification of fuzzy model," *Fuzzy Sets and Systems*, vol. 28, pp. 12–33, Oct. 1988.

[Tak85] T. Takagi and M. Sugeno, "Fuzzy identification of systems and its application to modeling and control," *IEEE Transactions on Systems, Man, and Cybernetics*, vol. 15, no. 1, pp. 116–132, Jan.-Feb. 1985.

[Tan92] K. Tanaka and M. Sugeno, "Stability analysis and design of fuzzy control systems," *Fuzzy Sets and Systems*, vol. 45, no. 2, pp. 135–156, Jan. 1992.

[Tan01] K. Tanaka and H. O. Wang, *Fuzzy Control Systems Design and Analysis: A Linear Matrix Inequality Approach*, John Wiley & Sons, New York, 2001.

[Tom07] M. L. Tomescu, S. Preitl, R.-E. Precup and J. K. Tar, "Stability analysis method for fuzzy control systems dedicated controlling nonlinear processes," *Acta Polytechnica Hungarica*, vol. 4, no. 3, pp. 127–141, Sep. 2007.

[Tri20] S. Tripathi, A. Shrivastava and K. C. Jana, "Self-tuning fuzzy controller for sun-tracker system using Gray Wolf Optimization (GWO) technique," *ISA Transactions*, vol. 101, pp. 50–59, Jun. 2020.

[Wai19] R.-J. Wai and A. S. Prasetia, "Adaptive neural network control and optimal path planning of UAV surveillance system with energy consumption prediction," *IEEE Access*, vol. 7, pp. 126137–126153, Aug. 2019.

[Wan96] H. O. Wang, K. Tanaka and M. Griffin, "An approach to fuzzy control of nonlinear systems: Stability and design issues," IEEE Transactions on Fuzzy Systems, vol. 4, no. 1, pp. 14–23, Feb. 1996.

[Wan19] Z.-H. Wang, Z. Liu, C. L. P. Chen and Y. Zhang, "Fuzzy adaptive compensation control of uncertain stochastic nonlinear systems with actuator failures and input hysteresis," *IEEE Transactions on Cybernetics*, vol. 49, no. 1, pp. 2–13, Jan. 2019.

[Wei20] Z.-X. Wei, F. Doctor, Y.-X. Liu, S.-Z. Fan and J.-S. Shieh, "An optimized type-2 self-organizing fuzzy logic controller applied in anesthesia for Propofol dosing to regulate BIS," *IEEE Transactions on Fuzzy Systems*, vol. 28, no. 6, pp. 1062–1072, Jun. 2020.

[Wu12] D.-R. Wu, "On the fundamental differences between interval type-2 and type-1 fuzzy logic controllers," *IEEE Transactions on Fuzzy Systems*, vol. 20, no. 5, pp. 832–848, Oct. 2012.

[Wu18] H.-N. Wu and S. Feng, "Mixed fuzzy/boundary control design for nonlinear coupled systems of ODE and boundary-disturbed uncertain beam," *IEEE Transactions on Fuzzy Systems*, vol. 26, no. 6, pp. 3379–3390, Dec. 2018.

[Wu19] T.-Z. Wu, W.-S. Yu and L.-X. Guo, "A study on use of hybrid energy storage system along with variable filter time constant to smooth DC power fluctuation in microgrid," *IEEE Access*, vol. 7, pp. 175377–175385, Nov. 2019.

[Wu20] D.-R. Wu and X.-F. Tan, "Multitasking Genetic Algorithm (MTGA) for fuzzy system optimization," *IEEE Transactions on Fuzzy Systems*, vol. 28, no. 6, pp. 1050–1061, Jun. 2020.

[Wu21] H.-N. Wu, X.-M. Zhang, J.-W. Wang and H.-Y. Zhu, "Observer-based output feedback fuzzy control for nonlinear parabolic PDE-ODE coupled systems," *Fuzzy Sets and Systems*, vol. 402, pp. 105–123, Jan. 2021.

[Xia20a] Y. Xia, J. Wang, B. Meng and X.-Y. Chen, "Further results on fuzzy sampled-data stabilization of chaotic nonlinear systems," *Applied Mathematics and Computation*, vol. 379, 125225, Aug. 2020.

[Xia20b] B. Xiao, H.-K. Lam, Y. Yu and Y.-D. Li, "Sampled-data output-feedback tracking control for interval type-2 polynomial fuzzy systems," *IEEE Transactions on Fuzzy Systems*, vol. 28, no. 3, pp. 424–433, Mar. 2020.

[Xie19] S.-W. Xie, Y.-F. Xie, F.-B. Li, Z.-H. Jiang and W.-H. Gui, "Hybrid fuzzy control for the goethite process in zinc production plant combining type-1 and type-2 fuzzy logics," *Neurocomputing*, vol. 366, pp. 170–177, Nov. 2019.

[Yan19] X.-Z. Yang, H.-K. Lam and L.-G. Wu, "Membership-dependent stability conditions for type-1 and interval type-2 T-S fuzzy systems," *Fuzzy Sets and Systems*, vol. 356, pp. 44–62, Feb. 2019.

[Yon17] Y. Yoneyama, "New conditions for stability and stabilization of Takagi-Sugeno fuzzy systems," in *Proceedings of 2017 Asian Control Conference*, Gold Coast, Australia, 2017, pp. 2154–2159.

[Yu18] G.-R. Yu, Y.-C. Huang and C.-Y. Cheng, "Sum-of-squares-based robust H_∞ controller design for discrete-time polynomial fuzzy systems," *Journal of the Franklin Institute*, vol. 355, no. 1, pp. 177–196, Jan. 2018.

[Zad65] L. A. Zadeh, "Fuzzy sets," *Information and Control*, vol. 8, no. 3, pp. 338–353, Jun. 1965.

[Zha19] Y.-X. Zhao, Y.-X. He, Z.-G. Feng, P Shi and X. Du, "Relaxed sum-of-squares based stabilization conditions for polynomial fuzzy-model-based control systems," *IEEE Transactions on Fuzzy Systems*, vol. 27, no. 9, pp. 1767–1778, Sep. 2019.

[Zim01] H.-J. Zimmermann, *Fuzzy Set Theory-and Its Applications,* 4th Edition, Kluwer Academic Publishers, Springer Netherlands, New York, 2001.

8 Cooperative Model-Free Adaptive Controllers for Multiagent Systems

8.1 INTRODUCTION

Multiagent systems are dynamic systems that include multiple individual dynamic systems (named as *agents*) that are connected together using a network called *communication graph*. In the network, there are several communication links between agents that provide means for transmitting data among them. By utilizing the communication graph, *distributed* and *cooperative* controllers and observer algorithms can be designed for regulating the performance of the multiagent system, without having any external unit for commanding and controlling the system. The initial concepts of control algorithms for multiagent systems are presented in [Tan03a, Tan03b, Olf06, Olf07]. For multiagent systems, a very basic problem is the *consensus* problem, which means reaching at a certain value for the states or outputs of each agents in the network [Li15]. The consensus problem can be extended to the *formation-tracking* control problem, in which a desired relative configuration between the states or outputs of the agents would be achieved in the network [Lew14].

Most of the proposed methods in the literature of cooperative control algorithms for multiagent systems are model-based control algorithms or model-free control algorithms with model-based parameter estimations. For example, *reinforcement learning* algorithms are utilized in [Mod16] and a cooperative control solution is provided. The current chapter brings the concept of model-free control with model-free parameter estimation into the consensus and formation-tracking problems for multiagent dynamic systems. Here, the proposed cooperative model-free control algorithms are not incorporating model-baed parameter estimators into the design, so as to remove the requirement of persistent excitation for the input signals.

8.2 THEORY

8.2.1 THE GENERIC STRUCTURE OF A NONLINEAR MULTIAGENT DYNAMIC SYSTEM

Definition 8.1

Let us consider a network of multiple dynamic agents with the generic dynamics proposed in *Definition 5.4* and *Definition 5.5* for the *i*-th agent of the network as follows [Saf21]:

$$\dot{\mathbf{x}}^i = \mathbf{A}^i \mathbf{x}^i + \mathbf{g}^i + \mathbf{u}^i,$$
$$\mathbf{y}^i = \mathbf{x}^i. \tag{8.1}$$

DOI: 10.1201/9781003143444-8

Here, $\mathbf{A}^i$ and $\mathbf{g}^i$ are unknown linear and nonlinear parameters at agent i. Assuming there are N agents in the network, we can integrate the dynamics of all agents into a lumped representation as follows [Saf21]:

$$\dot{\mathbf{x}}_t = \mathbf{A}_t \mathbf{x}_t + \mathbf{g}_t + \mathbf{u}_t,$$

$$\mathbf{y}_t = \mathbf{x}_t, \tag{8.2}$$

where $\mathbf{x}_t = [(\mathbf{x}^1)^T \quad (\mathbf{x}^2)^T \quad (\mathbf{x}^N)^T]^T \in \Re^{Nn\times 1}$ is the combined state vector, $\mathbf{u}_t = [(\mathbf{u}^1)^T \quad (\mathbf{u}^2)^T \quad (\mathbf{u}^N)^T]^T \in \Re^{Nn\times 1}$ is the combined system input (or control signal) vector, $\mathbf{y}_t = [(\mathbf{y}^1)^T \quad (\mathbf{y}^2)^T \quad \ldots \quad (\mathbf{y}^N)^T]^T \in \Re^{Nn\times 1}$ is the combined system (controlled) output vector, $\mathbf{g}_t = [(\mathbf{g}^1)^T \quad (\mathbf{g}^2)^T \quad \ldots \quad (\mathbf{g}^N)^T]^T \in \Re^{Nn\times 1}$ is the combined nonlinear function and $\mathbf{A}_t = \text{diag}(\mathbf{A}^1, \mathbf{A}^2, \ldots, \mathbf{A}^N) \in \Re^{Nn\times Nn}$ is the combined system matrix of the network.

Remark 8.1

The dynamics in *Definition 8.1* can be considered for networks of both single-integrator and double-integrator dynamic agents.

Remark 8.2

Similar to *Remark 5.7*, for the agents of the network defined in *Definition 8.1*, the first time derivatives of $\mathbf{A}^i$ and $\mathbf{g}^i$ are bounded with uknown values for the boundaries. In this regard, we define $|\mathbf{A}_t| \le \mathbf{U}_{A_t} \in \Re^{Nn\times Nn}$, $|\dot{\mathbf{A}}_t| \le \mathbf{U}_{\dot{A}_t} \in \Re^{Nn\times Nn}$, $|\mathbf{g}_t| \le \mathbf{U}_{g_t} \in \Re^{Nn\times 1}$ and $|\dot{\mathbf{g}}_t| \le \mathbf{U}_{\dot{g}_t} \in \Re^{Nn\times 1}$.

Definition 8.2

Considering the network of dynamic systems defined in *Definition 8.1*, we can define a communication graph for representing the available lines of data transmission among the agents, with a *Laplacian* matrix as follows ([Lew14] and [Li15]):

$$\mathbf{L}_N = \mathbf{D}_N - \mathbf{A}_N, \tag{8.3}$$

in which $\mathbf{A}_N = [a_{ij}] \in \Re^{N\times N}$ is the adjacency matrix of the network, with $a_{ij} = 1$ if there is a communication link between the i-th and j-th agents of the network and $a_{ij} = 0$ if not. Here, $\mathbf{D}_N = \text{diag}(d_1, d_2, \ldots, d_N) \in \Re^{N\times N}$, where $d_i = \sum_{j=1}^{N} a_{ij}$. Moreover, assuming the existance of a virtual leader for the network, a pinning-gain matrix is defined as $\mathbf{B}_N = \text{diag}(b_1, b_2, \ldots, b_N) \in \Re^{N\times N}$, with $b_i = 1$ if the i-th agent is connected to the virtual leader and $b_i = 0$ if not. Finally, we define ([Lew14] and [Li15])

$$\mathbf{H}_N = \mathbf{B}_N + \mathbf{L}_N. \tag{8.4}$$

Proposition 8.1

The virtual leader in the network, defined in *Definitions 8.1* and *8.2*, is a virtual agent that has access to all configuration of the network including the laplacian matrix. Moreover, the desired trajectory for the states of the entire network as well as the desired formation or consensus variables for the states of agents in the network are all avilable to that virtual leader. In this regard, only the connected agents of network to the virtual leader (i.e., agents with $b_i = 1$) have access to the mentioned information.

Remark 8.3

Here, it is assumed that all the communication links in the network are bidirectional, and there is only one spanning tree in the network rooted at the virtual leader. Moreover, it is assumed that at least there is one agent in the network that connects to the virtual leader, i.e., there is at least one $b_i = 1$.

Definition 8.3

The desired trajectory for the entire network defined in *Definitions 8.1* and *8.2* is defined as the state vector of the virtual leader $\mathbf{x}^0 \in \Re^{n \times 1}$ in a global frame named $\mathbf{F_g}$.

Definition 8.4

For the network of dynamic agents defined in *Definitions 8.1* and *8.2*, a set of desired formation among the states of all agents in the network is defined as follows [Saf21]:

$$\Delta = [(\eta^0)^T \quad (\eta^1)^T \quad (\eta^2)^T \quad \ldots \quad (\eta^N)^T] \in \Re^{(N+1) \times n}, \tag{8.5}$$

where for $i = 1 \ldots N$, $\eta^i \in \Re^{n \times 1}$ is the set-point (or the reference input or the desired state) vector for the state of i-th agent in a local frame attached to the virtual leader, named $\mathbf{F_v}$. In addition, $\eta^0 \in \Re^{n \times 1}$ is the desired state vector of the virtual leader in $\mathbf{F_v}$.

Proposition 8.2

According to *Definitions 8.3* and *8.4*, the formation-tracking problem is defined as satisfying the desired trajectory for the entire network, while the desired formation variables are maintained among the states of agents [Lew14]. In a particular case, if $\eta^0 = \eta^1 = \eta^2 = \ldots = \eta^N = \mathbf{0}$, where $\mathbf{0} = [0 \quad 0 \quad \ldots \quad 0]^T \in \Re^{n \times 1}$, then the problem would be a consensus problem in the network, where the objective is to just converge the states of all agents into the states of virtual leader, which in turn is the desired trajectory [Lew14].

8.2.2   COOPERATIVE MODEL-FREE ADAPTIVE CONTROLLER WITHOUT RELATIVE STATE MEASUREMENTS

Proposition 8.3

Recalling *Definition 8.2*, there is a minimal number of communication links in the network. It means that most of the agents are not connected to the virtual leader and hence do not have access to the desired trajectory of the network, i.e., $\mathbf{x}^0$, and the formation variables among the states of agents, i.e., Δ. In this regard, two cooperative observers are designed in this subsection to estimate the values of the above two parameters at each agent by considering the communication graph of the network.

Definition 8.5

For estimating the states of virtual leader at agent i of the network (i.e., $\hat{\mathbf{x}}_0^i \in \mathfrak{R}^{n \times 1}$), the following consensus error is defined:

$$\tilde{\mathbf{x}}_0^i = \sum_{j=1}^{N} a_{ij}(\hat{\mathbf{x}}_0^i - \hat{\mathbf{x}}_0^j) + b_i(\hat{\mathbf{x}}_0^i - \mathbf{x}_0). \tag{8.6}$$

Moreover, a lumped term including all of $\tilde{x}_0^i$ is defined in terms of

$$\tilde{\mathbf{x}}_{0t} = (\mathbf{H}_N \otimes \mathbf{I}_n)\hat{\mathbf{x}}_{0t} - (\mathbf{B}_N \otimes \mathbf{x}_0)\mathbf{1}, \tag{8.7}$$

where $\hat{\mathbf{x}}_{0t} = [(\hat{\mathbf{x}}_0^1)^T \ (\hat{\mathbf{x}}_0^2)^T \ (\hat{\mathbf{x}}_0^N)^T]^T \in \mathfrak{R}^{Nn \times 1}$, $\otimes$ is the Kronecker product [Kha02] and $\mathbf{1} = [1 \ 1 \ \cdots \ 1]^T \in \mathfrak{R}^{N \times 1}$. The objective of the cooperative observer is to eliminate $\tilde{\mathbf{x}}_0^i$ as time goes to infinity.

Theorem 8.1

Recalling *Definition 8.5*, a cooperative observer is defined as follows [Saf18]:

$$\Pi_m \dot{\hat{\mathbf{x}}}_0^i = -\lambda \tilde{\mathbf{x}}_0^i - \left(\mathbf{M} \left(\operatorname{sgn} \sum_{j=1}^{N} \mathbf{H}_N(i,j) \tilde{\mathbf{x}}_0^j \right) \right), \tag{8.8}$$

where $\lambda > 0$ is a tuning parameter, $\mathbf{M}(.)$ is a matrix generating function defined in *Definition 5.7* and sgn$(.)$ is the sign function. Moreover, $\Pi_m \in \mathfrak{R}^{n \times 1}$ is the upper bound for the absolute time derivative of the virtual leader's state, i.e., $|\dot{\mathbf{x}}_0| < \Pi_m$.

Proof (the proof provided here is following the proof presented in [Saf18]): Recalling the definition of $\tilde{x}_{0t}$ in (8.7), we define the following Lyapunov function:

$$V = \frac{1}{2}\tilde{\mathbf{x}}_{0t}^T \tilde{\mathbf{x}}_{0t}. \tag{8.9}$$

Then, its first time derivative is

$$\dot{V} = \tilde{\mathbf{x}}_{0t}^T((\mathbf{H}_N \otimes \mathbf{I}_n)\dot{\hat{\mathbf{x}}}_{0t} - (\mathbf{B}_N \otimes \dot{\mathbf{x}}_0)\mathbf{1}). \tag{8.10}$$

As the sum of elements on each row of $\mathbf{L}_N$ is equal to zero, one can propose

$$(\mathbf{B}_N \otimes \dot{x}_0)\mathbf{1} = ((\mathbf{B}_N \otimes \dot{\mathbf{x}}_0) + (\mathbf{L}_N \otimes \dot{\mathbf{x}}_0))\mathbf{1} = (\mathbf{H}_N \otimes \dot{\mathbf{x}}_0)\mathbf{1}. \tag{8.11}$$

Hence, we have

$$\dot{V} = \tilde{\mathbf{x}}_{0t}^T((\mathbf{H}_N \otimes \mathbf{I}_n)\dot{\hat{\mathbf{x}}}_{0t} - (\mathbf{H}_N \otimes \dot{\mathbf{x}}_0)\mathbf{1}). \tag{8.12}$$

At this point, let us assume that $\dot{\hat{\mathbf{x}}}_{0t}$ be presented in the following form:

$$\dot{\hat{\mathbf{x}}}_{0t} = -\lambda \tilde{\mathbf{x}}_{0t} + \hat{\alpha}, \tag{8.13}$$

where

$$\hat{\alpha} = -\mathbf{M}(\text{sgn}\{\tilde{\mathbf{x}}_{0t}^T(\mathbf{H}_N \otimes \mathbf{I}_n)\})(\mathbf{I}_N \otimes \Pi_m)\mathbf{1}. \tag{8.14}$$

In this regard, (8.12) is rewritten as follows:

$$\dot{V} = -\lambda \tilde{\mathbf{x}}_{0t}^T(\mathbf{H}_N \otimes \mathbf{I}_n)\tilde{\mathbf{x}}_{0t} + \tilde{\mathbf{x}}_{0t}^T(\mathbf{H}_N \otimes \mathbf{I}_n)\hat{\alpha} - \tilde{\mathbf{x}}_{0t}^T(\mathbf{H}_N \otimes \dot{\mathbf{x}}_0)\mathbf{1}. \tag{8.15}$$

Let us consider the scalar term $s_1 = -\lambda \tilde{\mathbf{x}}_{0t}^T(\mathbf{H}_N \otimes \mathbf{I}_n)\tilde{\mathbf{x}}_{0t}$. Recalling *Definition 8.2* and *Remark 8.4*, the matrix $\mathbf{H}_N$ and consequently the matrix $(\mathbf{H}_N \otimes \mathbf{I}_n)$ are symmetric matrices with positive diagonal and nonpositive off-diagonal elements. Thus, $(\mathbf{H}_N \otimes \mathbf{I}_n)$ has a positive determinant and eigenvalues. As a result, it is a *nonsingular M-matrix* [Tan03b], and hence, $(\mathbf{H}_N \otimes \mathbf{I}_n) > 0$. Accordingly, this leads to

$$s_1 = -\lambda \tilde{\mathbf{x}}_{0t}^T(\mathbf{H}_N \otimes \mathbf{I}_n)\tilde{\mathbf{x}}_{0t} < 0. \tag{8.16}$$

Now, consider the remaining terms in (8.15) with the notation

$$s_2 = \tilde{\mathbf{x}}_{0t}^T(\mathbf{H}_N \otimes \mathbf{I}_n)\hat{\alpha} - \tilde{\mathbf{x}}_{0t}^T(\mathbf{H}_N \otimes \dot{\mathbf{x}}_0)\mathbf{1}. \tag{8.17}$$

Using the *mixed-product* property of the Kronecker product [Lew14] leads to

$$(\mathbf{H}_N \otimes \dot{\mathbf{x}}_0) = (\mathbf{H}_N \otimes \mathbf{I}_n)(\mathbf{I}_N \otimes \dot{\mathbf{x}}_0), \tag{8.18}$$

and thus,

$$s_2 = \tilde{\mathbf{x}}_{0t}^T(\mathbf{H}_N \otimes \mathbf{I}_n)\hat{\alpha} - \tilde{\mathbf{x}}_{0t}^T(\mathbf{H}_N \otimes \mathbf{I}_n)(\mathbf{I}_N \otimes \dot{\mathbf{x}}_0)\mathbf{1}. \tag{8.19}$$

Then, by defining an absolute function $|\mathbf{v}|$ for any vector $\mathbf{v} \in \Re^{Nn \times 1}$, as follows:

$$|\mathbf{v}| = [\mathbf{v}(1) \quad \mathbf{v}(2) \quad \dots \quad \mathbf{v}(Nn)]^T, \tag{8.20}$$

the following inequality is expressed:

$$s_2 \leq \tilde{\mathbf{x}}_{0t}^T (\mathbf{H}_N \otimes \mathbf{I}_n) \hat{\boldsymbol{\alpha}} + |\tilde{\mathbf{x}}_{0t}^T (\mathbf{H}_N \otimes \mathbf{I}_n)| (\mathbf{I}_N \otimes \Pi_m) \mathbf{1} = s_3. \tag{8.21}$$

Now, by replacing $\hat{\boldsymbol{\alpha}}$ from (8.14) into (8.21), we have

$$s_3 = -\tilde{\mathbf{x}}_{0t}^T (\mathbf{H}_N \otimes \mathbf{I}_n) \mathbf{M}(\text{sgn}\{\tilde{\mathbf{x}}_{0t}^T (\mathbf{H}_N \otimes \mathbf{I}_n)\})(\mathbf{I}_N \otimes \Pi_m) \mathbf{1}$$
$$+ |\tilde{\mathbf{x}}_{0t}^T (\mathbf{H}_N \otimes \mathbf{I}_n)| (\mathbf{I}_N \otimes \Pi_m) \mathbf{1}, \tag{8.22}$$

and next by recalling

$$\tilde{\mathbf{x}}_{0t}^T (\mathbf{H}_N \otimes \mathbf{I}_n) \mathbf{M}(\text{sgn}\{\tilde{\mathbf{x}}_{0t}^T (\mathbf{H}_N \otimes \mathbf{I}_n)\}) = |\tilde{\mathbf{x}}_{0t}^T (\mathbf{H}_N \otimes \mathbf{I}_n)|, \tag{8.23}$$

we reach at $s_3 = 0$. Finally, according to (8.16) and (8.17), the concluding result is

$$\dot{V} = s_1 + s_2 \leq 0. \tag{8.24}$$

This shows that based on the *Lyapunov* stability theorem [Kha02], the vector $\tilde{\mathbf{x}}_{0t}$ is asymptotically stable and converges to zero. Moreover, by rewriting each row of the matrices defined in (8.13) and (8.14), for $\dot{\tilde{\mathbf{x}}}_{0t}$, the convergence of the cooperative observer proposed in (8.8) will be achieved. This completes the proof.

Definition 8.6

To estimate the formation variables Δ at agent i in the network, i.e., $\hat{\Delta}^i \in \Re^{(N+1) \times n}$, a consensus error $\tilde{\Delta}^i$ is defined

$$\tilde{\Delta}^i = \sum_{j=1}^{N} a_{ij} (\hat{\Delta}^i - \hat{\Delta}^j) + b_i (\hat{\Delta}^i - \Delta). \tag{8.25}$$

The objective of a cooperative observer should be converging $\tilde{\Delta}^i$ to zero.

Theorem 8.2

Recalling *Definition 8.6*, a cooperative observer is proposed for estimating the formation variables among the states of agents in the network as follows [Saf18]:

$$\dot{\hat{\Delta}}^i = -\mu \tilde{\Delta}^i - \left(\text{sgn}\left(\sum_{j=1}^{N} \mathbf{H}_N(i, j) \hat{\Delta}^i \right) \right) \mathbf{M}(\mathbf{Y}_m), \tag{8.26}$$

where $\mu > 0$ is a tuning parameter and $|\dot{\eta}_0| < \mathbf{Y}_m \in \Re^{n \times 1}$.

Proof: The proof for this theorem is completely following the method presented as the proof for *Theorem 8.1*.

Lemma 8.1

Having $\hat{\mathbf{x}}_0^i$ and $\hat{\Delta}^i$ at agent i of the network, one can determine the desired trajectory for system output of this agent as follows [Saf18]:

$$\mathbf{y}_d^i = \hat{\mathbf{x}}_0^i + \hat{\Delta}^i(1,:)^T + \hat{\Delta}^i(i,:)^T. \tag{8.27}$$

In this regard, the tracking error $\mathbf{e}^i$ at agent i is defined

$$\mathbf{e}^i = \mathbf{y}_d^i - \mathbf{y}^i. \tag{8.28}$$

The objective of model-free controller would be to eliminate this tracking error. This can be achieved by implementing the model-free adaptive controller proposed in *Theorem 5.3* on $\mathbf{e}^i$. Note that by utilizing *Definition 5.7*, the parameters $\xi^i = \int \mathbf{e}^i \, dt$ and $\sigma^i = \mathbf{e}^i + \xi^i$ are defined.

Remark 8.4

According to *Lemma 8.1*, for the implementation of model-free adaptive controller at agent i, the measurements on absolute states of this agent, i.e., $\mathbf{x}^i$, are required. No relative measurement on states of agents, i.e., $(\mathbf{x}^i - \mathbf{x}^j)$, in the network is required.

8.2.3 COOPERATIVE MODEL-FREE ADAPTIVE CONTROLLER WITH RELATIVE STATE MEASUREMENTS

Definition 8.7

For the network of dynamic agents defined in *Definitions 8.1* and *8.2*, a *consensus error* is defined as follows [Saf21]:

$$\mathbf{e}^i = \sum_{j=1}^{N} a_{ij}((\eta^i - \eta^j) - (\mathbf{x}^i - \mathbf{x}^j)) + b_i((\eta^i - \eta^0) - (\mathbf{x}^i - \mathbf{x}^0)). \tag{8.29}$$

Furthermore, we have

$$\mathbf{e}^i = \sum_{j=1}^{N} a_{ij}(\mathbf{z}^i - \mathbf{z}^j) + b_i(\mathbf{z}^i - \mathbf{z}^0), \tag{8.30}$$

where $\mathbf{z}^i = \eta^i - \mathbf{x}^i$ and $\mathbf{z}^0 = \eta^0 - \mathbf{x}^0$. In addition, the consensus errors at all agents of the network can be represented in the following lumped format [Saf21]:

$$\mathbf{e}_t = (\mathbf{H}_N \otimes \mathbf{I}_n)\mathbf{z}_t - (\mathbf{B}_N \otimes \mathbf{z}^0)\mathbf{1}, \tag{8.31}$$

where $\mathbf{e}_t = [(\mathbf{e}^1)^T \ (\mathbf{e}^2)^T \ \cdots \ (\mathbf{e}^N)^T]^T$ and $\mathbf{z}_t = [(\mathbf{z}^1)^T \ (\mathbf{z}^2)^T \ \cdots \ (\mathbf{z}^N)^T]^T$, both defined in $\mathfrak{R}^{Nn \times 1}$. Here, The objective of the cooperative model-free control algorithm is to converge e_t to zero as time goes to enfinity. Following *Remark 8.4*, this is a definition of a formation-tracking problem, and it can be reduced to a consensus problem.

Theorem 8.3

Considering the problem defined in *Definition 8.7*, the cooperative model-free adaptive controller at the i-th agent of the network is proposed as follows [Saf21]:

$$\mathbf{u}^i = \frac{1}{2}\mathbf{P}^i\mathbf{e}^i + \hat{\mathbf{A}}^i\mathbf{c}^i - \hat{\mathbf{g}}^i - \dot{\hat{\mathbf{c}}}^i, \tag{8.32}$$

where

$$\hat{\mathbf{c}}^i = \hat{\boldsymbol{\eta}}^0 - \hat{\boldsymbol{\eta}}^i - \hat{\mathbf{x}}_0^i, \tag{8.33}$$

while $\hat{\mathbf{A}}^i$ and $\hat{\mathbf{g}}^i$ are computed by the following adaptive laws (referring to *Definition 5.7* for $\mathbf{v}_x$ and $\mathbf{M}_x$) [Saf21]

$$\dot{\hat{\mathbf{g}}}^i = -\Gamma_1 \left(\sum_{j=1}^{N} \mathbf{H}_N(i,j)\mathbf{P}^j\mathbf{e}^j - \rho_1\Gamma_1\hat{\mathbf{g}}^i \right),$$

$$\mathbf{v}_{\hat{\mathbf{A}}} = \Gamma_2\mathbf{M}_{c^i} \left(\sum_{j=1}^{N} \mathbf{H}_N(i,j)\mathbf{P}^j\mathbf{e}^j \right) - \rho_2\Gamma_2\mathbf{v}_{\hat{\mathbf{A}}^i}, \tag{8.34}$$

where Γ_1 and Γ_2 defined in $\mathfrak{R}^{n \times n}$ are the learning rates, and ρ_1 and ρ_2 are two positive leakage gains. Moreover, the main controller gain $\mathbf{P}^i = (\mathbf{P}^i)^T \in \mathfrak{R}^{n \times n}$, which is a positive definite matrix, is updated online using the following DRE [Saf21]:

$$\dot{\mathbf{P}}^i = \mathbf{H}_N(i,i)\mathbf{P}^i\mathbf{P}^i - 2\hat{\mathbf{A}}^i\mathbf{P}^i - 2\mathbf{Q}, \tag{8.35}$$

with $\mathbf{Q} = \mathbf{Q}^T \in \mathfrak{R}^{n \times n}$ –a positive definite diagonal matrix.

Proof (the proof presented here is following the proof provided in [Saf21]): First, for a network defined in *Definition 8.1* and *Definition 8.1*, let us define lumped terms for the parameter estimation errors of all agents as $\tilde{\mathbf{A}}_t = \mathbf{A}_t - \hat{\mathbf{A}}_t$ and $\tilde{\mathbf{g}}_t = \mathbf{g}_t - \hat{\mathbf{g}}_t$, where $\hat{\mathbf{A}}_t = \mathrm{diag}(\hat{\mathbf{A}}^1, \hat{\mathbf{A}}^2, \ldots, \hat{\mathbf{A}}^N) \in \mathfrak{R}^{Nn \times Nn}$ and $\hat{\mathbf{g}}_t = [(\hat{\mathbf{g}}^1)^T \ (\hat{\mathbf{g}}^2)^T \ \cdots \ (\hat{\mathbf{g}}^N)^T]^T \in \mathfrak{R}^{Nn \times 1}$ are the lumped variables for estimated linear and nonlinear parameters of the network, respectively. Then, we can define the following Lyapunov function:

$$V = \frac{1}{2}\mathbf{e}_t^T\mathbf{P}_t\mathbf{e}_t + \frac{1}{2}\tilde{\mathbf{g}}_t^T\Gamma_{1t}^{-1}\tilde{\mathbf{g}}_t + \frac{1}{2}\mathbf{v}_{\hat{\mathbf{A}}_t}^T\Gamma_{2t}^{-1}\mathbf{v}_{\hat{\mathbf{A}}_t}, \tag{8.36}$$

where $\mathbf{P}_t = \mathrm{diag}(\mathbf{P}^1, \mathbf{P}^2, \dots, \mathbf{P}^N) \in \mathfrak{R}^{Nn \times Nn}$, $\boldsymbol{\Gamma}_{1t} = \mathrm{diag}(\boldsymbol{\Gamma}_1^1, \boldsymbol{\Gamma}_1^2, \dots, \boldsymbol{\Gamma}_1^N) \in \mathfrak{R}^{Nn \times Nn}$ and $\boldsymbol{\Gamma}_{2t} = \mathrm{diag}(\boldsymbol{\Gamma}_2^1, \boldsymbol{\Gamma}_2^2, \dots, \boldsymbol{\Gamma}_2^N) \in \mathfrak{R}^{Nn \times Nn}$ are the lumped variables for the time-varying main controller gains as well as the learning rates at all agents in the network. Furthermore, the first time derivative of V is

$$\dot{V} = \mathbf{e}_t^T \mathbf{P}_t \dot{\mathbf{e}}_t + \frac{1}{2} \mathbf{e}_t^T \dot{\mathbf{P}}_t \mathbf{e}_t + \tilde{\mathbf{g}}_t^T \boldsymbol{\Gamma}_{1t}^{-1} \dot{\tilde{\mathbf{g}}}_t + \mathbf{v}_{\tilde{\mathbf{A}}_t}^T \boldsymbol{\Gamma}_{2t}^{-1} \dot{\mathbf{v}}_{\tilde{\mathbf{A}}_t}. \tag{8.37}$$

Besides, from (8.31), we have

$$\dot{\mathbf{e}}_t = (\mathbf{H}_N \otimes \mathbf{I}_n)\dot{\mathbf{z}}_t - (\mathbf{B}_N \otimes \dot{\mathbf{z}}^0)\mathbf{1}, \tag{8.38}$$

and consequently by noting $\mathbf{z}_t = \boldsymbol{\eta}_t - \mathbf{x}_t$, where $\boldsymbol{\eta}_t = [(\boldsymbol{\eta}^1)^T \; (\boldsymbol{\eta}^2)^T \; \dots \; (\boldsymbol{\eta}^N)^T] \in \mathfrak{R}^{Nn \times 1}$, and recalling (8.2), we lead to

$$\dot{\mathbf{e}}_t = (\mathbf{H}_N \otimes \mathbf{I}_n)\dot{\boldsymbol{\eta}}_t - (\mathbf{H}_N \otimes \mathbf{I}_n)\mathbf{A}_t \mathbf{x}_t - (\mathbf{H}_N \otimes \mathbf{I}_n)\mathbf{u}_t - (\mathbf{H}_N \otimes \mathbf{I}_n)\mathbf{g}_t - (\mathbf{B}_N \otimes \dot{\mathbf{z}}^0)\mathbf{1}. \tag{8.39}$$

Thus, by replacing $\dot{\mathbf{e}}_t$ from (8.39) into (8.37), the result will be

$$\dot{V} = \mathbf{e}_t^T \mathbf{P}_t ((\mathbf{H}_N \otimes \mathbf{I}_n)\dot{\boldsymbol{\eta}}_t - (\mathbf{H}_N \otimes \mathbf{I}_n)\mathbf{A}_t \mathbf{x}_t - (\mathbf{H}_N \otimes \mathbf{I}_n)\mathbf{u}_t - (\mathbf{H}_N \otimes \mathbf{I}_n)\mathbf{g}_t$$
$$- (\mathbf{B}_N \otimes \dot{\mathbf{z}}^0)\mathbf{1}) + \frac{1}{2} \mathbf{e}_t^T \dot{\mathbf{P}}_t \mathbf{e}_t + \tilde{\mathbf{g}}^T \boldsymbol{\Gamma}_{1t}^{-1} \dot{\tilde{\mathbf{g}}} + \mathbf{v}_{\tilde{\mathbf{A}}_t}^T \boldsymbol{\Gamma}_{2t}^{-1} \mathbf{v}_{\tilde{\mathbf{A}}_t}. \tag{8.40}$$

Recalling *Definition 8.2*, we can observe that some of values in each row of $\mathbf{L}_N$ is zero. In this regard, one can write

$$(\mathbf{L}_N \otimes \dot{\mathbf{z}}^0)\mathbf{1} = \mathbf{0}, \tag{8.41}$$

for $\mathbf{0} \in \mathfrak{R}^{Nn \times 1}$ is a vector having zero elements. Also, by recalling (8.4), the following result will be obtained:

$$(\mathbf{B}_N \otimes \dot{\mathbf{z}}^0)\mathbf{1} = (\mathbf{H}_N \otimes \dot{\mathbf{z}}^0)\mathbf{1}. \tag{8.42}$$

Moreover, according to the *mixed-product* feature of the Kronecker product [Li15], the equation in (8.42) is expressed as follows:

$$(\mathbf{B}_N \otimes \dot{\mathbf{z}}^0)\mathbf{1} = (\mathbf{H}_N \otimes \mathbf{I}_n)(\mathbf{I}_N \otimes \dot{\mathbf{z}}^0)\mathbf{1}, \tag{8.43}$$

where $\mathbf{I}_N \in \mathfrak{R}^{N \times N}$ is an identity matrix. Now, by replacing (8.43) in (8.40), we reach at

$$\dot{V} = \mathbf{e}_t^T \mathbf{P}_t ((\mathbf{H}_N \otimes \mathbf{I}_n)\dot{\boldsymbol{\eta}}_t - (\mathbf{H}_N \otimes \mathbf{I}_n)\mathbf{A}_t \mathbf{x}_t - (\mathbf{H}_N \otimes \mathbf{I}_n)\mathbf{u}_t - (\mathbf{H}_N \otimes \mathbf{I}_n)\mathbf{g}_t$$
$$- (\mathbf{H}_N \otimes \mathbf{I}_n)(\mathbf{I}_N \otimes \dot{\mathbf{z}}^0)\mathbf{1}) + \frac{1}{2} \mathbf{e}_t^T \dot{\mathbf{P}}_t \mathbf{e}_t + \tilde{\mathbf{g}}^T \boldsymbol{\Gamma}_{1t}^{-1} \dot{\tilde{\mathbf{g}}} + \mathbf{v}_{\tilde{\mathbf{A}}_t}^T \boldsymbol{\Gamma}_{2t}^{-1} \mathbf{v}_{\tilde{\mathbf{A}}_t}. \tag{8.44}$$

In addition, as $(\mathbf{H}_N \otimes \mathbf{I}_n)$ and $\mathbf{A}_t$ are both symmetric matrices, one can have

$$(\mathbf{H}_N \otimes \mathbf{I}_n)\mathbf{A}_t\mathbf{x}_t = \mathbf{A}_t(\mathbf{H}_N \otimes \mathbf{I}_n)\mathbf{x}_t, \tag{8.45}$$

and hence,

$$\dot{V} = \mathbf{e}_t^T \mathbf{P}_t((\mathbf{H}_N \otimes \mathbf{I}_n)\dot{\boldsymbol{\eta}}_t - \mathbf{A}_t(\mathbf{H}_N \otimes \mathbf{I}_n)\mathbf{x}_t - (\mathbf{H}_N \otimes \mathbf{I}_n)\mathbf{u}_t - (\mathbf{H}_N \otimes \mathbf{I}_n)\mathbf{g}_t$$
$$- (\mathbf{H}_N \otimes \mathbf{I}_n)(\mathbf{I}_N \otimes \dot{\mathbf{z}}^0)\mathbf{1}) + \frac{1}{2}\mathbf{e}_t^T \dot{\mathbf{P}}_t\mathbf{e}_t + \tilde{\mathbf{g}}^T \Gamma_{1t}^{-1}\dot{\tilde{\mathbf{g}}} + \mathbf{v}_{\tilde{\mathbf{A}}_t}^T \Gamma_{2t}^{-1}\mathbf{v}_{\tilde{\mathbf{A}}_t}. \tag{8.46}$$

At this point, we want to construct a term including $\mathbf{A}_t\mathbf{e}_t$ within the paranthesis of the first term on the right-hand side of (8.46). In this regard, let us add and subtract $\mathbf{e}_t^T \mathbf{P}_t\mathbf{A}_t((\mathbf{H}_N \otimes \mathbf{I}_n)\boldsymbol{\eta}_t + (\mathbf{B}_N \otimes \mathbf{z}^0)\mathbf{1})$ in this equation. Thus, by recalling $\mathbf{z}_t = \boldsymbol{\eta}_t - \mathbf{x}_t$ as well as (8.30), we have

$$\dot{V} = \mathbf{e}_t^T \mathbf{P}_t((\mathbf{H}_N \otimes \mathbf{I}_n)\dot{\boldsymbol{\eta}}_t + \mathbf{A}_t\mathbf{e}_t + \mathbf{A}_t(\mathbf{B}_N \otimes \mathbf{z}^0)\mathbf{1} - \mathbf{A}_t(\mathbf{H}_N \otimes \mathbf{I}_n)\boldsymbol{\eta}_t - (\mathbf{H}_N \otimes \mathbf{I}_n)\mathbf{u}_t$$
$$- (\mathbf{H}_N \otimes \mathbf{I}_n)\mathbf{g}_t - (\mathbf{H}_N \otimes \mathbf{I}_n)(\mathbf{I}_N \otimes \dot{\mathbf{z}}^0)\mathbf{1}) + \frac{1}{2}\mathbf{e}_t^T \dot{\mathbf{P}}_t\mathbf{e}_t + \tilde{\mathbf{g}}^T \Gamma_{1t}^{-1}\dot{\tilde{\mathbf{g}}} + \mathbf{v}_{\tilde{\mathbf{A}}_t}^T \Gamma_{2t}^{-1}\mathbf{v}_{\tilde{\mathbf{A}}_t}. \tag{8.47}$$

Moreover, according to the discussion presented earlier above, one can have

$$\mathbf{A}_t(\mathbf{B}_N \otimes \mathbf{z}^0)\mathbf{1} - \mathbf{A}_t(\mathbf{H}_N \otimes \mathbf{I}_n)\boldsymbol{\eta}_t = (\mathbf{H}_N \otimes \mathbf{I}_n)\mathbf{A}_t((\mathbf{I}_N \otimes \mathbf{z}^0)\mathbf{1} - \boldsymbol{\eta}). \tag{8.48}$$

Next, inserting (8.47) into (8.46) leads to

$$\dot{V} = \mathbf{e}_t^T \mathbf{P}_t(\mathbf{A}_t\mathbf{e}_t + (\mathbf{H}_N \otimes \mathbf{I}_n)\dot{\boldsymbol{\eta}}_t + (\mathbf{H}_N \otimes \mathbf{I}_n)\mathbf{A}_t((\mathbf{I}_N \otimes \mathbf{z}^0)\mathbf{1} - \boldsymbol{\eta}) - (\mathbf{H}_N \otimes \mathbf{I}_n)\mathbf{u}_t$$
$$- (\mathbf{H}_N \otimes \mathbf{I}_n)\mathbf{g}_t - (\mathbf{H}_N \otimes \mathbf{I}_n)(\mathbf{I}_N \otimes \dot{\mathbf{z}}^0)\mathbf{1}) + \frac{1}{2}\mathbf{e}_t^T \dot{\mathbf{P}}_t\mathbf{e}_t + \tilde{\mathbf{g}}^T \Gamma_{1t}^{-1}\dot{\tilde{\mathbf{g}}} + \mathbf{v}_{\tilde{\mathbf{A}}_t}^T \Gamma_{2t}^{-1}\mathbf{v}_{\tilde{\mathbf{A}}_t}. \tag{8.49}$$

Next, we intend to include $\hat{\mathbf{A}}_t$ and $\hat{\mathbf{g}}_t$ into the first term of the right-hand side of (8.49); let us add and subtract the following term in (8.49):

$$\mathbf{e}_t^T \mathbf{P}_t(\mathbf{H}_N \otimes \mathbf{I}_n)(\hat{\mathbf{A}}_t((\mathbf{I}_N \otimes \mathbf{z}^0)\mathbf{1} - \boldsymbol{\eta}_t) + \hat{\mathbf{g}}). \tag{8.50}$$

Hence, we have

$$\dot{V} = \mathbf{e}_t^T \mathbf{P}_t(\mathbf{A}_t\mathbf{e}_t + (\mathbf{H}_N \otimes \mathbf{I}_n)\dot{\boldsymbol{\eta}}_t + (\mathbf{H}_N \otimes \mathbf{I}_n)\hat{\mathbf{A}}_t((\mathbf{I}_N \otimes \mathbf{z}^0)\mathbf{1} - \boldsymbol{\eta}_t) - (\mathbf{H}_N \otimes \mathbf{I}_n)\mathbf{u}_t$$
$$- (\mathbf{H}_N \otimes \mathbf{I}_n)\hat{\mathbf{g}}_t - (\mathbf{H}_N \otimes \mathbf{I}_n)(\mathbf{I}_N \otimes \dot{\mathbf{z}}^0)\mathbf{1}) + \tilde{\mathbf{g}}^T \Gamma_{1t}^{-1}\dot{\tilde{\mathbf{g}}} - \mathbf{e}_t^T \mathbf{P}_t(\mathbf{H}_N \otimes \mathbf{I}_n)\tilde{\mathbf{g}}_t \tag{8.51}$$
$$+ \mathbf{v}_{\tilde{\mathbf{A}}_t}^T \Gamma_{2t}^{-1}\mathbf{v}_{\tilde{\mathbf{A}}_t} + \mathbf{e}_t^T \mathbf{P}_t(\mathbf{H}_N \otimes \mathbf{I}_n)\tilde{\mathbf{A}}_t((\mathbf{I}_N \otimes \mathbf{z}^0)\mathbf{1} - \boldsymbol{\eta}_t) + \frac{1}{2}\mathbf{e}_t^T \dot{\mathbf{P}}_t\mathbf{e}_t.$$

Besides, recalling $\mathbf{z}^0 = \boldsymbol{\eta}^0 - \mathbf{x}^0$ and defining $\mathbf{c}^i = \boldsymbol{\eta}^0 - \boldsymbol{\eta}^i - \mathbf{x}^0 = \mathbf{z}^0 - \boldsymbol{\eta}^i$ for $i = 1...N$, one can have

$$(\mathbf{I}_N \otimes \mathbf{z}^0)\mathbf{1} - \boldsymbol{\eta}_t = \mathbf{c}_t, \tag{8.52}$$

where $\mathbf{c}_t = [(\mathbf{c}^1)^T \quad (\mathbf{c}^2)^T \quad \cdots \quad (\mathbf{c}^N)^T]^T \in \mathfrak{R}^{Nn \times 1}$. Also, we have

$$(\mathbf{I}_N \otimes \dot{\mathbf{z}}^0)\mathbf{1} - \dot{\boldsymbol{\eta}}_t = \dot{\mathbf{c}}_t. \tag{8.53}$$

Hence, by replacing (8.52) and (8.53) into (8.51), we have

$$\dot{V} = \mathbf{e}_t^T \mathbf{P}_t(\mathbf{A}_t \mathbf{e}_t - (\mathbf{H}_N \otimes \mathbf{I}_n)\dot{\mathbf{c}}_t + (\mathbf{H}_N \otimes \mathbf{I}_n)\hat{\mathbf{A}}_t \mathbf{c}_t - (\mathbf{H}_N \otimes \mathbf{I}_n)\mathbf{u}_t - (\mathbf{H}_N \otimes \mathbf{I}_n)\hat{\mathbf{g}}_t)$$
$$+ \tilde{\mathbf{g}}^T \boldsymbol{\Gamma}_{1t}^{-1} \dot{\tilde{\mathbf{g}}} - \mathbf{e}_t^T \mathbf{P}_t(\mathbf{H}_N \otimes \mathbf{I}_n)\tilde{\mathbf{g}}_t + \mathbf{v}_{\tilde{\mathbf{A}}_t}^T \boldsymbol{\Gamma}_{2t}^{-1} \mathbf{v}_{\tilde{\mathbf{A}}_t} + \mathbf{e}_t^T \mathbf{P}_t(\mathbf{H}_N \otimes \mathbf{I}_n)\tilde{\mathbf{A}}_t \mathbf{c}_t + \frac{1}{2}\mathbf{e}_t^T \dot{\mathbf{P}}_t \mathbf{e}_t. \tag{8.54}$$

Now, let us focus on the estimation of the nonlinear terms $\hat{\mathbf{g}}_t$. In this regard, we define

$$s_1 = \tilde{\mathbf{g}}^T \boldsymbol{\Gamma}_{1t}^{-1} \dot{\tilde{\mathbf{g}}} - \mathbf{e}_t^T \mathbf{P}_t(\mathbf{H}_N \otimes \mathbf{I}_n)\tilde{\mathbf{g}}. \tag{8.55}$$

Then, by adding and subtracting the following term from s_1:

$$\frac{1}{4\rho_1}(\boldsymbol{\Gamma}_{1t}^{-1}\dot{\mathbf{g}}_t + \rho_1 \mathbf{g}_t)^T(\boldsymbol{\Gamma}_{1t}^{-1}\dot{\mathbf{g}}_t + \rho_1 \mathbf{g}_t) + \rho\rho_1\tilde{\mathbf{g}}^T(\tilde{\mathbf{g}} + \hat{\mathbf{g}}), \tag{8.56}$$

s_1 will result as follows:

$$s_1 = \tilde{\mathbf{g}}_t^T(-\boldsymbol{\Gamma}_{1t}^{-1}\dot{\hat{\mathbf{g}}}_t - \rho_1\hat{\mathbf{g}}_t - (\mathbf{H}_N \otimes \mathbf{I}_n)\mathbf{P}_t\mathbf{e}_t) - s_3 + s_4. \tag{8.57}$$

In (8.57), we have

$$s_3 = \frac{1}{4\rho_1}\left(\boldsymbol{\Gamma}_{1t}^{-1}\dot{\mathbf{g}}_t + \rho_1 \mathbf{g}_t\right)^T(\boldsymbol{\Gamma}_{1t}^{-1}\dot{\mathbf{g}}_t + \rho_1 \mathbf{g}_t) + \rho_1\tilde{\mathbf{g}}_t^T\tilde{\mathbf{g}}_t$$
$$- 2\sqrt{\rho_1}\tilde{\mathbf{g}}_t^T\frac{1}{2\sqrt{\rho_1}}\left(\rho_1\tilde{\mathbf{g}}_t + \rho_1\hat{\mathbf{g}}_t + \boldsymbol{\Gamma}_{1t}^{-1}\dot{\mathbf{g}}_t\right) \tag{8.58}$$
$$= \left(\frac{1}{2\sqrt{\rho_1}}(\boldsymbol{\Gamma}_{1t}^{-1}\dot{\mathbf{g}}_t + \rho_1 \mathbf{g}_t) - \sqrt{\rho_1}\tilde{\mathbf{g}}_t\right)^T\left(\frac{1}{2\sqrt{\rho_1}}(\boldsymbol{\Gamma}_{1t}^{-1}\dot{\mathbf{g}}_t + \rho_1 \mathbf{g}_t) - \sqrt{\rho_1}\tilde{\mathbf{g}}_t\right) \geq 0,$$

and

$$s_4 = \frac{1}{4\rho_1}(\boldsymbol{\Gamma}_{1t}^{-1}\dot{\mathbf{g}}_t + \rho_1 \mathbf{g}_t)^T(\boldsymbol{\Gamma}_{1t}^{-1}\dot{\mathbf{g}}_t + \rho_1 \mathbf{g}_t). \tag{8.59}$$

In addition, by recalling *Remark 8.2*, it can be shown that

$$s_4 \leq \frac{1}{4\rho_1}(\Gamma_{1t}^{-1}\mathbf{U}_{\dot{\mathbf{g}}_t} + \rho_1\mathbf{U}_{\mathbf{g}_t})^T(\Gamma_{1t}^{-1}\mathbf{U}_{\dot{\mathbf{g}}_t} + \rho_1\mathbf{U}_{\mathbf{g}_t}) = \beta_4. \tag{8.60}$$

Furthermore, by defining

$$\dot{\hat{\mathbf{g}}}_t = -\Gamma_{1t}(\mathbf{H}_N \otimes \mathbf{I}_n)\mathbf{P}_t\mathbf{e}_t - \rho_1\Gamma_{1t}\hat{\mathbf{g}}_t, \tag{8.61}$$

we can have

$$s_1 \leq -r_3 + \beta_4. \tag{8.62}$$

In a similar way, if we define

$$s_2 = \mathbf{v}_{\tilde{\mathbf{A}}_t}^T \Gamma_{2t}^{-1}\mathbf{v}_{\dot{\tilde{\mathbf{A}}}_t} + \mathbf{e}_t^T\mathbf{P}_t(\mathbf{H}_N \otimes \mathbf{I}_n)\mathbf{M}_c\mathbf{v}_{\tilde{\mathbf{A}}_t}, \tag{8.63}$$

for $\tilde{\mathbf{A}}_t\mathbf{c}_t = \mathbf{M}_c\mathbf{v}_{\tilde{\mathbf{A}}_t}$, then by adding and subtracting the following term from s_2:

$$\frac{1}{4\rho_1}(\Gamma_{2t}^{-1}\mathbf{v}_{\dot{\mathbf{A}}_t} + \rho_2\mathbf{v}_{\mathbf{A}_t})^T(\Gamma_{2t}^{-1}\mathbf{v}_{\dot{\mathbf{A}}_t} + \rho_2\mathbf{v}_{\mathbf{A}_t}) + \rho_2\mathbf{v}_{\tilde{\mathbf{A}}_t}^T(\mathbf{v}_{\tilde{\mathbf{A}}_t} + \mathbf{v}_{\dot{\mathbf{A}}_t}), \tag{8.64}$$

s_2 will result in terms of

$$s_2 = \mathbf{v}_{\tilde{\mathbf{A}}_t}^T(-\Gamma_{2t}^{-1}\mathbf{v}_{\dot{\mathbf{A}}} - \rho_2\mathbf{v}_{\hat{\mathbf{A}}} + \mathbf{M}_c(\mathbf{H}_N \otimes \mathbf{I}_n)\mathbf{P}_t\mathbf{e}_t) - s_5 + s_6, \tag{8.65}$$

where

$$
\begin{aligned}
s_5 &= \frac{1}{4\rho_2}(\Gamma_{2t}^{-1}\mathbf{v}_{\dot{\mathbf{A}}_t} + \rho_2\mathbf{v}_{\mathbf{A}_t})^T(\Gamma_{2t}^{-1}\mathbf{v}_{\dot{\mathbf{A}}_t} + \rho_2\mathbf{v}_{\mathbf{A}_t}) + \rho_2\mathbf{v}_{\tilde{\mathbf{A}}_t}^T\mathbf{v}_{\tilde{\mathbf{A}}_t} \\
&\quad - 2\sqrt{\rho_2}\mathbf{v}_{\tilde{\mathbf{A}}_t}^T\frac{1}{2\sqrt{\rho_2}}(\rho_2\mathbf{v}_{\tilde{\mathbf{A}}_t} + \rho_2\mathbf{v}_{\dot{\mathbf{A}}_t} + \Gamma_{2t}^{-1}\mathbf{v}_{\dot{\mathbf{A}}_t}) \\
&= \left(\frac{1}{2\sqrt{\rho_2}}(\Gamma_{2t}^{-1}\mathbf{v}_{\dot{\mathbf{A}}_t} + \rho_2\mathbf{v}_{\mathbf{A}_t}) - \sqrt{\rho_2}\mathbf{v}_{\tilde{\mathbf{A}}_t}\right)^T\left(\frac{1}{2\sqrt{\rho_2}}(\Gamma_{2t}^{-1}\mathbf{v}_{\dot{\mathbf{A}}_t} + \rho_2\mathbf{v}_{\mathbf{A}_t}) - \sqrt{\rho_2}\mathbf{v}_{\tilde{\mathbf{A}}_t}\right) \geq 0,
\end{aligned}
\tag{8.66}
$$

and

$$s_6 = \frac{1}{4\rho_2}(\Gamma_{2t}^{-1}\mathbf{v}_{\dot{\mathbf{A}}_t} + \rho_2\mathbf{v}_{\mathbf{A}_t})^T(\Gamma_{2t}^{-1}\mathbf{v}_{\dot{\mathbf{A}}_t} + \rho_2\mathbf{v}_{\mathbf{A}_t}). \tag{8.67}$$

Also, according to *Remark 8.2*, the following inequality is obtained:

$$s_6 \leq \frac{1}{4\rho_2}(\Gamma_{2t}^{-1}\mathbf{U}_{\dot{\mathbf{A}}_t} + \rho_2\mathbf{U}_{\mathbf{A}_t})^T(\Gamma_{2t}^{-1}\mathbf{U}_{\dot{\mathbf{A}}_t} + \rho_2\mathbf{U}_{\mathbf{A}_t}) = \beta_6. \tag{8.68}$$

Moreover, by defining

$$\mathbf{v}_{\hat{\mathbf{A}}} = +\Gamma_{2t}\mathbf{M}_c(\mathbf{H}_N \otimes \mathbf{I}_n)\mathbf{P}_t\mathbf{e}_t - \rho_2\Gamma_{2t}\mathbf{v}_{\hat{\mathbf{A}}}, \tag{8.69}$$

this leads to

$$s_2 \leq -r_5 + \beta_6. \tag{8.70}$$

Finally, by substituting the inequalities (8.62) and (8.70) into (8.54) and rearranging, we have

$$\dot{V} \leq \mathbf{e}_t^T \mathbf{P}_t(\mathbf{A}_t\mathbf{e}_t + (\mathbf{H}_N \otimes \mathbf{I}_n)(\hat{\mathbf{A}}_t\mathbf{c}_t - \dot{\mathbf{c}}_t - \mathbf{u}_t - \hat{\mathbf{g}}_t)) + \frac{1}{2}\mathbf{e}_t^T\dot{\mathbf{P}}_t\mathbf{e}_t - s_7 + \beta_7, \tag{8.71}$$

where $s_7 = s_3 + s_5$ and $\beta_7 = \min\{\beta_4, \beta_6\}$. Now, by defining the lumped control input (or control signal) as follows:

$$\mathbf{u}_t = \frac{1}{2}\mathbf{P}_t\mathbf{e}_t + \hat{\mathbf{A}}_t\mathbf{c}_t - \hat{\mathbf{g}}_t - \dot{\mathbf{c}}_t, \tag{8.72}$$

the inequality in (8.71) is represented as follows:

$$\dot{V} \leq \frac{1}{2}\mathbf{e}_t^T(2\mathbf{P}_t\mathbf{A}_t - \mathbf{P}_t(\mathbf{H}_N \otimes \mathbf{I}_n)\mathbf{P}_t + \dot{\mathbf{P}}_t)\mathbf{e}_t - s_7 + \beta_7. \tag{8.73}$$

In this regard, by having a lumped DRE as follows:

$$\dot{\mathbf{P}}_t = 2\mathbf{P}_t\mathbf{A}_t - \mathbf{P}_t(\mathbf{H}_N \otimes \mathbf{I}_n)\mathbf{P}_t - 2\mathbf{Q}_t, \tag{8.74}$$

for $\mathbf{Q}_t = (\mathbf{I}_N \otimes \mathbf{Q}) \in \mathfrak{R}^{Nn \times Nn}$, then we reach at

$$\dot{V} \leq -(\mathbf{e}_t^T\mathbf{Q}_t\mathbf{e}_t + s_7) + \beta_7. \tag{8.75}$$

At last, according to the *LaSalle-Yoshizawa* theorem presented in *Lemma 3.3*, we can say that V is UUB, and consequently, $\mathbf{e}_t$, $\tilde{\mathbf{g}}_t$ and $\tilde{\mathbf{A}}_t$ converge to small bounded sets around origin. At this point, we should only note that the equations proposed in (8.32–8.35) are achieved by formulating each row of the equations in (8.61), (8.69), (8.72) and (8.74), respectively. This completes the proof.

Remark 8.5

Recalling *Definition 8.7* and *Theorem 8.3*, the realtive measurements on the states of neighboring agents, i.e., $(\mathbf{x}^i - \mathbf{x}^j)$ should be available for implementation of the cooperative model-free adaptive controller. But, no absolute measurements on the state of each agent in the network is needed.

Remark 8.6

For tuning $\Gamma_1, \Gamma_2, \rho_1, \rho_2$, the same procedure presented in *Remark 5.9* should be utilized.

Remark 8.7

According to (8.32) and (8.33), the parameters $\hat{\eta}^0$, $\hat{\eta}^i$ and $\hat{x}_0^i$ are required to determine the value of $\mathbf{u}^i$ of the proposed cooperative model-free adaptive controller in *Theorem 8.3*. These parameters are provided using the cooperative observers presented in *Theorems 8.1* and *8.2*.

Remark 8.8

According to *Theorem 8.3*, $\mathbf{P}^i$ should be a positive definite matrix. The following lemma is proposed to show that by updating $\mathbf{P}^i$ using (8.35), this requirement is satisfied.

Lemma 8.2

In (8.35), the value for all diagonal elements of $\mathbf{P}^i$ would be positive, if the diagonal elements of $\mathbf{Q}$ are chosen large enough and if the following solution is utilized [Saf21]:

$$\mathbf{P}^i(k,k) = \mathbf{P}_0^k + \frac{1}{w^k}, \quad k = 1...n, \tag{8.76}$$

where

$$\mathbf{P}_0^k = \frac{\hat{\mathbf{A}}(k,k) - \sqrt{\tau}}{\mathbf{H}_N(i,i)},$$

$$w^k = \frac{\mathbf{H}_N(i,i)}{2\sqrt{\tau}} + \left(w_0^k - \frac{\mathbf{H}_N(i,i)}{2\sqrt{\tau}} \right) \exp\left(-\int_0^t 2\sqrt{\tau}\, dr \right),$$

$$\tau = (\hat{\mathbf{A}}^i(k,k))^2 + 2\mathbf{Q}(k,k)\mathbf{H}_N(i,i),$$

$$w_0^k = \frac{1}{1 - \mathbf{P}_0^k}.$$

$\tag{8.77}$

Proof (the proof presented here is following the proof provided in [Saf21]: Let us write the equation corresponding to the kth diagonal element of $\dot{\mathbf{P}}^i$ in (8.35) as follows:

$$\dot{\mathbf{P}}^i(k,k) = -2\hat{\mathbf{A}}^i(k,k)\mathbf{P}^i(k,k) + \mathbf{H}_N(i,i)(\mathbf{P}^i(k,k))^2 - 2\mathbf{Q}(k,k). \tag{8.78}$$

Note that the off-diagonal elements of $\dot{\mathbf{P}}^i$ in (8.35) would be always zero by choosing $\mathbf{Q}$ to have zero off-diagonal elements. Moreover, by letting $p = \mathbf{P}^i(k,k)$, $a = \hat{\mathbf{A}}^i(k,k)$, $h = \mathbf{H}_N(i,i)$ and $q = \mathbf{Q}(k,k)$, (8.78) is simplified as follows:

$$\dot{p} = hp^2 - 2ap - 2q. \tag{8.79}$$

This is a nonhomogeneous second-order differential equation and has the general solution as

$$p = p_0 + \frac{1}{w}, \tag{8.80}$$

where $p_0 = \mathbf{P}_0^k$ and $w = w^k$. Here, p_0 is the homogeneous solution to (8.79) and is proposed as follows:

$$hp_0^2 - 2ap_0 - 2q = 0. \tag{8.81}$$

The definition of $\mathbf{P}_0^k$ in (8.77) is a solution to (8.81) as

$$p_0 = \frac{a - \sqrt{\Delta}}{h}, \quad \Delta = a^2 + 2qh. \tag{8.82}$$

Note that according to *Definition 8.2*, each diagonal element of $\mathbf{H}_N$ is positive, i.e., $h > 0$. Furthermore, by substituting w from (8.77) into (8.80), we have

$$p = \frac{a - \sqrt{\Delta}}{h} + \frac{1}{\tau + \gamma}, \tag{8.83}$$

where

$$\tau = \frac{h}{2\sqrt{\Delta}}(1 - \mu), \quad \gamma = \frac{\mu}{1 - \dfrac{a - \sqrt{\Delta}}{h}}. \tag{8.84}$$

Since $h > 0$ and $0 < \mu \leq 1$, we have $\tau \geq 0$. In addition, by rearranging γ as follows:

$$\gamma = \frac{h\mu}{h - a + \sqrt{\Delta}}, \tag{8.85}$$

we reach at

$$\frac{1}{\tau + \gamma} = \frac{h - a + \sqrt{\Delta}}{\tau(h - a + \sqrt{\Delta}) + h\mu} = \frac{h - (a - \sqrt{\Delta})}{h(\tau + \mu) - \tau(a - \sqrt{\Delta})}. \tag{8.86}$$

Moreover, by replacing (8.86) into (8.83), we have

$$p = \frac{h^2 - (a - \sqrt{\Delta})p_1}{h(h(\tau + \mu) - \tau(a - \sqrt{\Delta}))}, \tag{8.87}$$

where

$$p_1 = h(1 - \tau - \mu) + \tau(a - \sqrt{\Delta}). \tag{8.88}$$

According to the definition of Δ in (8.82), we know that $\sqrt{\Delta} > a$, and hence, $(a - \sqrt{\Delta}) < 0$. So, by recalling $\tau > 0$, $h > 0$ and $0 < \mu < 1$, we deduce that the denominator of p presented in (8.87) is always positive. For the nominator of p, let us rephrase p_1 by replacing τ from (8.84) as follows:

$$p_1 = h(1 - \frac{h}{2\sqrt{\Delta}} + \frac{\mu h}{2\sqrt{\Delta}} - \mu) + \frac{h}{2\sqrt{\Delta}}(1 - \mu)(a - \sqrt{\Delta})$$

$$= \frac{h}{2\sqrt{\Delta}}((1 - \mu)(2\sqrt{\Delta} + a - \sqrt{\Delta}) - (1 - \mu)h) \tag{8.89}$$

$$= \frac{h}{2\sqrt{\Delta}}(1 - \mu)(\sqrt{\Delta} + a - h).$$

Finally, since $h^2 > 0$ and $(a - \sqrt{\Delta}) < 0$, then by choosing q large enough in order to have $p_1 > 0$ in (8.89), we can guarantee that the nominator of p in (8.87) will be positive. This completes the proof.

8.2.4 OPERATING PRINCIPLES OF THE COOPERATIVE MODEL-FREE ADAPTIVE CONTROLLERS

The operating principles of the proposed cooperative MFAC algorithms, also referred to as the cooperative adaptive model-free control algorithms CAMFC-1 and CAMFC-2, are synthesized in Figures 8.1 and 8.2, respectively. Additional details are given in [Saf18] and [Saf21].

8.3 SIMULATION RESULTS

Here, a team of four quadrotors is considered for implementing the CAMFC-1 and CAMFC-2 algorithms. The objective is to let the team follow a target trajectory while maintaining a desired formation topology. In this regard, a communication graph as depicted in Figure 8.3 is utilized between the quadrotors in the team. This communication graph has only one *spanning tree* rooted at agent-1. Note that agent-0 is a virtual leader that has information on the target trajectory and the desired formation topology. This information is only shared with agent-1. Using MATLAB R2018b is recommended to run all Simulink models presented in this chapter.

8.3.1 SIMULATION OF CAMFC-1 ALGORITHM

The schematic for the implementation of CAMFC-1 on each of the quadrotors is presented in Figure 8.4. As it is observed, two cooperative observers are utilized for determining the desired position set-point at each quadrotor. Then, the MFAC-1 unit (incorporating the MFAC algorithm proposed in Section 5.2.3) receives the difference

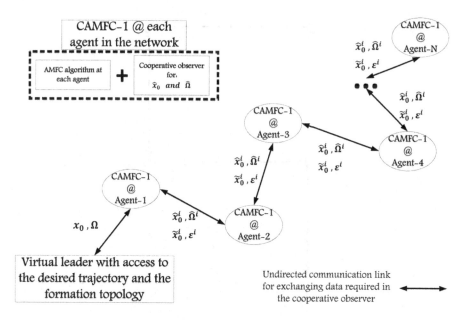

FIGURE 8.1 Operating princple of CAMFC-1 algorithm.

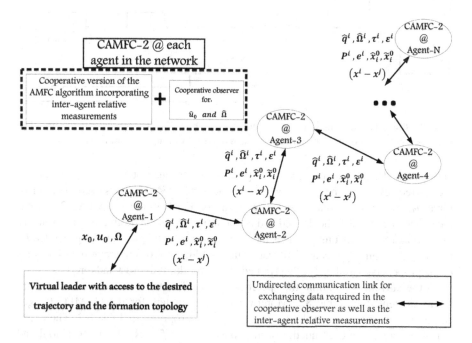

FIGURE 8.2 Operating princple of CAMFC-2 algorithm.

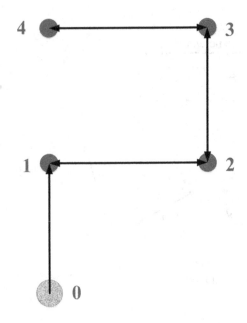

FIGURE 8.3 The communication graph between the quadrotors.

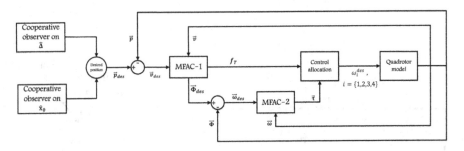

FIGURE 8.4 The schematic of implementation of the CAMFC-1 on each quadrotor. The measurement over the absolute position of each quadrotor in the network is required.

between the desired and actual absolute positions of the quadrotor and considers that as the desired velocity of the quadrotor. The output of the MFAC-1 unit is the desired thrust force, f_T, and the desired attitude of the quadrotor. Furthermore, the MFAC-2 unit has the difference between the desired and actual attitudes, as the desired angular velocity of the quadrotor. This unit provides the desired vector of torque, $\vec{\tau}$, to be generated by the propellers. Finally, a control allocation unit produces the desired rotational velocities of the propellers, based on the determined values for desired thrust and torque.

In the current simulation, the specifications of the model for each quadrotor are set as presented in Table 8.1. Moreover, the parameters for the CAMFC-1 algorithm at each quadrotor in the team are tuned as proposed in Table 8.2. The external disturbances on each quadrotor are considered as $\vec{f}_d = \mathbf{R}_r[0 \quad 0 \quad 0.1\sin(t)]$ and $\vec{\tau}_d = 0.1\sin(t)[1 \quad 1 \quad 1]$.

TABLE 8.1

The Specifications of the Quadrotor Model

Parameter	Value	Parameter	Values
m	2 kg	1	0.2 m
J	$1.25 \cdot 10^{-3} \cdot \mathrm{diag}(1,1,2)$ kg.m²	k_l	10^{-5} N.s²/rad²
k_f	0.01 N.s/m	k_t	10^{-7} N.m.s²/rad²
k_τ	0.01 N.m.s/rad		

TABLE 8.2

The Tuning Parameters for the CAMFC-1 Algorithm

Parameter	Value	Parameter	Value
Π_m	[1 1 1]	Y_m	[1 1 1]
λ	10	μ	10
	MFAC-1		**MFAC-2**
Q	$10 \times \mathrm{diag}(1,1,10)$	Q	$0.1 \times \mathrm{diag}(1,1,1)$
ρ_1	1.0	ρ_1	1.0
Γ_1	$100 \times \mathrm{diag}(1,1,1)$	Γ_1	$100 \times \mathrm{diag}(1,1,1)$
ρ_2	1.0	ρ_2	1.0
Γ_2	$100 \times \mathrm{diag}(1,1,1)$	Γ_2	$100 \times \mathrm{diag}(1,1,1)$

In addition, the desired trajectory of the virtual leader is considered as

$$x_0 = \left[4 \times \sin\left(0.2t + \frac{\pi}{2} \right) \quad 6 \times \cos(0.2t) \quad 0.1 \times t \right]^T,$$ where t is the time variable. The desired formation topology among the quadrotors is also defined as follows:

$$\Delta = \begin{bmatrix} 0 & 0 & 0 \\ r_x & r_y & r_z \\ -r_x & r_y & -r_z \\ -r_x & -r_y & 2r_z \\ r_x & -r_y & -2r_z \end{bmatrix}, \quad (8.90)$$

where

$$\begin{cases} 0 \le t < 120: \ r_x = +1, r_y = +0.5, r_z = +3 \\ 120 \le t < 240: \ r_x = -1, r_y = -0.5, r_z = +2 \end{cases}. \quad (8.91)$$

The simulation results for this case study are presented in Figures 8.5–8.16. These results have been obtained in terms of running the accompanying Matlab & Simulink programs and schemes. The readers are invited to test them. Programs and schemes in three Matlab & Simulink versions are offered in order to increase the availability for the readers and thus simplify the applicability.

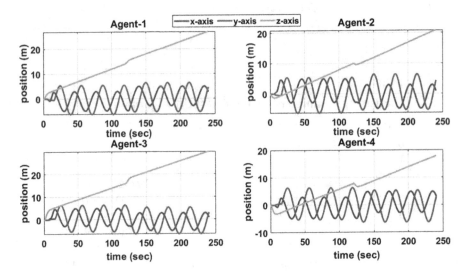

FIGURE 8.5   CAMFC-1: The positions of all quadrotors.

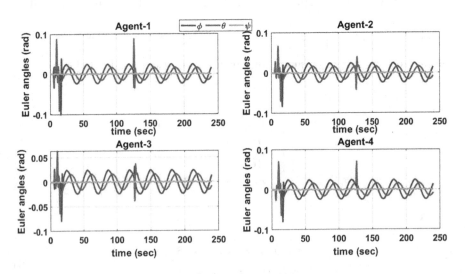

FIGURE 8.6   CAMFC-1: The Euler angles of all quadrotors.

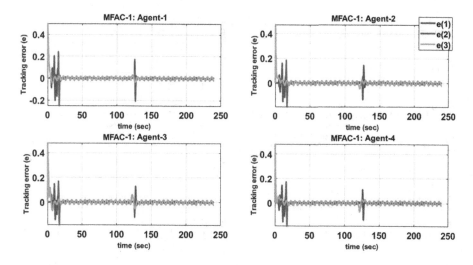

FIGURE 8.7   CAMFC-1: The tracking errors for MFAC-1 at all quadrotors.

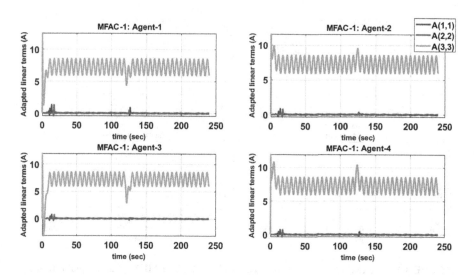

FIGURE 8.8   CAMFC-1: The estimated linear terms for MFAC-1 at all quadrotors.

8.3.2   SIMULATION OF CAMFC-2 ALGORITHM

The schematic of the solution for formation-tracking problem of a network of four quadrotors using the CAMFC-2 algorithm is provided in Figure 8.17. Here, the observed values for $\hat{\Delta}$ and $\hat{\mathbf{x}}_0$ along the measured relative position of neighboring agents are considered as the required input variables of the CAMFC-2 algorithm. This algorithm is implemented distributed at each quadrotor and generates the

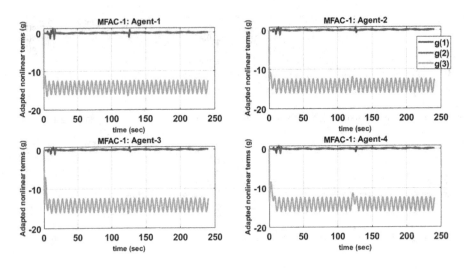

FIGURE 8.9   CAMFC-1: The estimated nonlinear terms for MFAC-1 at all quadrotors.

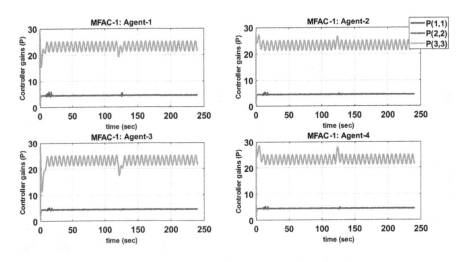

FIGURE 8.10   CAMFC-1: The estimated controller gains for MFAC-1 at all quadrotors.

set-points for the linear velocity of it. Then, similar to the simulation study performed for the CAMFC-1 algorithm, the MFAC-1 unit (incorporating the MFAC algorithm proposed in Section 5.2.3) is utilized for the velocity control of the quadrotor, while the MFAC-2 unit is controlling the angular velocity of the drone. Note that here there is no requirement for measuring the absolute position of all quadrotors. Only agent-1 needs to have its own measured absolute position, since, according to Figure 8.1, it is the only agent connected to the virtual leader.

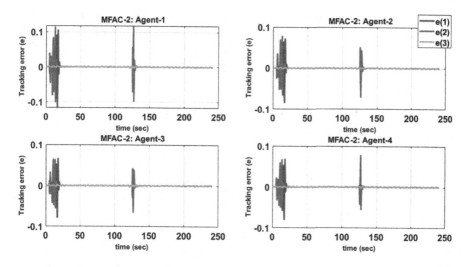

FIGURE 8.11 CAMFC-1: The tracking errors for MFAC-2 at all quadrotors.

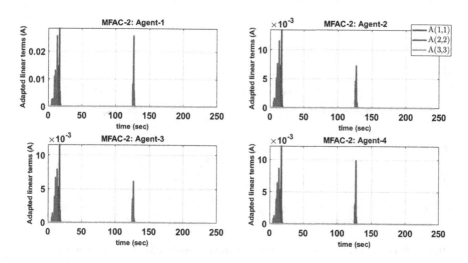

FIGURE 8.12 CAMFC-1: The estimated linear terms for MFAC-2 at all quadrotors.

In the current simulation, the specifications of the model for each quadrotor are set the same as the ones presented in Table 8.1. The tuning parameters for the CAMFC-2 algorithm are provided in Table 8.3. The external disturbances on each quadrotor are considered as $\vec{f}_d = \mathbf{R}_r[0 \quad 0 \quad 0.1\sin(t)]$ and $\vec{\tau}_d = 0.1\sin(t)[1 \quad 1 \quad 1]$.

In addition, the desired trajectory of the virtual leader is defined as follows (t is the time variable):

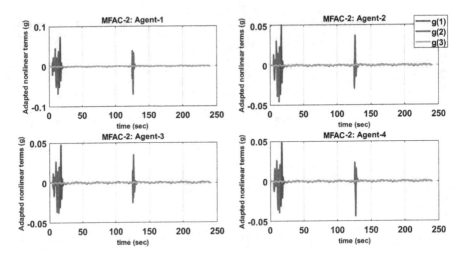

FIGURE 8.13   CAMFC-1: The estimated nonlinear terms for MFAC-2 at all quadrotors.

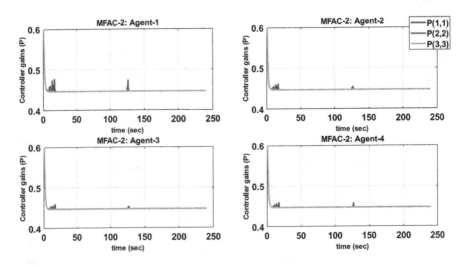

FIGURE 8.14   CAMFC-1: The estimated controller gains for MFAC-2 at all quadrotors.

$$
\left\{
\begin{aligned}
&0 \le t < 60: \ r_x = 0, \ r_y = 0, \ r_z = +5 \\
&60 \le t < 120: \ r_x = +1, \ r_y = 0, \ r_z = +5 \\
&120 \le t < 180: \ r_x = +1, \ r_y = +1, \ r_z = +5 \\
&180 \le t < 240: \ r_x = 0, \ r_y = +1, \ r_z = +5 \\
&240 \le t < 360: \ r_x = 0, \ r_y = 0, \ r_z = +5 \\
&360 \le t < 400: \ r_x = 0, \ r_y = 0, \ r_z = 0
\end{aligned}
\right\} .
\tag{8.92}
$$

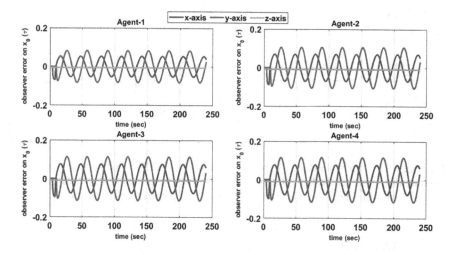

FIGURE 8.15 CAMFC-1: The consensus errors for cooperative observer on $\mathbf{x}_0$ at all quadrotors.

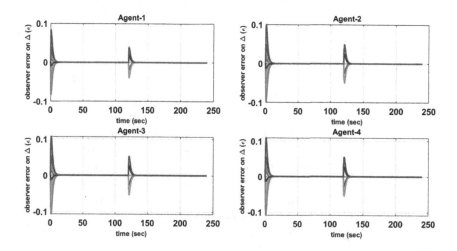

FIGURE 8.16 CAMFC-1: The consensus errors for cooperative observer on Δ at all quadrotors.

The desired formation topology among the quadrotors is also defined as follows:

$$\Delta = \begin{bmatrix} 0 & 0 & 0 \\ r_x & r_y & r_z \\ -r_x & r_y & 2r_z \\ -r_x & -r_y & 3r_z \\ r_x & -r_y & 4r_z \end{bmatrix}, \tag{8.93}$$

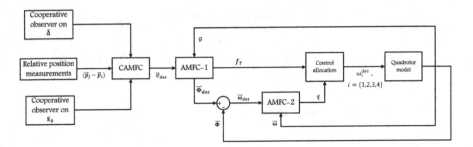

FIGURE 8.17 The schematic of implementation of the CAMFC-2 on each quadrotor. The measurement over the absolute position of all quadrotors in the network is revoked; instead, the relative position estimation between the quadrotors in the network is needed.

TABLE 8.3
The Tuning Parameters for the CAMFC-2 Algorithm

Parameter	Value	Parameter	Value
Π_m	[1 1 1]	Y_m	[1 1 1]
λ	10	μ	10
CAMFC			
Q	$10 \times \mathrm{diag}(1,1,10)$	Γ_1	$100 \times \mathrm{diag}(1,1,1)$
ρ_1	1.0	Γ_2	$\mathrm{diag}(1,1,1)$
ρ_2	1.0		
MFAC-1		**MFAC-2**	
Q	$10 \times \mathrm{diag}(1,1,10)$	Q	$0.1 \times \mathrm{diag}(1,1,1)$
ρ_1	1.0	ρ_1	1.0
Γ_1	$100 \times \mathrm{diag}(1,1,1)$	Γ_1	$100 \times \mathrm{diag}(1,1,1)$
ρ_2	0.1	ρ_2	0.1
Γ_2	$\mathrm{diag}(1,1,1)$	Γ_2	$\mathrm{diag}(1,1,1)$

where

$$\left.\begin{array}{l} 0 \le t < 60: \ r_x = 0, \ r_y = 0, \ r_z = +0.5 \\ 60 \le t < 120: \ r_x = +1, \ r_y = 0, \ r_z = +0.5 \\ 120 \le t < 180: \ r_x = +1, \ r_y = +1, \ r_z = +0.5 \\ 180 \le t < 240: \ r_x = 0, \ r_y = +1, \ r_z = +0.5 \\ 240 \le t < 400: \ r_x = 0, \ r_y = 0, \ r_z = +0.5 \end{array}\right\}. \tag{8.94}$$

The simulation results for this case study are presented in Figures 8.18–8.33. They have been obtained in terms of running the accompanying Matlab & Simulink programs and schemes.

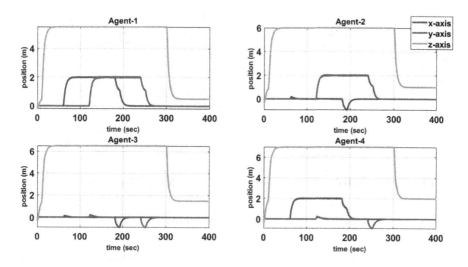

FIGURE 8.18 CAMFC-2: The positions of all quadrotors.

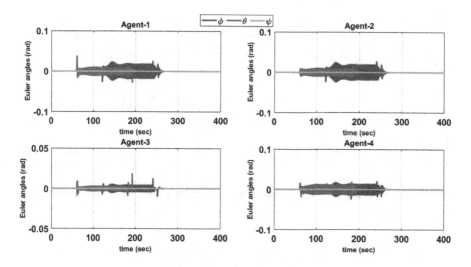

FIGURE 8.19 CAMFC-2: The Euler angles of all quadrotors.

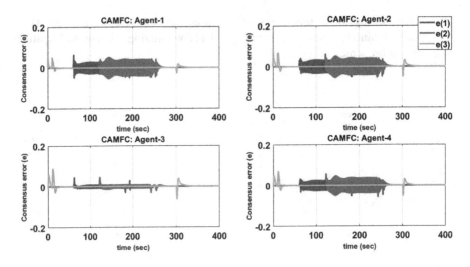

FIGURE 8.20   CAMFC-2: The consensus errors for CAMFC at all quadrotors.

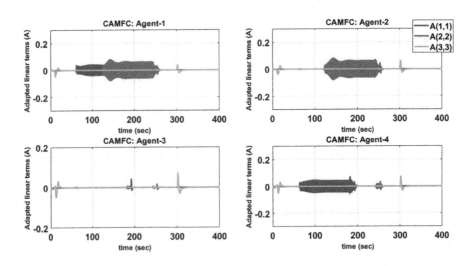

FIGURE 8.21   CAMFC-2: The estimated linear terms for CAMFC at all quadrotors.

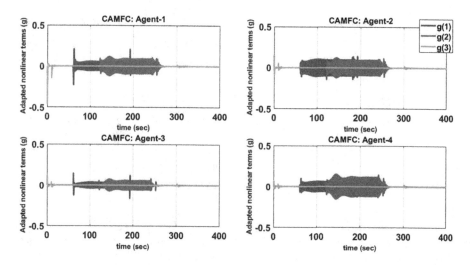

FIGURE 8.22 CAMFC-2: The estimated nonlinear terms for CAMFC at all quadrotors.

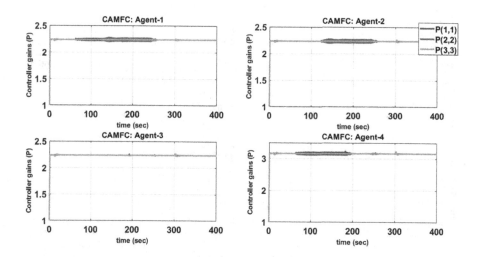

FIGURE 8.23 CAMFC-2: The estimated controller gains for CAMFC at all quadrotors.

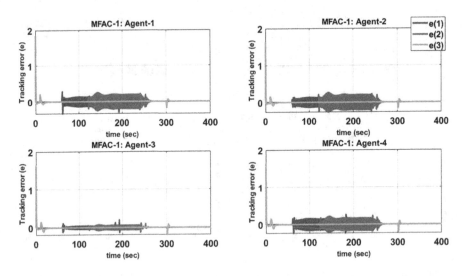

FIGURE 8.24   CAMFC-2: The tracking errors for MFAC-1 at all quadrotors.

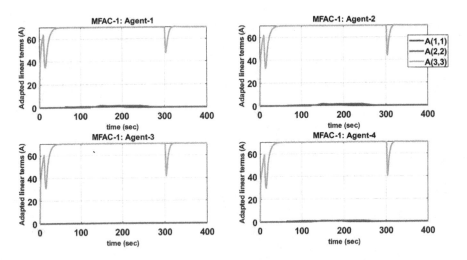

FIGURE 8.25   CAMFC-2: The estimated linear terms for MFAC-1 at all quadrotors.

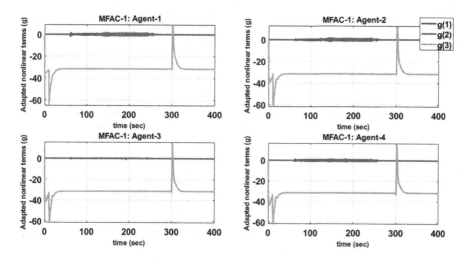

FIGURE 8.26 CAMFC-2: The estimated nonlinear terms for MFAC-1 at all quadrotors.

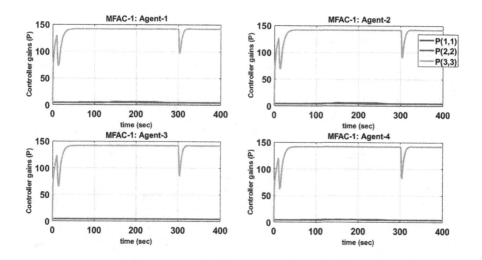

FIGURE 8.27 CAMFC-2: The estimated controller gains for MFAC-1 at all quadrotors.

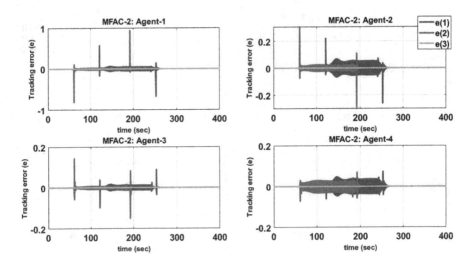

FIGURE 8.28   CAMFC-2: The tracking errors for MFAC-2 at all quadrotors.

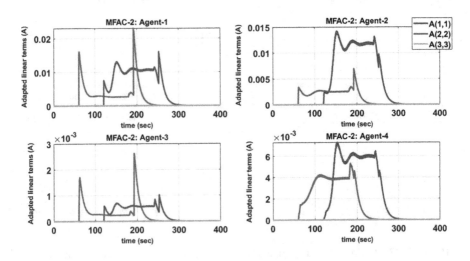

FIGURE 8.29   CAMFC-2: The estimated linear terms for MFAC-2 at all quadrotors.

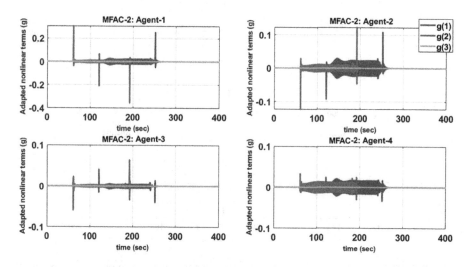

FIGURE 8.30 CAMFC-2: The estimated nonlinear terms for MFAC-2 at all quadrotors.

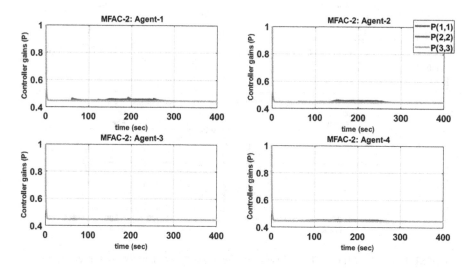

FIGURE 8.31 CAMFC-2: The estimated controller gains for MFAC-2 at all quadrotors.

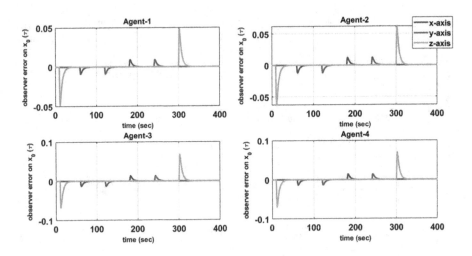

FIGURE 8.32   CAMFC-2: The consensus errors for the cooperative observer on $\mathbf{x}_0$ at all quadrotors.

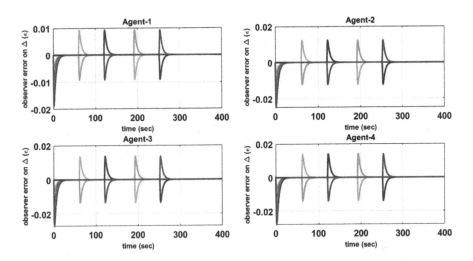

FIGURE 8.33   CAMFC-2: The consensus errors for the cooperative observer on Δ at all quadrotors.

REFERENCES

[Kha02] H. Khalil, *Nonlinear Systems*, Prentice Hall, Upper Saddle River, NJ, 2002.

[Lew14] F. L. Lewis, H. Zhang, K. Hengster-Movric and A. Das, *Cooperative Control of Multi-Agent Systems: Optimal and Adaptive Design Approaches*, Springer-Verlag, London, 2014.

[Li15] Z. Li and Z. Duan, *Cooperative Control of Multi-Agent Systems: A Consensus Region Approach*, CRC Press, Taylor & Francis, Boca Raton, FL, 2015.

[Mod16] H. Modares, S. P. Nageshrao, G. A. Delgado Lopes, R. Babuska and F. L. Lewis, "Optimal model-free output synchronization of heterogeneous systems using off-policy reinforcement learning," *Automatica,* vol. 71, pp. 334–341, Sep. 2016.

[Olf06] R. Olfati-Saber, "Flocking for multi-agent dynamic systems: algorithms and theory," *IEEE Transactions on Automatic Control,* vol. 51, no. 3, pp. 401–420, Mar. 2006.

[Olf07] R. Olfati-Saber, J. A. Fax and R. M. Murray, "Consensus and cooperation in networked multi-agent systems," *Proceedings of the IEEE,* vol. 95, no. 1, pp. 215–233, Jan. 2007.

[Saf18] A. Safaei and M. N. Mahyuddin, "A solution for the cooperative formation-tracking problem in a network of completely unknown nonlinear dynamic systems without relative position information," *International Journal of Systems Science,* vol. 49, no. 16, pp. 3459–3475, Dec. 2018.

[Saf21] A. Safaei, "Cooperative adaptive model-free control with model-free parameter estimation and online gain tuning," *IEEE Transactions on Cybernetics,* pp. 1–13, Mar. 2021. doi: 10.1109/TCYB.2021.3059200

[Tan03a] H. G. Tanner, A. Jadbabaie and G. J. Pappas, "Stable flocking of mobile agents, part I: fixed topology," in *Proceedings of 42nd IEEE International Conference on Decision and Control*, Maui, HI, 2003, pp. 2010–2015.

[Tan03b] H. G. Tanner, A. Jadbabaie and G. J. Pappas, "Stable flocking of mobile agents, part II: dynamic topology," in *Proceedings of 42nd IEEE International Conference on Decision and Control*, Maui, HI, 2003, pp. 2016–2021.

Appendix
Simulation Results for Implementation of Model-Free Adaptive Controller on a Differential-Drive Ground Mobile Robot

Here, the dynamic system for nonholonomic wheeled mobile robot (WMR) presented in Section 1.2.2 is utilized for implementing the MFAC algorithm proposed in Section 5.2.3. In this regard, a cascade control scheme is utilized as depicted in Figure A.1. According to this figure, a switching controller is used for triggering either the distance controller or heading controller, based on the magnitude of distance tracking error e_{dis} and heading tracking error e_{hed}. Then, two proportional controllers are utillized for converting the distance and heading tracking errors into the desired linear and angular velocities of the robot (i.e., v_{des} and ω_{des}), respectively. After that, two MFAC modules are implemented to determine the desired force and torque at the robot, by considering the tracking errors for linear and angular velocities. Here, MFAC-1 is dedicated for the distance controller and MFAC-2 unit is used for controlling the heading angle. Finally, an allocation unit is used for defining the desired rotational speed of right and left electric motors on the robot.

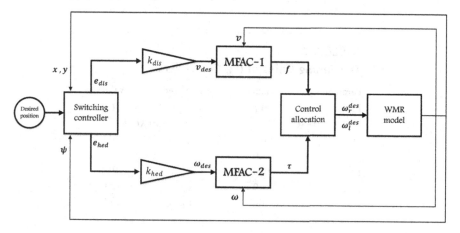

FIGURE A.1   The proposed control scheme for position tracking of a nonholonomic WMR.

This case study is implemented in AMFC_Nonholonom_Ground_Rover.mdl Simulink model (using MATLAB R2018b is recommended). For this simulation, the specifications of the WMR model are considered as in Table A.1. Moreover, the proportional controllers and the MFAC modules are tuned with the gains presented in Table A.2. Here, the external disturbances acting on the WMR model are considered as $f_d = 0.001 \times \sin t$ and $\tau_d = 0.0001 \times \sin t$.

In addition, the desired trajectory for the robot in current simulation is defined as follows

$$
\left\{
\begin{array}{l}
0 \leq t < 100 : x_{des} = 0, y_{des} = 0 \\
100 \leq t < 200 : x_{des} = 10, y_{des} = 0 \\
200 \leq t < 300 : x_{des} = 10, y_{des} = 10 \\
300 \leq t < 400 : x_{des} = 0, y_{des} = 10 \\
400 \leq t < 500 : x_{des} = 0, y_{des} = 0
\end{array}
\right\}
$$

The simulation results for the current implementation, which is done in Matlab & Simulink, are presented in Figures A.2–A.12.

TABLE A.1

The Specifications of the WMR Model

Parameter	Value	Parameter	Values
M	10 kg	d	0. 2 m
J	0.1 kg.m^2	r	0.02 m
k_f	0.1 N.s/m	k_v	0.01 N.m.s/rad
k_τ	0.1 N.m.s/rad		

TABLE A.2

The Tuning Gains of the Controllers

Parameter	Value	Parameter	Values
k_{dis}	0.1	k_{hed}	0.9
	MFAC-1		MFAC-2
Q	0.1	Q	0.1
ρ_1	1.0	ρ_1	1.0
Γ_1	10.0	Γ_1	10.0
ρ_2	1.0	ρ_2	1.0
Γ_2	1.0	Γ_2	1.0

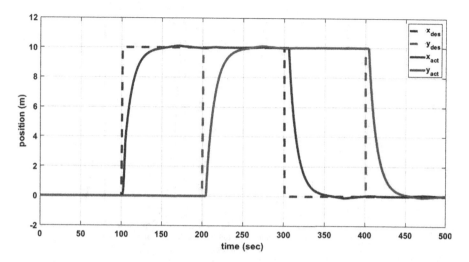

FIGURE A.2 The position of WMR in 2D space.

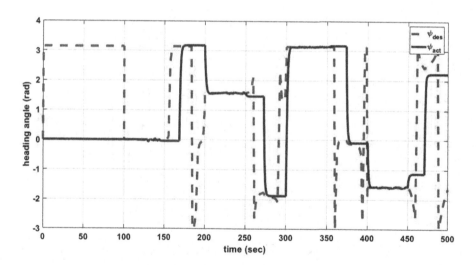

FIGURE A.3 The heading angle of WMR.

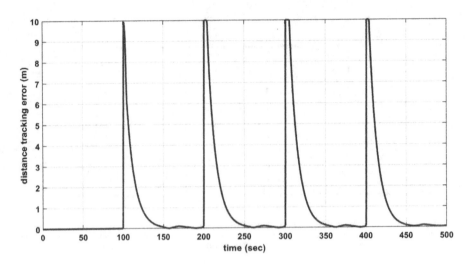

FIGURE A.4   The distance tracking error of WMR.

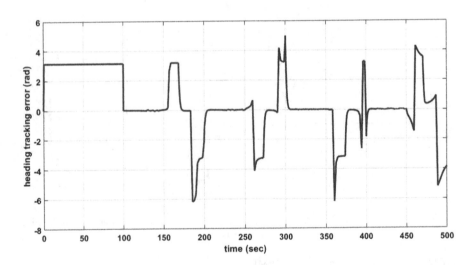

FIGURE A.5   The heading tracking error of WMR.

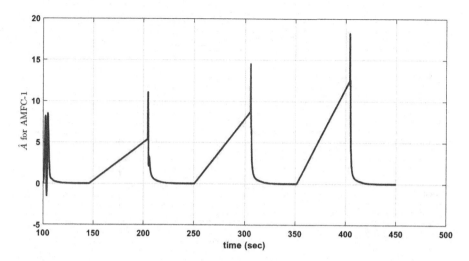

FIGURE A.6   The estimated linear term in MFAC-1.

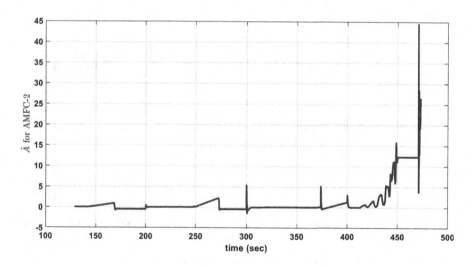

FIGURE A.7   The estimated linear term in MFAC-2.

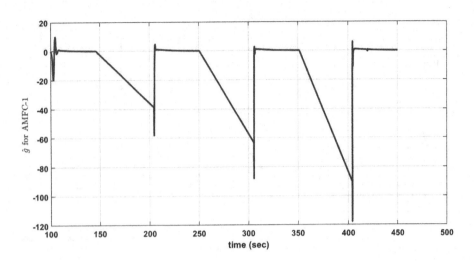

FIGURE A.8   The estimated nonlinear term in MFAC-1.

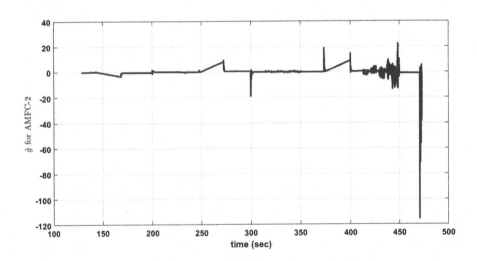

FIGURE A.9   The estimated nonlinear term in MFAC-2.

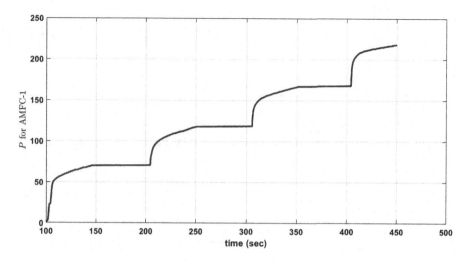

FIGURE A.10   The controller gain P in MFAC-1.

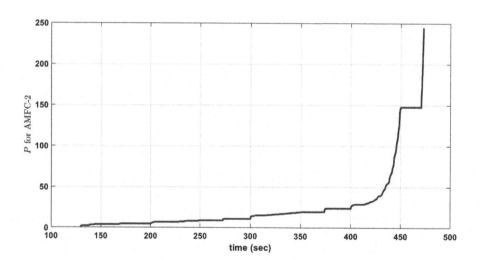

FIGURE A.11   The controller gain P in MFAC-2.

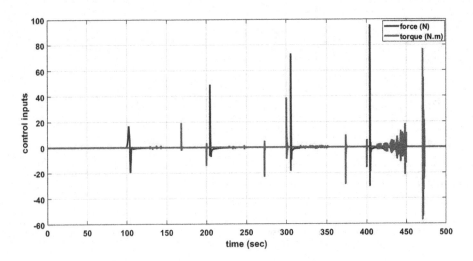

FIGURE A.12   The control inputs (ordered force and torque) of WMR.

Index

Printed in the United States
by Baker & Taylor Publisher Services